Lecture Notes in Networks and Systems

Volume 269

The series "Lecture Notes in Networks and Systems" publishes the latest developments in Networks and Systems—quickly, informally and with high quality. Original research reported in proceedings and post-proceedings represents the core of LNNS.

Volumes published in LNNS embrace all aspects and subfields of, as well as new challenges in, Networks and Systems.

The series contains proceedings and edited volumes in systems and networks, spanning the areas of Cyber-Physical Systems, Autonomous Systems, Sensor Networks, Control Systems, Energy Systems, Automotive Systems, Biological Systems, Vehicular Networking and Connected Vehicles, Aerospace Systems, Automation, Manufacturing, Smart Grids, Nonlinear Systems, Power Systems, Robotics, Social Systems, Economic Systems and other. Of particular value to both the contributors and the readership are the short publication timeframe and the world-wide distribution and exposure which enable both a wide and rapid dissemination of research output.

The series covers the theory, applications, and perspectives on the state of the art and future developments relevant to systems and networks, decision making, control, complex processes and related areas, as embedded in the fields of interdisciplinary and applied sciences, engineering, computer science, physics, economics, social, and life sciences, as well as the paradigms and methodologies behind them.

Indexed by SCOPUS, INSPEC, WTI Frankfurt eG, zbMATH, SCImago.

All books published in the series are submitted for consideration in Web of Science.

More information about this series at http://www.springer.com/series/15179

Salman Nazir · Tareq Z. Ahram ·
Waldemar Karwowski
Editors

Advances in Human Factors in Training, Education, and Learning Sciences

Proceedings of the AHFE 2021 Virtual Conference on Human Factors in Training, Education, and Learning Sciences, July 25–29, 2021, USA

Springer

Editors
Salman Nazir
Institute of Maritime Operations
University of South-Eastern Norway
Borre, Norway

Tareq Z. Ahram
Institute for Advanced Systems Engineering
University of Central Florida
Orlando, FL, USA

Waldemar Karwowski
Department of Industrial Engineering
and Management System
University of Central Florida
Orlando, FL, USA

ISSN 2367-3370 ISSN 2367-3389 (electronic)
Lecture Notes in Networks and Systems
ISBN 978-3-030-79999-1 ISBN 978-3-030-80000-0 (eBook)
https://doi.org/10.1007/978-3-030-80000-0

This Springer imprint is published by the registered company Springer Nature Switzerland AG
The registered company address is: Gewerbestrasse 11, 6330 Cham, Switzerland

Advances in Human Factors and Ergonomics 2021

AHFE 2021 Series Editors

Tareq Z. Ahram, Florida, USA
Waldemar Karwowski, Florida, USA

12th International Conference on Applied Human Factors and Ergonomics and the Affiliated Conferences (AHFE 2021)

Proceedings of the AHFE 2021 Virtual Conference on Human Factors in Training, Education, and Learning Sciences, July 25–29, 2021, USA.

Advances in Neuroergonomics and Cognitive Engineering	Hasan Ayaz, Umer Asgher and Lucas Paletta
Advances in Industrial Design	Cliff Sungsoo Shin, Giuseppe Di Bucchianico, Shuichi Fukuda, Yong-Gyun Ghim, Gianni Montagna and Cristina Carvalho
Advances in Ergonomics in Design	Francisco Rebelo
Advances in Safety Management and Human Performance	Pedro M. Arezes and Ronald L. Boring
Advances in Human Factors and Ergonomics in Healthcare and Medical Devices	Jay Kalra, Nancy J. Lightner and Redha Taiar
Advances in Simulation and Digital Human Modeling	Julia L. Wright, Daniel Barber, Sofia Scataglin and Sudhakar L. Rajulu
Advances in Human Factors and System Interactions	Isabel L. Nunes
Advances in the Human Side of Service Engineering	Christine Leitner, Walter Ganz, Debra Satterfield and Clara Bassano
Advances in Human Factors, Business Management and Leadership	Jussi Ilari Kantola, Salman Nazir and Vesa Salminen
Advances in Human Factors in Robots, Unmanned Systems and Cybersecurity	Matteo Zallio, Carlos Raymundo Ibañez and Jesus Hechavarria Hernandez
Advances in Human Factors in Training, Education, and Learning Sciences	Salman Nazir, Tareq Z. Ahram and Waldemar Karwowski

(continued)

(continued)

Advances in Human Aspects of Transportation	Neville Stanton
Advances in Artificial Intelligence, Software and Systems Engineering	Tareq Z. Ahram, Waldemar Karwowski and Jay Kalra
Advances in Human Factors in Architecture, Sustainable Urban Planning and Infrastructure	Jerzy Charytonowicz, Alicja Maciejko and Christianne S. Falcão
Advances in Physical, Social & Occupational Ergonomics	Ravindra S. Goonetilleke, Shuping Xiong, Henrijs Kalkis, Zenija Roja, Waldemar Karwowski and Atsuo Murata
Advances in Manufacturing, Production Management and Process Control	Stefan Trzcielinski, Beata Mrugalska, Waldemar Karwowski, Emilio Rossi and Massimo Di Nicolantonio
Advances in Usability, User Experience, Wearable and Assistive Technology	Tareq Z. Ahram and Christianne S. Falcão
Advances in Creativity, Innovation, Entrepreneurship and Communication of Design	Evangelos Markopoulos, Ravindra S. Goonetilleke, Amic G. Ho and Yan Luximon
Advances in Human Dynamics for the Development of Contemporary Societies	Daniel Raposo, Nuno Martins and Daniel Brandão

Preface

This book provides researchers and practitioners a forum to share research and best practices in the application of human factors to training, education, and learning sciences. Just as human factors' discipline has been applied to hardware, software, and the built environment, there is now a growing interest in the optimal design of training, education, and learning experiences. Principles of behavioral and cognitive science are extremely relevant to the design of instructional content and the effective application of technology to deliver the appropriate learning experience. These principles and best practices are important in corporate, higher education, and military training environments.

This book also aims to share and transfer not just knowledge, learning experiences, and best training approaches that are of real value in practical terms, a value that can help leaders ensure their organizations stay ahead of the competition through continued innovation, strong competitive advantage, and inspired leadership.

This book is organized into nine sections that contain the following subject areas:

1. Advanced Learning Technologies
2. Implementation of Learning Technologies in Education
3. Novel Technology and Methods for Training and Certification
4. Interactive Multimedia and Gamification
5. Student Training and Environment Interaction
6. Analysis of Learning Strategies and Learning Performance
7. Teaching and Learning
8. Distance Learning during COVID-19 Era
9. Competency and Skill Management in Education

Each section contains research papers that have been reviewed by the members of the International Editorial Board. Our sincere thanks and appreciation to the board members as listed below:

S. Abramovich, USA
D. Al Thani, Qatar
T. Alexander, Germany
T. Barath, Hungary
J. Bartnicka, Poland
J. Castro-Alonso, Chile
J. Elliott, USA
M. Freeman, USA
Gronstedt, USA
Madinger, USA
S. Mallam, Norway
Mansoor, Qatar
McClernon, USA
K. Moore, USA
K. Orvis, USA
B. Pokorny, USA
R. Roper, USA
Sampson, Greece
J. Smith, Canada
J. Syversen, Norway

July 2021

Salman Nazir
Tareq Z. Ahram
Waldemar Karwowski

Contents

Student Training and Environment Interaction

Analysis of Learning Strategies and Learning Performance

Teaching and Learning

Advanced Learning Technologies

Structure of a Socio-Technical Learning and Innovation Factory

Barbara Tropschuh(✉), Fabian Dillinger, Quirin Gärtner, Svenja Korder, Harald Bauer, and Moritz Kagerer

Institute for Machine Tools and Industrial Management (*iwb*), Technical University of Munich, 85748 Garching b. Munich, Germany
barbara.tropschuh@iwb.tum.de

Abstract. Nowadays, the socio-technical manufacturing environment faces numerous challenges due to digitalization and demographic change. Testing new technologies or applying them into ongoing manufacturing systems without reducing productivity is very difficult, but it is crucial to stay competitive. Learning factories are particularly suitable in such volatile and challenging times, as they offer a protected environment for applying new ideas and methods. Since learning factories today are mainly focused on technology and organization, they can no longer meet the aforementioned challenges. Therefore, this paper presents a new framework for learning factories that also considers human aspects using the example of the InnoLab at the Institute for Machine Tools and Industrial Management (*iwb*).

Keywords: Digitalization · Assistance systems · Laboratory · Learning factory

1 Introduction – Manufacturing Environment

The highly competitive market environment of manufacturing companies is characterized by steadily increasing quality requirements, demographic change, mass personalization, and cost pressure on manufacturing processes along the entire value chain [1]. One approach to counteract this turbulent environment is the transformation of manufacturing in the direction of Industry 4.0, which is defined as a "real-time, intelligent, horizontal and vertical networking of people, machines, objects and information and communication technologies for the dynamic management of complex systems" [2]. Therefore, Industry 4.0 includes a wide range of smart and digital technologies to support employees in their daily activities and maintain their well-being. This still involves many unknowns for companies' organizational forms and employees and is not yet sufficiently advanced [1]. Kollmann & Schmidt [3] see strengthening the innovative capacity of companies as an essential prerequisite for promoting the implementation of new manufacturing approaches and technologies. This is usually associated with a considerable expenditure of resources (e.g. human) and economic risk, which companies are often unwilling or unable to take. It therefore appears desirable to strengthen the innovative capacity of companies and thus prepare them for the technological challenges of the

S. Nazir et al. (Eds.): AHFE 2021, LNNS 269, pp. 3–11, 2021.
https://doi.org/10.1007/978-3-030-80000-0_1

future. In this context, learning factories play a crucial role to explore and evaluate new technologies in a protected production environment. According to Abele [4], a *learning factory* is an environment reproducing authentic production processes on a technical and organizational level, resembling a real value chain, changeable in its setting, producing a real product, and pursuing a didactical concept with different learning methods within its on-site learning approach. In this paper, the concept of a socio-technical manufacturing learning factory with its technology fields and organizational approaches is described using the example of the InnoLab at the Institute for Machine Tools and Industrial Management (*iwb*).

2 Theoretical Background

This chapter sets up the reference frame of the basic principles addressed in this paper and presents an introduction on the concepts of Ambidexterity (2.1), Innovation Management (2.2), and Learning Factory Fields (2.3). Moreover, Sect. 2.4 describes the research gap regarding existing learning factories.

2.1 Ambidexterity

An innovation is defined as the successful implementation of an idea improving a procedure, an organization or a process [5]. Scientific literature distinguishes between five different types of innovation, relevant in manufacturing: product/service, process, organization, production, and management innovation [6]. Innovations can be further differentiated into radical and incremental innovations. Radical innovations increase the performance of existing processes significantly, while incremental innovations serve successive improvements of performance and competitiveness [7]. Still, radical and incremental innovations require fundamentally opposite organizational behaviors. A behavioral distinction is made between explorative and exploitative patterns. Explorative behavior is fostering radical innovation through experimentation, flexibility, divergent thinking, improvisation, organic structures, autonomy or adaptation. Exploitative behavior is promoting incremental innovation via efficiency, refinement, focus, mechanistic structures or routinisation [8, 9].

As resources within innovation management are limited, explorative and exploitative behaviors cannot be addressed equally and simultaneously [10]. Therefore, ambidexterity management addresses an organizational balance between these two behavioral patterns in order to enable both types of innovations [9]. One approach to ambidexterity management is structural ambidexterity, proposing a separate organizational unit for explorative organizational behavior and divergent activities. Recently, academic institutions and companies have set up structural ambidexterity using so-called *Innovation Labs* in order to devise radical innovation in parallel to the existing structures and processes [11].

2.2 Innovation Management

There is no blueprint to successfully implement innovation. The types of innovation projects and the circumstances of their implementation are too diverse. However, various approaches have dealt with identifying the most important elements characterizing successful and comprehensive innovation management. By analyzing and comparing these approaches, two things emerge. The first common aspect is that the innovation management process is generally formalized as a sequential procedure triggered by a potential or need and followed by the steps idea generation, evaluation, selection, production, introduction, and utilization [12, 13]. Secondly, it can be stated that almost all approaches describe so-called organizational design dimensions of innovation management, which enable a consistent innovation management. These dimensions are formed by the elements innovation strategy, culture and innovation processes and methods [14]. Transferred to manufacturing, this leads to a cyclic innovation management framework for manufacturing, consisting of the process steps impulse, ideation, concept, change and serial production [15], which in turn are supported by the organizational design dimensions innovation strategy, culture, processes and methods.

2.3 Learning Factory Fields: Human – Technology – Organization

In order to drive forward the digitalization in manufacturing, a holistic transformation process into an open socio-technical system is necessary [16]. Dombrowski & Mielke [17] note, with regard to manufacturing systems in general, that these would not be limited to technologies alone, but would rather encompass "the interplay of technology, organization and human[s] in production". Accordingly, most of the change processes affecting the manufacturing system also have an impact on functional, organizational, and technological aspects [18]. The digitalization in manufacturing in the sense of Industry 4.0 represents such a holistic corporate transformation [19] and, pursuant to modeling by Oks et al. [20], concerns all three key dimensions – human, technology, and organization. Neglecting only one of these dimensions or intermediate areas could trigger the failure of an entire Industry 4.0 implementation project. Figure 1 provides a representation of the key dimensions of a digital business transformation as formulated by Oks et al. [20]. In order to support manufacturing companies in their digital transformation process, learning factories should cover all areas of the HTO-model. Relevant human topics could for example be strain-oriented workplace design or individual work scheduling [21]. Furthermore, new technologies should be part of learning factories. Rüssmann et al. [22] list nine fundamental technologies, which are composed of (1) Autonomous Robots, (2) Industrial Internet of Things, (3) Additive Manufacturing (AM) Technologies, (4) Cloud, (5) Augmented Reality and (6) Simulations, (7) Big Data and Analytics, (8) Horizontal and Vertical System Integration, and finally (9) measures to strengthen Cyber-Security. Barig & Balzereit [23] add the field Assistance Systems to these nine main technologies. These ten technology fields can only successfully implement the essence of Industry 4.0 in interaction and are therefore important elements of learning factories. Furthermore, the introduction of new technologies has an impact on the organization [20]. The use of new assistance systems could lead to changed leadership styles, as employees take on more responsibility and receive information through assistance systems rather than their supervisor [24]. It therefore is important to strengthen openness to new organizational forms, such as lean culture, flat hierarchies or the mentor-mentee principle.

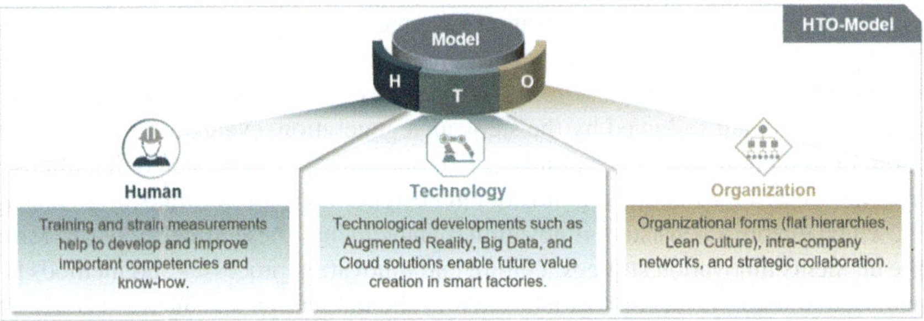

Fig. 1. HTO-model (based on [20])

2.4 Research Gap

The innovation management and the described HTO-model show how important the triad between human, technology, and organization is. In order to maintain a balance between efficiency and proactive promotion of innovation within this field of tension, ambidexterity management is an essential management tool. A structural ambidexterity in the form of a learning factory is often used to transfer the abstract concept of ambidexterity management into practice and makes its effects tangible on the shop floor. In the volatile industrial manufacturing environment, employees are confronted with new devices and organizational changes. Therefore, it is even more important for companies to know the state of research in these fields and to explore the organizational and technical solutions by themselves in a realistic but protected manufacturing environment. Renowned learning factories demonstrate different smart devices, production processes, and lean methods as well as new technologies like augmented reality [25]. Nevertheless, the focus on human-centered innovations, such as strain measurement methods for manufacturing, is not sufficiently taken into account. Humans play a central role in transformation and in innovation processes [17]. In order to address this topic comprehensively, the learning factory of the *iwb*, the InnoLab, aims to integrate the human-centered approach in all three focus areas human, technology, and organization.

3 Structure of a Socio-technical Innovation Lab

In order to explain the elements and the structure of a socio-technical innovation lab, a framework has been created, which contains the design dimensions, the focus areas according to the HTO-model as well as the overarching vision of an innovation lab (see Fig. 2). The individual elements are described in more detail as follows.

Design Dimensions

The design dimensions of the innovation lab framework consist of innovative strategies, methods, and an innovation culture. These three elements interact and support each other. In the area of **strategy**, the innovation lab supports the participants to develop visions and strategies for different time horizons, various topics, and future scenarios. For this purpose, an external business perspective (e.g. PESTEL analysis) as well as an internal business perspective (e.g. value stream mapping) are analyzed in an use case. Subsequently, the strengths, weaknesses, opportunities, and threats are identified, for example with the help of a SWOT-analysis.

To be able to develop and implement the strategy as well as to foster innovation transformations in companies, people are guided and supported by **methods** including the following three categories. ***Methods of technology management*** support the early recognition of emerging technologies, their accessibility for companies, and the further implementation of existing solutions. The methods in this area can be assigned to the five areas of technology management: identification (e.g. benchmark), selection (e.g. road mapping), acquisition (e.g. make-or-buy), exploitation (e.g. incremental development), and protection (e.g. patent analysis) [26]. ***Creativity methods*** are defined techniques for generating ideas or solving problems that stimulate the inventiveness of the participants. They support the development of innovations as well as continuous improvements and are thus an essential component for ensuring the long-term competitiveness of companies. Thereby, intuitive methods (e.g. brainstorming) and discursive, systematic methods (e.g. morphological box) as well as combined methods (e.g. 6-thinking-hats) can be distinguished [27]. ***Project management methods*** are supporting tools and techniques applied to plan, develop, and control projects efficiently and effectively [28]. These methods can be categorized into classic sequential approaches (e.g. waterfall), agile customer-oriented methods (e.g. Scrum), and hybrid approaches (e.g. V-Scrum-Modell) [29].

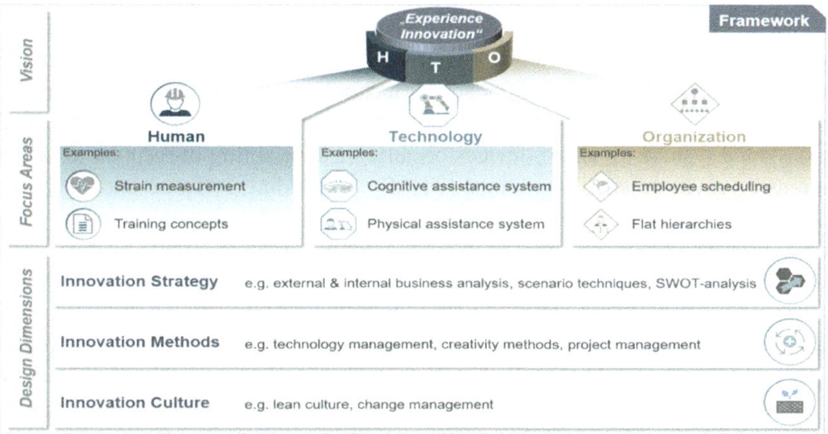

Fig. 2. Innovation lab framework

The **culture** of a company strongly influences the employees and their actions. This far-reaching influence can be seen especially in the lean culture, where the method of continuous improvement is integrated in the mindset of the employees. To strengthen the lean culture and to foster a culture of innovation, it is particularly important to involve employees in new ideas and changes. Therefore, the methods of change management, failure acceptance, and employee engagement are also included. Leaders can step into different roles and experience the impact of several strategies and methods for operative employees. This change of perspective supports the implementation, as possible challenges are experienced first hand.

Focus Areas

The three focus areas human, technology, and organization represent the areas that can be developed or experimented within the innovation lab framework. In the socio-technical system, **humans** are characterized by physical and mental performance requirements as well as competencies. Individual physical and mental strain can be measured by analyzing vital signs, e.g. heart rate or muscle activity, or by using questionnaires, e.g. NASA-Task Load Index. Learning concepts for individual employee development and employee scheduling according to individual characteristics are part of this focus area. In the InnoLab, for example, the individual strain of participants can be measured during the performance of assembly work. Vital signs are recorded by using various smart devices and assisting systems like muscle shirts, fitness tracker, chest straps or ear sensors. The individual strain depends on the performance requirements of the test person as well as the duration and intensity of the applied workload, which has to be identified and adjusted individually. Due to innovative training concepts in the innovation lab, it is possible to explore different methods and to experience individually adaptable workplaces. In the area of **technology**, the focus lies on assistance systems as part of the socio-technical design of the framework. Thereby, each system that supports an employee in an industrial environment in the execution of work can be described as an assistance system [30]. A fundamental distinction can be made between the support of cognitive work tasks and executive, physical activities. Innovative assistance systems additionally embed a software unit or sensor technology that enables digitalization and connectivity with regard to the changes due to Industry 4.0. They are referred to as digital assistance systems or smart devices because of their ability to respond to different situations or environmental conditions [31]. By embedding software and sensors, the scope of assistance systems is expanded. They can not only support the employee, but in some cases also record or measure the individually perceived strain and thus enable a quick reaction when employees are experiencing overload. For example, a collaborative robot can support employees when needed, relieve them of specific work tasks and thus, reduce physical and psychological strain. Numerous worker assistance systems and tools for data analysis are provided in the InnoLab. The area of **organization** can be divided into structure and process. New forms of cooperation, novel work, and team forms are considered in the context of the structure in and between different organizations. In the area of organizational procedures, topics such as employee scheduling, innovative working time models, flat hierarchies or organizational processes are highlighted. The flexible room design of an innovation lab allows, for example, novel forms of collaboration to be tested. It is important that creative areas are located adjacent to the assembly environment in order to test new ideas in practice, which in turn fosters the experience of a culture of innovation. In the InnoLab, a flexible open-space environment with creative tools such as digital boards or convertible assembly stations for individual design and use is provided.

Vision
The top of the framework reflects the overarching vision of an innovation lab. The special feature lies in a method-supported, structured continuous learning and innovation process, in which innovations can be experienced. In contrast to a classic creative workshop, the innovation lab focuses on the contribution of methods and technologies to increasing the value of processes. Participants can test their ideas and potential solutions directly in a realistic, haptic simulation environment. The goal is to create iterative, short-cycle learning experiences. The developed concepts are immediately validated with the help of prototypes. The identified improvement potentials can thus be used directly for the further iteration stages. This approach leads to rapid learning progress in all fields of the innovation lab framework (see Fig. 2), solves practical challenges and increases the problem-solving skills of the participants.

4 Conclusion and Outlook

This paper presents a framework for a socio-technical innovation lab using the example of the InnoLab at the *iwb*. In view of the challenges in the manufacturing environment, focusing on organizational and technical aspects is no longer sufficient for learning factories. The human component must be included and further considered. This is realized in the presented framework based on the HTO-model and the innovation management process. The elements of the framework are now being implemented on an ongoing basis with various practice-oriented workshops in cooperation with different companies. In addition, further research is being conducted to determine whether other aspects need to be considered and supplemented within the framework of manufacturing innovation management.

Acknowledgements. The German Research Foundation (DFG) funds this research and development project 431696521 "Bewertung des Wettbewerbspotenzials von Produktionstechnologien". We extent our sincere thanks to the DFG for the generous support.

References

1. Wagner, R.M.: Industrie 4.0 für die Praxis: Mit realen Fallbeispielen aus mittelständischen Unternehmen und vielen umsetzbaren Tipps. Springer Gabler, Wiesbaden (2018)
2. Bauer, W., Schlund, S., Marrenbach, D., Ganschar, O.: Industrie 4.0 – Volkswirtschaftliche Potentiale für Deutschland (2014)
3. Kollmann, T., Schmidt, H.: Deutschland 4.0: Wie die digitale Transformation gelingt. Springer Gabler, Wiesbaden (2016)
4. Abele, E.: Learning factory. In: The International Academy for Produ, Laperrière, L., Reinhart, G. (eds.) CIRP Encyclopedia of Production Engineering, p. 112. Springer, Heidelberg (2016)
5. Schumpeter, J.: Theorie der wirtschaftlichen Entwicklung: Eine Untersuchung über Unternehmergewinn, Kapital, Kredit, Zins und den Konjunkturzyklus. Duncker & Humblot, Berlin (2013)

6. Trott, P.: Innovation Management and New Product Development. Pearson, Harlow (2017)
7. Olivan, P.: Methode zur organisatorischen Gestaltung radikaler Technologieentwicklungen unter Berücksichtigung der Ambidextrie, Stuttgart (2019)
8. Koryak, O., Lockett, A., Hayton, J., Nicolaou, N., Mole, K.: Disentangling the antecedents of ambidexterity: exploration and exploitation. Res. Policy **47**, 413–427 (2018)
9. Fourné, S.P., Rosenbusch, N., Heyden, M.L., Jansen, J.J.: Structural and contextual approaches to ambidexterity: a metaanalysis of organizational and environmental contingencies. Eur. Manag. J. **37**, 564–576 (2019)
10. Ardito, L., Besson, E., Petruzzelli, A.M., Gregori, G.L.: The influence of production, IT and logistics process innovations on ambidexterity performance. Bus. Process Manag. J. **24**, 1271–1284 (2018)
11. Niewöhner, N., Asmar, L., Wortmann, F., Röltgen, D., Kühn, A., Dumitrescu, R.: Design fields of agile innovation management in small and medium sized enterprises. Procedia CIRP **84**, 826–831 (2019)
12. Langer, S., Lindemann, U.: Managing cycles in development process: analysis and classification of external context, vol. 1, pp. 539–550 (2009)
13. Vahs, D., Brem, A.: Innovationsmanagement: Von der Idee zur erfolgreichen Vermarktung. Schäffer Poeschel, Stuttgart (2015)
14. Mugge, P., Markham, S.K.: An innovation management framework: a model for managers who want to grow their businesses. In: The PDMA Handbook of New Product Development, pp. 35–50. Wiley, Hoboken (2013)
15. Hofer, A., Brandl, F., Bauer, H., Haghi, S., Reinhart, G.: A framework for managing innovation cycles in manufacturing systems. Procedia CIRP **93**, 771–776 (2020)
16. Horvat, D., Stahlecker, T., Zenker, A., Lerch, C., Mladineo, M.: A conceptual approach to analysing manufacturing companies' profiles concerning Industry 4.0 in emerging economies. Procedia Manuf. **17**, 419–426 (2018)
17. Dombrowski, U., Mielke, T.: Ganzheitliche Produktionssysteme: Aktueller Stand und zukünftige Entwicklungen. Springer Vieweg, Heidelberg (2015)
18. Alter, S.: 18 Reasons Why IT-Reliant Work Systems Should Replace "The IT Artifact" as the Core Subject Matter of the IS Field 12 (2003)
19. Thoben, K.-D., Wiesner, S., Wuest, T.: "Industrie 4.0" and smart manufacturing – a review of research issues and application examples. Int. J. Autom. Technol. **11**(1), 4–16 (2017)
20. Oks, S.J., Fritzsche, A., Möslein, K.M.: An application map for industrial cyber-physical systems. In: Jeschke, S., Brecher, C., Song, H., Rawat, D.B. (eds.) Industrial Internet of Things. SSWT, pp. 21–46. Springer, Cham (2017). https://doi.org/10.1007/978-3-319-42559-7_2
21. Abele, E., et al.: Innovative Lernmodule und -fabriken - Validierung und Weiterentwicklung einer neuartigen Wissensplattform für die Produktionsexzellenz von morgen. TU Darmstadt (2015)
22. Rüßmann, M., Lorenz, M., Waldner, M., Engel, P., Harnisch, M., Justus, J.: The Future of Productivity and Growth in Manufacturing Industries. https://www.bcg.com/de-de/publications/2015/engineered_products_project_business_industry_4_future_productivity_growth_manufacturing_industries
23. Barig, B., Balzereit, K.: Assistance systems for industry 4.0 environments (2019)
24. Schwarzmüller, T., Brosi, P., Duman, D., Welpe, I.M.: How does the digital transformation affect organizations? Key themes of change in work design and leadership. MREV Manag. Revue **29**, 114–138 (2018)
25. Abele, E., Metternich, J., Tisch, M.: Learning Factories: Concepts, Guidelines Best-Practice Examples. Springer, Cham (2019)
26. Gregory, M.J.: Technology management: a process approach. Proc. Instit. Mech. Eng. Part B J. Eng. Manuf. **209**, 347–356 (1995)

27. Drews, G., Hillebrand, N.: Lexikon der Projektmanagement-Methoden: Die besten Methoden für jede Situation; der GPM-Werkzeugkasten für effizientes Projektmanagement. Haufe, Freiburg (2007)
28. PMI: A guide to the project management body of knowledge: PMBOK guide, Newtown Square, Pennsylvania, USA (2017)
29. Masciadra, E.: Traditional project management. In: Handzic, M., Bassi, A. (eds.) Knowledge and Project Management. KMOL, vol. 5, pp. 3–23. Springer, Cham (2017). https://doi.org/10.1007/978-3-319-51067-5_1
30. Reinhart, G.: Handbuch Industrie 4.0: Geschäftsmodelle, Prozesse, Technik. Hanser, München (2017)
31. Apt, W., Bovenschulte, M., Priesack, K., Weiß, C., Hartmann, E.A.: Forschungsbericht: Einsatz von digitalen Assistenzsystemen im Betrieb, Berlin (2018)

Teaching Robotics with Virtual Reality: Developing Curriculum for the 21st Century Workforce

Eric Peterson(✉), Biayna Bogosian, Jorge Tubella, and Shahin Vassigh

Architecture Department, Florida International University, 11200 SW 8th Street, Miami, FL 33199, USA
{eric.peterson,bbogosia,jtubella,svassigh}@fiu.edu

Abstract. The nature of work is changing. The ongoing shift to automated manufacturing and fabrication in the 21st Century will require a workforce with a radically different set of skills focused less on manual dexterity for the performance of repetitive tasks and more on interaction with robotic systems. It is expected that both Industrial Manufacturing and the Architecture Engineering and Construction (AEC) sector will begin to adopt more automation technology for the production of manufactured goods as well as the design and construction of the built environment. Teaching the fundamental skills required for 21st century workers in the Industrial Manufacturing and AEC sectors may no longer be feasible or efficient using on the job training or apprenticeship as a primary educational modality. Virtual Reality (VR) environments offer a possible solution to train workers in an immersive, interactive, and safe environment where they are less likely to be injured, slow down production, or cause unintentional damage to products or projects in construction. Developing curriculum for these workers requires a modular approach to pedagogy that draws on skills and strategies of game design. Assessment tools for evaluating knowledge and skills acquisition can be structured using a reward system drawn from game design and game theory. Pedagogic sequence can thus be more self-directed as a learner explores their developing knowledge, skills, and mastery of specific tasks. The addition of biometric sensors integrated with the VR Headset promises to add a powerful tool for testing curriculum in real time and providing feedback that can influence the duration, sequence, and progression of a learner's exploration of the learning environment. In this paper we will describe the development of our own AI-powered VR learning environment for teaching fundamentals of robotics, discuss the curriculum and assessment tools we have developed for in-person workshops and VR environment training, and present our aspirations for the further development of this educational tool.

Keywords: Pedagogy · Curriculum · Robotics · Experiential learning theory · Immersive learning · Interactive learning · Game-based learning · Virtual Reality · User Interface · User Experience · Artificial Intelligence · Adaptive Learning Systems · Machine Learning · Natural Language Processing

S. Nazir et al. (Eds.): AHFE 2021, LNNS 269, pp. 12–18, 2021.
https://doi.org/10.1007/978-3-030-80000-0_2

1 Introduction

The project described here outlines the efforts of an interdisciplinary group of faculty from Florida International University who received funding from the National Science Foundation's (NSF) 2019 Convergence Accelerator program (OIA-1937019). Funding was awarded for developing a comprehensive plan and a roadmap for creation of an immersive training software and a support system for the AEC industry provisionally titled the Robotics Academy. The project was successful in developing a curriculum prototype for a Virtual Reality (VR) learning environment that is currently being tested with a range of students from different backgrounds and different skill levels; additional Artificial Intelligence (AI) components remain in the development stages with limited application and testing.

2 A Robotics Curriculum for the 21st Century Workforce

Sophisticated robots that enhance human dexterity, visual perception, speed, and strength are creating rapid changes in the global economy. Robotic technologies are transforming job markets and altering the requisite training and skill sets needed for employment. Preparing the workforce for an economy that is increasingly defined by these technologies is imperative. Robotics automation is transforming jobs in Industrial Manufacturing and the Architecture, Engineering, and Construction (AEC) sectors at a speed and scale never experienced before. While the manufacturing sector has adopted automation technologies quite readily, AEC industries have been slow to adjust to robotic production technologies in spite of evidence that,"The ability of robot systems has grown, allowing them to work more and more in comparably unstructured environments as well as to be deployed in numerous and diverse fields." [1].

The AEC sector accounts for 23% of all jobs in the US, driving the design, construction, and operation of the built environment, and impacting nearly every industry, trade, labor, and employment market across the economy. The AEC sector impacts a broad range of commodity markets including steel, copper, aluminum, wood, minerals, and petrochemicals as well as manufactured products including concrete, glass, gypsum board, paper, plastics, mechanical equipment, electrical equipment, electronics, fabric, tile, carpeting, paint, sealants, and fasteners. Despite its crucial role in the national economy, the AEC industry is not prepared to meet critical national needs. Most buildings and infrastructure continue to be built on-site using centuries-old materials and techniques with labor productivity rates that are lower than those reported over 50 years ago (US BLS Labor Productivity, 2020). At the same time, training and apprenticeship models in the building trades have remained largely constrained to on the job and workplace training with little or no focus on the high-tech training and knowledge required for working with robotic tools and automation processes.

In November 2019, we held a Summit addressing the future of automation and AI across the global economy with industry leaders and analysts within the AEC industry. Sixty representatives of the region's leading AEC firms attended the event, including two of the nation's largest housing developers and builders. The Summit featured presentations from each of the keynote speakers and an extended focus group roundtable involving invited business representatives, Summit keynote speakers and the Robotics Academy project team. The Summit revealed a number of concerns: the AEC industry faces a growing labor shortage in the US, the construction industry must dramatically improve its productivity and building performance in order to meet demand. While prefabricated automated construction of buildings and infrastructure is a clear necessity, only a small percentage of the US construction industry is currently geared toward prefabrication. At a regional level, investment in automated, manufactured construction is almost non-existent; firms are aware of new technology applications but are adopting modular and automated prefabricated construction techniques at a very slow rate. Meanwhile, prefabrication in infrastructure construction is expanding indicating a potential growth sector that will need support for the adoption of robotic automation technology. A transition to automated engineering, design and construction will have severely disruptive impacts on the AEC sector with its massive potential for increased productivity leading to new jobs in robotic programming and factory-based prefabrication. Addressing these challenges will require educational tools for job training that can adapt to innovative robotic automation technologies.

In order to determine the significance of anticipated changes due to automation in the AEC industry we conducted a series of interviews with employers and employees in the building industry, including automation engineers, software developers, roboticists, designers, educators, and system integrators as well as product design specialists. Conclusions drawn from employer interviews reinforce our conclusion that new applications for robotics automation are rapidly increasing; demand for AEC workers with technical proficiency in automation is high while supply remains low; and the adoption of automation and robotics in the building industry is creating new jobs, entirely new job classes, and broad opportunities for entrepreneurship. Meanwhile, employees in the AEC sector complained of inadequate access to training in robotic technology, reporting that current available resources do not meet their needs. Overall, the lack of standardization for robotic tools and programming procedures is a major hurdle for workers tasked with deploying automation technology; employees need some level of in-person training and interaction with robots in order to gain knowledge and experience; employee programs that include certificates and credentials lead to confidence and job security. The Robotics Academy's training and resources are targeted to address these issues with an accessible AEC robotics curriculum, including a broad range of automation technologies. We have developed a certification path using a stackable micro-credential *badging* system similar to existing certification systems that can be easily adopted by the AEC industry.

Addressing the learning needs of AEC students, future professionals, and industry workers are critical for ensuring the competitiveness of a large portion of the US workforce. As the construction industry historically has invested far less than other industries for training, it is critical that higher education plays a larger role in skill-development in robotics automation for workers in these industries [3]. It is our belief that the ideal medium for disseminating this knowledge is immersive learning environments that draw on techniques used by game designers to engage a broad range of learners.

3 Immersive and Interactive Learning with Virtual Reality

Drawing on Constructivist and Experiential Learning Theories that suggest knowledge is actively produced from experience [4], and that knowledge is acquired in the act of learning by doing [5], we have developed a VR environment that situates learners in a series of task-driven activities that simulate real-world scenarios. Immersive technologies, simulation, visualization, and geospatial datasets are ideal tools to enhance learning. VR provides computer-generated simulations of the real or an imagined world that can serve as a rich and engaging context for learning [6–8]. Research shows that these environments facilitate learning and the assessment of learning by providing safe and low-cost settings for practice and rehearsal of real-world tasks [9, 10].

The qualities of immersion and interactivity that these technologies offer can promote active experiential learning (Bogosian et al. 2020). Immersion in a VR environment can be designed to facilitate knowledge as a product of experience. At the same time, interactivity can provide targeted feedback to facilitate experimentation and exploration [12]. An effective game design strategy to increase player motivation is to enhance the User Interface (UI) and User Experience (UX) capabilities with entertaining and engaging features to reinforce a sense of embodiment and improve interaction with the learning content. These strategies include introducing a compelling storyline, customizable virtual environments, high-quality graphics, and game characters. Responsive virtual environments that allow learners to navigate simulated scenarios using spatial cues and modify the learning context based on their own needs offer further enhancements to immersive and interactive experiential learning.

To design the training curriculum, we studied existing AEC and robotics courses, interviewed academics and professionals, consulted with STEM curriculum developers and evaluators, and worked with AI experts. Our immersive and interactive curriculum offers a secure space for conducting exercises that might otherwise be costly and potentially dangerous in real life. Each lesson is developed around scenarios to provide real-world context typically encountered in construction, manufacturing, or material handling. This use case-centered approach offers learners the flexibility to develop skills with the equipment, procedures, and tasks most relevant to their individual needs. We designed and tested three VR mini-courses in 2020. Each course included 24 h of combined in-person and online training offered to AEC professionals and University students representing a total of 39 students, 35 of whom received micro-credential badges. The three courses that we designed and tested include 1) *Computational Thinking for Robotics Processes*, 2) *Fundamentals of Robotic Arms*, 3) *Parametric Design for Robotics Processes.*

For the NSF C-Accel program, we developed a VR Minimum Viable Product (MVP) for testing our curriculum by building a narrative to demonstrate various aspects of the Robotics Academy as a VR experience. For this task, we researched UI design to accommodate trainees with diverse backgrounds and technical abilities, and we conducted UX research to develop a comfortable and engaging VR experience. We then built 3D assets for various aspects of the Robotics Academy platform and established a workflow for translating traditional in-person lessons to VR lessons, implementing Microsoft's Azure cognitive services for voice recognition and text-to-speech. Finally, we enabled customized curricular sequences and evaluation by using Deep Learning frameworks such as PyTorch and TensorFlow allowing AI to estimate user learning progress based on data gathered from profile settings, biometrics, and UI interactions during the lessons.

We tested our MVP with trainees of different backgrounds and have planned MVP revisions for further playtesting and evaluation. To train our AI algorithms for a customized experience, we learned that the user should have the option to utilize different profile setting modes, such as uploading an existing CV or extracting data from existing social media platforms such as LinkedIn. We have determined that optional voice input and output is preferred by almost all play-testers; thus, it will become a permanent feature with the ability to modify voice output. The ability of users to provide feedback on the lessons reinforces their sense of community and engagement with the platform; thus, we will enable this feature through user forums in future testing. We are now poised for phase 2 testing with a modified curriculum in a broader pool of up to 500 students beginning in Spring 2021.

Curriculum testing has significantly informed the design and structure of our training platform. The pilot courses allowed us to identify curriculum areas that are best suited for VR integration. In our experience, using practical AEC applications for teaching content in simulation format increases user interest and learning performance, and we have found that activities that allow a student to manipulate an actual robotic arm from the VR environment both increases hands-on experience with robots and bridges the divide between virtual and physical worlds. Participant feedback indicated that a mixture of content delivery methods was ideal for supporting learner needs and preferences. In response, our course content platform is being expanded to include multiple formats that encourage independent and supported student tasks, including further integration of VR and Augmented Reality (AR), videos, animations, audiovisual guides, discussion boards, and online communities.

4 Guides to Further Research

A fundamental aim of our research and development of robotics curriculum and a VR learning environment is to Incorporate Artificial Intelligence (AI) more comprehensively, including Machine Learning (ML) and Natural Language Processing (NLP), to monitor and respond to progress toward learning objectives of each individual learner. To develop an AI-assisted Adaptive Learning Systems (ALS), our curriculum is designed as a series of modular microlessons allowing learning content to be strategically rearranged based on learner performance. Each microlesson provides the potential for granular assessment that can be aggregated into personalized lesson plans. As an ALS is trained by incorporating learner performance data from the lessons, it can reconfigure lesson plans and suggest individualized learning experiences for each student to improve learning outcomes.

Recent advances in data-driven AI models offer the possibility to transform the efficacy of ALS by gathering and dynamically responding to learner input as well as biometric feedback. An ALS typically uses static rules crafted by researchers and learning specialists to adjust and individualize learning content based on pre-defined objectives, preferences, and knowledge in order to provide an efficient, effective, and personalized learning experience. In our next phase of developing the curriculum, we will deploy novel ML models, including deep neural networks to dynamically respond to learner status and performance, and individualize lesson plans to improve the learning experience. In order to effectively train ML models, the ALS monitors learners in the virtual environment and integrates static data such as personal information and learning status with dynamic data, including biometrics, performance data, and verbalization of specific concepts. The ALS can use these data to provide more targeted feedback and recommendations for individualized learning plans.

Until recently, customization of immersive environments for individual learning patterns has been quite limited. Many existing VR applications have not provided enough scenarios and interactivity for learning complex tasks, relying on the same routines and simulations for all learners [13]. Integration of AI with immersive technologies provides the opportunity for developing intelligent environments with high levels of realism and interactivity to support the quality of perception, learning, and communication through natural language and reasoning [14]. Educational games exhibit qualities that can facilitate adaptive learning by personalizing the environment.

Building on the inherent capacity of game engines for immersive interaction design and ALS, we will closely monitor learner progress by collecting and analyzing three types of data: 1) *Personal Data* including information about the personal and professional background, learning objectives, learning preference, skill level, and experience of each learner; 2) *Performance Data* including activity completion time, number of attempts, test scores, level of engagement with course content to complete a task, and conceptual progress; 3) *Biometric Data* including information extracted from head-mounted displays as well as add-on sensors to track eye-gaze, heartbeat, body movement, and other haptic interactions. By creating a workflow for high-resolution temporal data collection and processing of these three data types, we will apply ML and NLP algorithms for ALS evaluation to generate an accurate user profile for each learner. During the learning process, behavior and status can be monitored and processed by the ALS, allowing for customized feedback and automated lesson planning.

By deploying these emerging technologies to the project of creating and delivering automation and robotics curriculum in a VR environment, we hope to not only improve access to automation technology knowledge but also to contribute to the science of learning and knowledge production. At the same time, we hope that our efforts to improve educational tools for 21st Century workers in robotics can help to provide the necessary momentum for the AEC industry sector to modernize its approach to designing, building, and managing the built environment.

Acknowledgments. This material is based upon work supported by the National Science Foundation under Grant No. OIA-1937019. Any opinions, findings, and conclusions or recommendations expressed in this material are those of the author(s) and do not necessarily reflect the views of the National Science Foundation.

References

1. Bock, T.: The future of construction automation: technological disruption and the upcoming ubiquity of robotics. Autom. Constr. **59**, 113–121 (2015)

2. United States Bureau of Labor Statistics (2020). https://www.bls.gov
3. Chui, M., Mischke, J.: The Impact and Opportunities of Automation in Construction. McKinsey and Company (2019). https://www.mckinsey.com
4. Jonassen, D.: Evaluating constructivist learning. In: Educational Technology, vol. 28 (1991)
5. Dewey, J.: Experience and Education. Kappa Delta Pi, New York (1938)
6. Davis, D., Chen, G., Hauff, C., Houben, G.: Activating learning at scale: a review of innovations in online learning strategies. Comput. Educ. **125**, 327–344 (2018)
7. Martirosov, S., Kopecek, P.: Virtual reality and its influence on training and education. In: Katalinic, B. (ed.) Proceedings of the 28th DAAAM International Symposium, pp. 708—717. DAAAM International, Vienna (2017)
8. Mantovani, F.: 12 VR learning: potential and challenges for the use of 3D environments in education and training. In: Riva, G., Galimberti, C. (eds.) Towards Cyberpsychology: Mind, Cognition, and Society in the Internet Age, pp. 207–224 (2001)
9. Beck, D.: Augmented and virtual reality in education: immersive learning research. J. Educ. Comput. Res. **57**(7), 1619–1625 (2019)
10. Dede, C.: Immersive interfaces for engagement and learning. Science **323**, 66–69 (2009)
11. Bogosian, B., Bobadilla, L., Alonso, M., Elias, A., Perez, G., Alhaffar, H., Vassigh, S.: Towards an immersive robotics training for the future of architecture, engineering, and construction workforce. In: IEEE World Conference on Engineering Education (EDUNINE) (2020)
12. Jarmon, L.: Virtual world teaching, experiential learning, and assessment: an interdisciplinary communication course in second life. Comput. Educ. **53**(1), 169–182 (2009)
13. Gonzalez-Franco, M., Cermeron, J., Li, K., Pizarro, R., Thorn, J., Hutabarat, W., Hannah, P., Tiwari, A., Bermell-Garcia, P.: Immersive augmented reality training for complex manufacturing scenarios arXiv:1602.01944 (2016)
14. Mateus, S., Branch, J.: Intelligent virtual environment using artificial neural networks. In: Lackey, S., Chen, J. (eds.) VAMR 2017. LNCS, vol. 10280, pp. 43–53. Springer, Cham (2017). https://doi.org/10.1007/978-3-319-57987-0_4

I Feel the Need for Speed: Empirical Evidence of the Effectiveness of VR Training Technology on Knowledge and Skill Acquisition

Victoria Trabysh[1](✉), T. J. Feldhacker[1], Cait Rizzardo[2], Jake Johnson[1], Lisa Tripp[1], Kent Halverson[2], and Antonio Gonzalez III[1]

[1] United States Air Force, Randolph Air Force Base, Randolph, TX, USA
{victoria.trabysh,theodore.feldhacker,jake.johnson.12, lisa.tripp.1,antonio.gonzalez.20}@us.af.mil
[2] Aptima, Inc., 12 Gill Street, Suite 1400, Woburn, MA 01801, USA
{crizzardo,khalverson}@aptima.com

Abstract. The U.S. Air Force (AF) is invested in improving the effectiveness of training. In this pursuit, the AF recently explored the application of augmented and virtual reality (AR/VR) for aircraft maintenance technical training via a prototype course. The present study evaluated the effectiveness of this prototype aircraft maintenance training curriculum and training environment that leveraged modern AR/VR capabilities, instructional methodologies, and assessment techniques. This paper describes a quasi-experimental study comparing the AR/VR-based course with the traditional aircraft maintenance course taught using conventional instructional methods, such as classroom lectures and hands-on demonstrations at part-task training stations. Results indicate that students in the AR/VR course completed the course in half the time while scoring slightly lower on knowledge and skill assessments. This study offers both exciting insights into the potential of VR utilization in the training environment and a better understanding of the challenges involved in future integration of VR technologies into training.

Keywords: U.S. Air Force · AR/VR technology · AR/VR-based course · Training environment · Aircraft maintenance · Knowledge · Skills

1 Introduction

"Our Air Force must accelerate change to control and exploit the air domain to the standard the Nation expects and requires from us. If we don't change – if we fail to adapt – we risk losing the certainty with which we have defended our national interests for decades. We risk losing a high-end fight." - CSAF General Charles Q. Brown

Training environments across government, industry, and academia are undergoing high levels of modernization due to continuous advancements in technological capabilities. The rapid evolution of computer-based learning technologies—specifically augmented and virtual reality (AR/VR)—has created exciting opportunities for enhancing

S. Nazir et al. (Eds.): AHFE 2021, LNNS 269, pp. 19–26, 2021.
https://doi.org/10.1007/978-3-030-80000-0_3

the training experience in various domains. In the medical field, the use of VR-integrated technology has generated evidence of effectiveness, reducing operating time and augmenting patient outcomes associated with laparoscopic surgeries [4]. In pilot training, full-motion simulators with integrated AR/VR capabilities are being used to teach pilots to fly, while significantly reducing the number of hours needed in a real aircraft [5]. Similar initiatives have reliably provided evidence of enhanced learning outcomes through increased interaction with representative environments via simulation [1]. This modernization effort has become a way of life in the U.S. Air Force (AF), which is committed to exploring the application of such technologies in re-envisioning training.

For the AF's Air Education and Training Command (AETC), this exploration includes endeavors such as the AR/VR-based aircraft maintenance technical training course model, one of several initiatives evaluating the effectiveness of training models that leverage virtual reality environments and the inherent changes in flexibility and scalability to the overall training curriculum. While traditional aircraft maintenance training has evolved, constraints associated with time, money, training devices, manpower, and available facilities still prove challenging in increasing production rates and quality of aircraft maintainers. In fact, a report by the Government Accounting Office in 2019 revealed a significant skill gap in experienced aircraft maintainers and emphasized that this problem cannot be quickly resolved due to the lengthy time it takes to train aircraft maintainers [2]. This suggests that training solutions that can accelerate knowledge and skill acquisition are desperately needed.

The traditional aircraft maintenance technical training course takes 21–23 days to complete and is delivered through conventional, lecture-based instructional methods. The lessons progress at a single, standardized pace, regardless of students' variable learning speeds, with little room for individualized instruction. Although given live demonstrations of hands-on tasks, students have minimal time to practice these hands-on skills. The shortcomings and challenges of the traditional course may be overcome by leveraging AR/VR and its attributes, as these technologies have proven promising in supporting knowledge acquisition and providing students with a platform to practice complex skills in high fidelity, low cost, and low risk learning environments [3]. By utilizing such technology to redesign the aircraft maintenance training model, the goal was to increase the rate of student learning, while still achieving an acceptable level of proficiency in both knowledge and hands-on skills. One concern was that students may, however, experience some short-term physiological effects, such as eye strain, due to prolonged exposure to AR/VR. This paper discusses these considerations and the impacts of the prototype training model for aircraft maintainer fundamentals: a single step in the ongoing efforts to integrate and test technologies for the benefit of training fundamentals and beyond.

2 AR/VR-Based Aircraft Maintenance Course

The AR/VR-based aircraft maintenance training initiative created a unique opportunity to study the effectiveness of utilizing AR/VR technology to support the aircraft maintenance technical training program. The prototype system completely redesigned the aircraft maintenance training curriculum and training environment, transforming the traditional lecture-based classroom environment into an AR/VR-equipped environment configured for more efficient training. The AR/VR-based course model utilized commercially available technologies to create an immersive learning environment where students could perform virtual task practice and unlimited rehearsal through a hands-on approach. This allowed students to accumulate knowledge rooted in application and in-context of the mission, such that they were "experiencing" training in both a visual and contextual format. A self-paced learning environment was also leveraged, freeing students from the traditional constraints of the classroom, and thereby creating greater access to learning at a time and place most advantageous to the students. Students were given tablets that contained all course reading materials, as well as desktop versions of the VR lessons to promote study outside of the classroom setting. Thus, students were able to move at their own pace within the learning environment and progress through the course at a rate commensurate with their learning speed.

As a result of this course model, the extent to which instructors could allocate individualized attention was enhanced. Instructors were liberated from lecturing for extended periods of time, allowing them to focus on individual student needs. This dramatically shifted the interactions between the student and the instructor. Rather than the traditional instructor-lecturer role, the AR/VR course employed instructors in a manner that more closely resembles the trainer and technical expert role they play in the field. As a means to easily identify where these student needs were present, a course dashboard with advanced student performance monitoring capabilities provided instructors real-time data on student learning progression, such as where they were at in the course and what grades they had received. Aside from course features, an efficient, rolling on-boarding process was implemented. This process enabled a greater number of technical training graduates in a shorter period of time; once a student graduated, the next student in the queue was notified of their start date and time. These processes and advanced technologies aimed at infusing the best techniques, available learning science, and student performance tracking to increase instructor awareness of the learner's status in real time, as well as catalyze individualized assistance when necessary.

The creation of the course followed an iterative approach to product development, applying a test-learn-adjust-expand-repeat pattern. The approach established a sequence of validation tests categorized into phases. From simply exploring the VR development process in the first phase, to later, more refined goals, the iteration process ultimately led to a full course prototype in Phase III. This piloting approach allowed course developers to detect limitations or failures early, learn from them, and improve the course prior to the next iteration. The biggest challenge in earlier phases was the traditional courses' inability to capture student performance in a way that allowed for comparative analyses. The traditional course relies on a pass/fail grading rubric that was discarded once the student progressed on to the next objective. This lack of performance knowledge led to the study teams' development of a procedure designed to compare all aspects of aircraft maintenance training.

3 Procedure

The AR/VR-based course was incorporated into the AF's operational training pipeline for 45 days during the fall of 2020. During this period, students attending aircraft maintenance training using the AR/VR course model were required to successfully complete learning assessments identical to those of students attending the traditional course. Study participants included 73 students (64 men and 9 women, with a mean age of 20.6 years and an age range of 17–30 years) attending Fundamentals of Aircraft Maintenance Technical Training. Participants were randomly selected based on their technical training start date and assigned to one of three groups: two control groups (1 and 2) enrolled in the traditional training course and the treatment group enrolled in the AR/VR-based training course. Control Group 2 was created to account for potential instructor influence; both trainees and instructors in Control Group 2 were unaware that they were part of the study until the conclusion of the course. Because this study took place in the field, with constraints beyond the control of the researchers, a design was required that could control for as many threats to validity as possible. The nature of the treatment (i.e., innovative training methods) made it impossible to prevent participating students from knowing that they were subjects to a new training approach. The design involved pre- and post-test observations in the form of self-report surveys. Throughout both training courses, student performance data and attitudinal data were systematically collected as part of the respective training curricula.

Student performance was measured for both knowledge and hands-on performance throughout the training program. The study team established two standardized assessment instruments: (1) a cumulative *end of course* (EoC) *exam* consisting of test questions from the approved exam test bank, and (2) the *job task evaluation rubric* (JTER) based on the approved AF Form 98 for assessing hands-on task proficiency. The rating scales for task performance included the following values: 0 = "Could not complete the task"; 1 = "Substantial assistance,"; 2 = "Moderate assistance," 3 = "Minimal assistance,"; and 4 = "No assistance needed." Rating scales for task safety included the following values: 0 = "No safety violations"; 1 = "Injury involving first aid and/or minimal threat to property"; 2 = "injury involving lost day and/or minor property damage"; 3 = "Permanent partial disability and/or major property damage"; and 4 = "Death, permanent total disability and/or loss of facility or asset." These newly established assessment instruments allowed numerous analyses, each providing insight into different aspects of the new training technologies and methodologies. The knowledge portion was broken into progress checks, block/module assessments, and an EoC exam. Hands-on skill performance was measured after students completed a module and observed a demonstration (live demonstration in the traditional course; virtual demonstration in the AR/VR course). Each student was then required to demonstrate his or her competence by completing a hands-on task at a part-task training station in the hangar bay.

Few studies, if any, have explored the effects of prolonged exposure to AR/VR, specifically in the capacity of a full course replacement (in this case, several hours per day, on average, for the duration of the course). As a preemptive mitigation strategy to avoid any physiological issues arising from prolonged exposure to VR, students were encouraged to monitor their time in the environment and take a minimum of a ten minute break for every 50 min spent in VR. Students also completed a symptoms survey at the conclusion of each training day, allowing researchers to monitor any adverse symptoms that students may be experiencing over time.

4 Results

The utilization of the self-paced, AR/VR course model in support of aircraft maintenance technical training resulted in students completing the course in an average of 12.4 academic days (with a range of 9 to 16 days), reducing training time by 46.1% on average.

The substantial time savings was associated with a mean difference of 7.7% on the EoC exam: traditional student scores (M = 73.9%, SD = 11.0%) and the self-paced, AR/VR student scores (M = 66.2%, SD = 11.9%). While this difference was not statistically significant with the current sample size ($t(56.2) = 2.63$, $p = 0.09$), the practical significance will be monitored as these students continue on to subsequent phases of training.

Hands-on task performance was evaluated using instructor observer ratings for skill and safety violations using the established AF Form 98. A ceiling effect was observed with both groups scoring at levels near the top of the rating scale (i.e., "No assistance needed"), suggesting both courses achieved acceptable student proficiency. As is common when measuring safety metrics, instructors observed a low incident rate (i.e., very few observed unsafe behaviors) during hands-on tasks, suggesting AR/VR students performed hands-on tasks with similar attention paid to safety considerations and procedures.

Early on, over 50% of students reported mild eye strain symptoms due to immersion in the virtual environment, however, this lessened to under 10% as the course progressed. Overall, there were very few instances in which other physiological effects, such as cyber sickness, headache, nausea, etc. were a factor in the study.

5 Conclusions

Overall, the AR/VR prototype course was a successful method for presenting learning material, such that 93.5% of the 32 students were able to meet or exceed the minimum requirements in the aircraft maintenance training course learning objectives in half the time of the traditional course students. The AR/VR-based course demonstrated that it is possible to fundamentally shift from an instructor-led teaching approach to a student-led learning approach and gain efficiencies. Although the AR/VR course students did not perform as well relative to the traditional course students, they were able to train to the minimum acceptable criteria in far less time, producing remarkably positive results when examining students' progress through the course.

Successful training transfer of skills learned in the AR/VR environment to the real-world environment is evident through hands-on task performance. AR/VR course students were able to achieve only slightly lower proficiency levels than traditional course students even without any prior live hands-on practice, deeming virtual task rehearsal effective. Additionally, the rolling onboarding process proved extremely successful in reducing time wasted through students waiting to begin the course; as one student graduated, another student began the course in their place. These efficiencies were gained through leveraging AR/VR, establishing hands-on evaluation kiosks, and allowing students to take control of their learning and progress through the course at a rate commensurate with their learning speed.

Although the self-paced nature of the course resulted in a significant decrease in total training time needed for technical training completion, the system did not reduce the instructor to student ratio. The traditional course effectively operates at a 16:1 student to instructor ratio. Due to the self-paced nature of the AR/VR course, the initial student to instructor ratio is significantly lower. During the study, the ratio was set at 2:1 (two primary instructors and six hands-on evaluators), however, this 2:1 ratio is not indicative of what the student-instructor ratio requirement would be if the AR/VR course was scaled. Unlike the set ratio in the traditional course, AR/VR becomes more economical as the student population increases; instructor requirements in the AR/VR course are based on the number of VR machines and hands-on evaluation devices rather than the number of students. Additionally, because instructors were no longer required to deliver lectured content, they were able to more effectively observe, assist, and coach individual students on an as needed basis in ways not permitted by the traditional course.

Lastly, student performance monitoring capabilities proved beneficial to the AR/VR course model, particularly regarding instructor awareness of each students' performance in real-time. These include, but are not limited to, decisional report capabilities and the ability to detect low performers. These capabilities are critical in ensuring students receive tailored learning experiences to overcome the absence of a conventional instructor. This component successfully empowered students to control their learning experience in a way that permitted them to advance when they were ready. However, a critical limitation in system control measures was exploited by lower performing students advancing through objectives without achieving the required learning outcome, ultimately allowing them to arrive at some hands-on and knowledge-based assessments unprepared.

6 Recommendations/Lessons Learned

Despite the success of utilizing the AR/VR prototype course in replacement of the traditional aircraft maintenance technical training course, consideration of the tradeoffs and areas requiring improvement are also critical. Further refinement in course content and continued development in AR/VR artificial intelligence to control student progress will be vital prior to full integration of this training platform. Although time efficiency and self-efficacy are important benefits resulting from a self-paced learning style, observation and student feedback suggested that many students felt compelled to finish the course as quickly as possible. The unrestricted self-paced nature of the course enabled students to rush or move through the material as quickly as possible, which lead students to miss key learning objectives and fail to perform repetitive practice when necessary. In the future, performance deficiencies could be mitigated through employment of additional progression "gates," preventing students from advancing further in the course until a specific level of proficiency has been achieved. In support of this strategy, additional gates were added to the Phase III iteration that were not present in Phase II, potentially one explanation for the 15% academic performance increase witnessed in Phase III.

In attempt to avoid any physiological issues arising from prolonged exposure to VR, students were encouraged to monitor their own time in the environment and take a minimum of a ten minute break for every 50 min spent in VR. Again, however, due to the self-paced nature of the course, some students did not diligently monitor their time spent in the VR and continued to press on with the course material. To address this, the system could incorporate additional features that restrict students to a certain duration in the environment, followed by a mandatory break before further progression is enabled. Although it is critical to consider the adverse physiological effects that prolonged AR/VR exposure can have and take appropriate precautions, results of the present study suggest that utilizing AR/VR to this capacity likely has benefits that far outweigh the risks of limited and mild symptoms, such as eye strain, early on in the course.

Given the significant amount of time saved through the utilization of AR/VR as a full course replacement, it is vital to consider how that time can be optimally reinvested. While AF senior leaders have a multitude of options to consider, there are two particular approaches that present themselves as most advantageous. 1) AR/VR training allows graduates to be produced at a faster rate, therefore a greater number of aircraft maintainers can be produced overall, and/or 2) data collected through the AR/VR course study and other similar initiatives can be used to identify and target gaps in the course. Identification of these gaps allows for further course refinement in order to maximize student proficiency levels. In addition to course refinements, individualized opportunities for further learning could exist based on student performance feedback, allowing them to use the additional time to cycle back through content in which they may have struggled with the most and/or delve deeper into a particular area.

Although average performance scores were not on par with those from traditional course students, the AR/VR-based course allowed for other beneficial learning methods to be utilized that are not generally present in a conventional classroom setting. For example, some students took advantage of additional opportunities to review course material with their instructor prior to completing a hands-on evaluation or a knowledge assessment. In order to capture such instances more accurately, a more comprehensive analysis of student-instructor interactions could be useful for pursuing more in depth analyses of the course and avoiding an overreliance on performance scores. In addition to this, further opportunity for student-student interactions could be beneficial. Student-instructor and student-student interactions were not mandated throughout the AR/VR course; however, glimpses of these student-initiated interactions highlighted the potential role they could play in further improving future iterations of the course.

Overall, results of this study indicate that there are significant efficiencies to be gained through employment of AR/VR training. By unhinging the learning curriculum, the prototype AR/VR course proved effective as an alternative to the traditional course. Even in prototype form, the self-paced, immersive approach generated the required proficiency to meet the current knowledge and skill requirements outlined in the AF's aircraft maintenance training curriculum. In its current form, there is no evidence that the prototype course produces higher performing Airmen, however, minor tweaks to the AR/VR course, as well as wisely reinvesting time could improve student knowledge and skill proficiency while retaining significant efficiency gains. Researchers recommend monitoring the long-term impact of the course in regards to both training transfer and return on investment, specifically with an eye toward the impact on time spent producing proficient aircraft maintainers.

References

1. Beckmann, J., Menke, K., Weber, P.: Holistic evaluation of AR/VR trainings in the ARSUL project. In: Proceedings of INTED2019 Conference, March 2019, Valencia, Spain (2019). ISBN 978-84-09-08619-1
2. Inhofe, J.M., Reed, J., Smith, A., Thornberry, M.: Military personnel: strategy needed to improve retention of experienced air force aircraft maintainers. United States Government Accountability Office (2019). https://www.gao.gov/assets/700/696747.pdf
3. Mayberry, C.R., Jaszlics, S., Stottlemyer, G., Fritz, G.: Augmented reality training application for C-130 aircrew training systems. M&S J. **35**, 36–46 (2014)
4. Nagendran, M., Gurusamy, K.S., Aggarwal, R., Loizidou, M., Davidson, B.R.: Virtual reality training for surgical trainees in laparoscopic surgery. Cochrane Database Syst. Rev. **2013**(8), CD006575 (2013). https://doi.org/10.1002/14651858.CD006575.pub3
5. Paur, J.: Learning to fly in a simulator first, airplane second. Wired (2011). https://www.wired.com/2011/11/learning-to-fly-in-a-simulator-first-airplane-second/

Digital Creative Abilities for Achieving Digital Maturity

Mara Rita Canina(✉) and Carmen Bruno

IDEActivity Center, Design Department, Politecnico di Milano, Via Durando 38/A, 20158 Milan, Italy
{marita.canina,carmen.bruno}@polimi.it

Abstract. The ongoing digital evolution is changing people's mindsets as well as behavioural and social attitudes, creating both significant opportunities and threats that need to be managed properly. In a near future, there will be a strong demand for Digital Creative Abilities (DCA) which include both digital skills along with human skills (e.g. problem solving, strategic and creative thinking, emotional intelligence, etc.). Their development allows achieving a Digital Maturity that means enabling people to continuously adapt to a changing digital landscape, learning how to use and collaborate with digital technologies to serve human needs. The paper presents the DC4DM model built to empower DCA to understand and anticipate the opportunities and threats offered by the digital evolution developing a strategic approach to the adoption and application of such technology. Training digital wise professionals to drive the digital transformation has become a mission for design education.

Keywords: Digital Maturity · Creativity · Design mindset · Future skills · Human skills

1 Introduction

Design and engineer education urgently need to co-evolve with the human, technological and cultural evolution of the digital era we are facing. Indeed, digital transformation is changing people's mindsets, behavioural and social attitudes, impacting every sector of our society [1]. Artificial Intelligence, Robotics, Virtual and Augmented Reality, and other emerging technologies, are impacting the process of creating and innovating, transforming the industrial economy and the associated job market, creating significant opportunities and threats that need to be properly understood, managed and guided.

According to the WEF's *Future of Jobs Report* (2020) [2], 50% of all employees will need reskilling by 2025, as the "double-disruption" of the economic impacts of the pandemic and adoption of technology increases, transforming jobs. At least 133 million new roles may emerge globally because of the new division of labour between humans, machines and algorithms. This will require people to develop new skills, and education to improve their training. In a near future, there will be a strong demand for digital skills (e.g. information skills, programming and app development) as declared in the

S. Nazir et al. (Eds.): AHFE 2021, LNNS 269, pp. 27–35, 2021.
https://doi.org/10.1007/978-3-030-80000-0_4

Digital Skills and Jobs Coalition launched by the European Commission [3], along with human skills that computers can't easily master such as complex problem solving, strategic and creative thinking, critical thinking, emotional intelligence, communication and negotiation, relationship and network building abilities. Critical thinking and problem-solving top the list of skills employees and the newly emerging this year are skills in self-management such as active learning, resilience, stress tolerance and flexibility.

This set of fundamental skills can be defined in a unique way as Digital Creative Abilities (DCA) and represent the levers to be activated and enhanced to allow individuals and teams to express their maximum creative potential. Their development allows managing the digital transition achieving a Digital Maturity (DM) [4] that means enabling people to continuously adapt to a changing digital landscape, learning how to collaborate with digital technologies and how to use them to serve the human needs in any field.

Leading the transformation and fostering the upgrade of such set of skills is, therefore, a fundamental requirement for digital transformation and for post-pandemic economic recovery in all sectors, as also proposed by the Reskilling Revolution[1]. Companies have a primary benefit in transforming the skills of their human capital, empowering them to activate and express their creative potential, to think and act in a non-predictable digital world.

The nurture and development of the DCA of the future professionals is a compulsory step for education that will allows companies to manage the digital transition achieving a DM that requires adjusting and implementing the entire organization to continuously adapt to an evolving digital landscape. People empowerment is intrinsic and transversal to the five key practices that companies should implement to become digitally mature [4]: (a) increasing collaboration organized in cross-functional teams, (b) stimulating workplace innovation, digitally-minded cultures, visions and experiences, supporting changes by committing resources (c) renewing their approach to talent, creating enjoyable environments where workers can continuously learn, (d) scaling small digital experiments into enterprise-wide initiatives that have business impact, and (e) planning a long term vision/strategy to face the changes emerging in the digital landscape. They are enabling elements that activate the process of digital transformation that leads to the increase of the digital maturity level.

Therefore, training and guiding the development of the DCA to manage the digital transition strategically and maturely with an approach centred on people's needs, is a fundamental requirement in this complex digital transformation and the focus on which the model, presented in this paper, is based.

Based on this knowledge, the paper presents the Digital Creativity for Digital Maturity (DC4DM) model aimed at enhancing people's creative and design skills for reaching a Digital Maturity. The DC4DM model is the core of the Erasmus + project, "Digital Creativity for developing Digital Maturity future skills". The project aims to implement and spread a new scenario of cross-functional teams' education for training future professionals to face the complex real-world challenges brought by the digital transformation.

[1] Reskilling Revolution initiative aims to provide one billion people with better education, skills and jobs by 2030 https://widgets.weforum.org/reskillingrevolution/index.html.

2 Skills Gap: The DC4DM Project

When we specifically refer to new emerging technologies it is largely recognized the need to strategically drive them. The DCA represents the levers to be activated and enhanced to allow teams to express their maximum creative potential to guide digital transition. This can be done through 4 main steps that (i) Explore how a specific digital technology is impacting the human being (ii) Understand how those impacts influence the DCA and the design process (iii) Define actions to enhance the DCA and the design process according to the identified needs for innovation (iv) Design the actions to empower them by exploiting the potentialities of digital tech.

The DC4DM project aim is to train digital wise professionals able to drive the digital transformation through an open learning approach based on real-world challenges, fostering sharing of cross-cultural knowledge from academia to industry and vice-versa.

Based on this ground, the project is implementing, applying, and spreading within a European network of Universities, SMEs and Startups, Business Incubators, a human-centred design model to facilitate and strategically guide the ongoing process of digital evolution for achieving DM through people creative enhancement.The operative application of the model and its tools facilitates and enhances the cognitive, emotional, and social skills that intervene in the creative design process to empower the generation of innovative technological ideas/solutions through a design approach and mindset.

Specifically, the project is targeting the following three main priorities for action. The first one is tackling skills gaps and mismatches promoting excellence in skills development by (i) *supporting the development of learning outcomes-oriented curricula* that better meet the learning needs of students, while also being relevant for the labour market and for the wider society; (ii) *promoting project-based activities that address real-world problems and applications*, offering international mobility and inquiry-based and ICT enriched learning, collaborative practices, including university-business cooperation. The second priority is supporting opportunities for all in acquiring and developing key competencies, especially the transversal skill needed in the fast-changing world such as entrepreneurial mindset, critical thinking, creativity, collaboration, digital literacy, etc. in order to foster employability, socio-educational and personal development, as well as participation in civic and social life. Finally, the third one is generating innovative practices in the digital age developing open educational resources, free and open textbook, and innovative digital tools for teaching.

3 DC4DM Model

Through basic research and co-design actions developed in the previous EU project Digital DIY [5, 6], the IDEActivity team from Politecnico di Milano has implemented the creativity-driven design model for empowering the DCA in the creative design process. Indeed, the model promotes the development of learners' creativity, design and entrepreneurial skills with tools and methods to adapt and advance new collaborative practices for reaching a DM. It integrates digital technologies, creative process and design methods to boost employability, companies' competitiveness and innovation potential in different contexts of application. The creativity-driven design model has been built (a) to empower DCA of cross-functional teams of design, engineers, and management students, (b) to transfer a design and creative mindset, and (c) to guide the design of innovative digital solutions in different fields.

The model background is based on Digital Creativity (DC), defined as the human ability to create an innovative and original digital outcome and to strategically exploiting the opportunity of digital technologies [7]. This ability is driven and empowered by the interconnection of different human factors that are shaped and influenced by the digital age and its digital technologies [8].

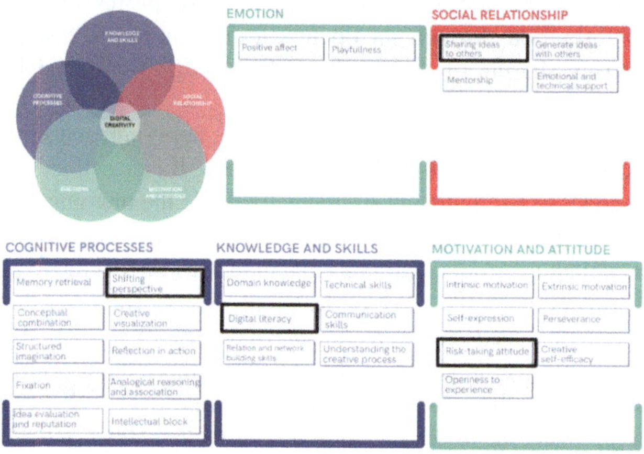

Fig. 1. Digital Creativity components and the factors of the human being selected for each component.

The DC Observatory tool [1, 9], defined and further evolved by the authors, allowed to understand the multiple dimensions of creativity in the digital age and to identify and map the most relevant factors of DC. These cognitive, attitudinal, emotional and social factors of the human being that influence individual DC and therefore the outcome of the creative process have been grouped into the human components essential for creativity [8].

The explorations by IDEActivity team of the influences brought by the digital transition on the human creative abilities, allowed to identify only the factors considered as the most relevant in this digital age. As shown in Fig. 1 the factors are clustered according to the human component they refer to.

For design research and training, this investigation becomes determining if it is structured within a theoretical frame that associates the impact identified and the influenced factors with the design process. Therefore, a DC Framework Tool [8, 10] has been built within the research, to be used as part of the final DC4DM model. The DC Framework Tool (Fig. 2) is developed to deeply understand how the digital transition is influencing the creative design process and its cognitive, emotional and social factors. The framework allowed to map the less rationale creativity factors emphasizing their interconnections with the design process deconstructing it in stages, steps, activities and thinking style.

This tool allows also to define design actions and tools to empower it, enhancing the DC factors that intervene in a specific moment of the process. The DC Observatory and Framework tools represent the fundamental knowledge needed to design actions to empower DCA in the digital age, exploiting the opportunities provided by the digital transition and digital technologies.

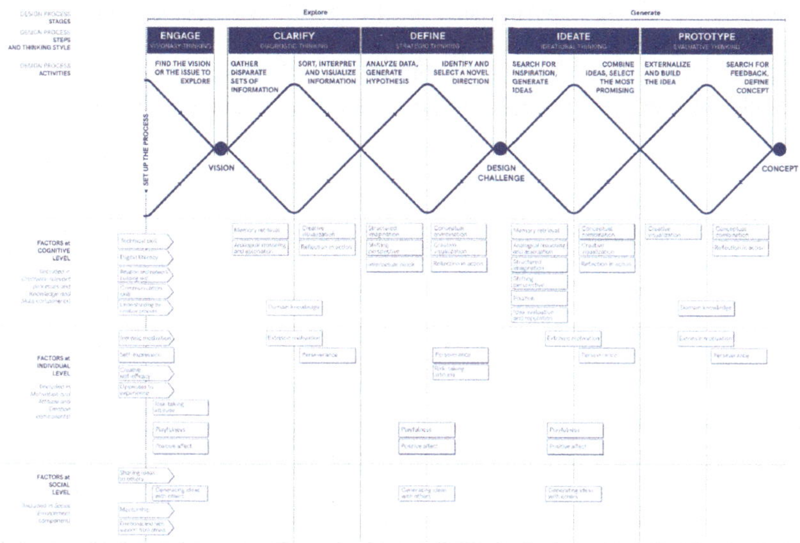

Fig. 2. Factors of the human being selected for each component.

Empowering Digital Creativity means training DC Factors and the Creative Design process to obtain innovative digital outcome for both individual and team.

Starting from the theoretical background it was necessary, for the implementation of the model, to prioritize the key practices adopted within this project that will be developed through different activities. The choice has been made by identifying the specific needs of companies, linking them to the training objectives that the model has to fulfil. For each identified needs the model has a specific objective. Firstly, companies should be able to apply digital tech within their organization to develop new business, to digitalize operation and processes, therefore the model has to enable students to acquire competencies and mindset to understand the potentialities of digital technologies and apply them to design digital solutions with a human-centred approach. Secondly, projects are becoming complex, requiring employees with different functions to work in a team, sharing their knowledge and solving projects together, working on collaborative digital platforms. In order to work in cross-functional teams students need to develop individual abilities of creative self-enhancement, and a digitally-minded culture, but also being able to communicate and share knowledge with others with a different background that means acquire team abilities. And finally, companies are facing a transitional time and they should be competitive even in an uncertain future, planning long term strategies, facing also sustainable and social challenges. The model should therefore allow students to help companies have a long-term strategic vision need to acquire skills in future and anticipatory thinking and develop a mindset helping them to face complex challenges by envisioning future scenarios. According to this analysis of companies' needs, it has been decided to work mainly on developing high-quality cross-functional team curricula integrating design, engineering and management to develop digital wise professional that can drive the digital maturity building strategical scenarios. Obviously, this also requires the integration of (i) scaling small digital experiments into enterprise-wide initiatives that have a business impact through Human-Centered projects addressing digital technologies and (ii) cultivating digitally-minded culture enhancing strategic vision, through the combination of design, technology and entrepreneur mentorship to support students in remaining human-centered while designing digital tech-driven, intelligent products, services, and systems.

Essentially, it requires to shift the concept of team creativity - a synergetic progression that occurs during a social process of sense-making and collaboration where one individual's actions may inspire the team to devise and follow a more creative process to address the problem with higher levels of creativity [11] - into a group of people with different functional specialties or skill sets, responsible for carrying out all phases of a process.

Team creativity is more than just the combined creativity of the individual team participants. There are other factors that influence the contribution of the participants and their interactions. According to Team Creativity Model (TCM) [11] - both individual creativity and shared mental models contribute to team creativity, and the latter act as a mediator between knowledge sharing and team creativity. It is clear that in order to enhance team creativity it is necessary to act in both the individual and social areas, i.e. in the sense of the relationship between collaborating individuals.

The individual creativity includes an individual's propensity to be creative and individual knowledge. Individual knowledge is also impacting knowledge sharing as an individual's propensity to share knowledge and trust within the team. In the digital context, both are affected by Digital Intelligence, the ability to acquire and apply new knowledge and skills related to digital technologies, to improve operational efficiency and outcomes quality [12].

As stated above the DC4DM model is focusing on cross-functional team creativity this implies to recognize the DC framework, named process, as part of the model. Indeed, individual digital creativity and digital knowledge sharing are antecedents of the process through them the digital team creativity is defined (Fig. 3).

Each dimension of the model was outlined and defined in detail to enable the mapping of the factors of creativity included in the 5 components of the DC. As shown in Fig. 3 the factors that were identified and defined through the DC Observatory tool are included. The intention of the paper is to show the process that led to the DC4DM model in relation to the elements that constitute it without going in the details of the factors' mapping. In the final configuration of the model, there are three main dimensions: individual, team and process in which are collected different types of skills useful to allow students to be aware of the digital maturity company needs and grasping the key practices (in the circles).

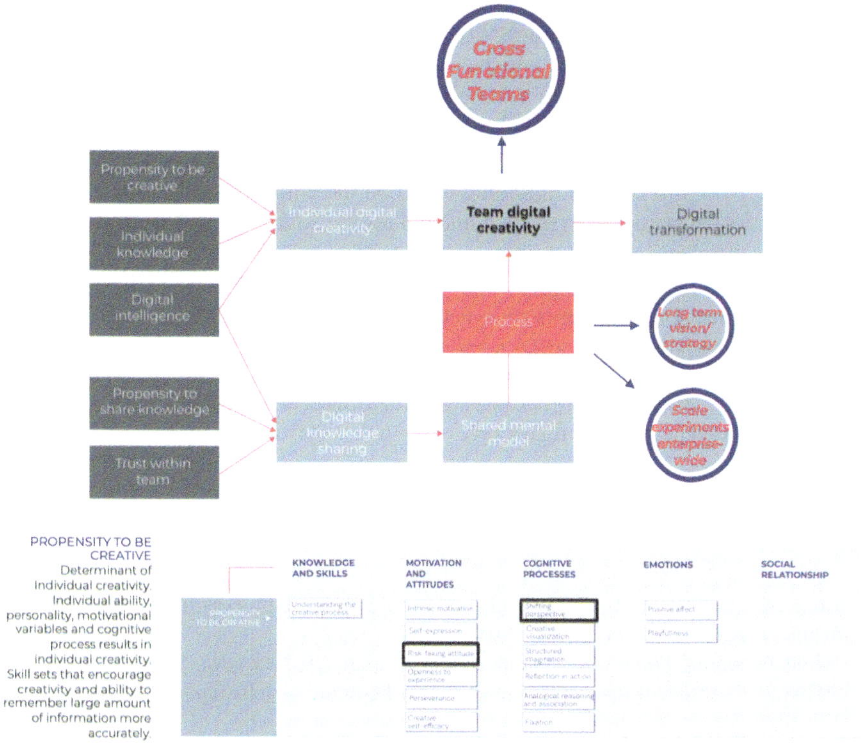

Fig. 3. DC4DM model (top). An example of DC4DM model dimension (down).

4 Reflection and Future Action

The research and the model allow to combine the fragmented data collected within the literature from different fields and disciplines and give them a new interpretation and a new meaning, providing a frame of reference for defining the boundaries of a complex phenomenon such as creativity in the digital age. The DC4DM model integrates the individual and team creativity with the various steps of the design process, enhancing the motivational, cognitive, and attitudinal, social and technological factors that facilitate and guide the ongoing process of digital evolution for supporting organizations in achieving DM. The model is not proposed to be exhaustive or concluded; it is a work in progress to outline the boundaries of cross-functional open innovative teaching method based on the evolving and emerging needs of students and industry, to address the development of DCA and train "digital wise" professional that can drive the DM by building strategical scenarios in the real-world environment.

The complexity of the model is evident, and it needs to be transferred to learners and companies in a simple and immediate way.

A new implementation is therefore envisaged which will work in two directions: the first is to create a new configuration which has a linear organization with respect to the process integrating a pre- and post-process moments; the second is the integration of a dimension, currently not present, which encompasses the skills of social, environmental and ethical responsibility through a strategic vision of the future.

Putting the process at the centre of the DC4DM model means shifting individual digital creativity and digital knowledge sharing to the pre-process phase, which thus become antecedents. This shift enables a training based on the empowerment of individual and team DCAs, which are considered fundamental for the "digital wise" professionals who will then put them into practice in the process itself.

Acknowledgments. This paper presents reflections from the research tasks carried out by the authors within the *Digital Creativity for developing Digital Maturity future skills* Project, which addresses the KA203 - Strategic Partnerships for higher education. This project has received funding from the EU Erasmus+ program under grant agreement No 2020-1-IT02-KA203-079913. The views expressed in this paper do not necessarily reflect the views of the EC.

References

1. Bruno, C., Canina M.: Creativity 4.0. empowering creative process for digitally enhanced people. Des. J. **22**(1), 2119–2131 (2019)
2. World Economic Forum: The Future of Jobs Report 2020. Geneva, Switzerland, October 2020 (2020). https://www.weforum.org/reports/the-future-of-jobs-report-2020. Accessed 10 Feb 2021
3. Shaping Europe's digital future. https://ec.europa.eu/digital-single-market/en. Accessed 10 Feb 2021
4. Kane, C.: Digital maturity, not digital transformation. MIT Sloan Manage. Rev. (2017). https://sloanreview.mit.edu. Accessed 11 Feb 2021
5. Digital Do It Yourself. http://www.didiy.eu/. Accessed 11 Feb 2021
6. Salvia, G., Bruno, C., Canina, M.: Digitally making as an opportunity for skilling and empowerment. In: This Place: Cumulus Association Biannual International Conference, Conference Proceedings, pp. 105–110 (2016)
7. Bruno, C.: Digital Creativity dimension: a new domain for creativity. In: Bruno, C. (ed.) Creativity in the Design Process. Exploring the Influences of the Digital Evolution, Springer Series in Design and Innovation (2021, authorised for publishing)
8. Bruno, C., Canina, M.R.: The Creativity 4.0 Framework. Outlining the influences on creativity in the digital era. Int. J. Bus. Syst. Res. (2021, authorised for publishing)
9. Bruno, C., Canina M.R: Creativity 4.0. empowering creativity in the digital era. In: DS 95: Proceedings of the 21st International Conference on Engineering and Product Design Education (E&PDE 2019), pp. 1–6 (2019)
10. Bruno, C., Canina M.R: Designing a framework to investigate creativity enablers and inhibitors in the digital era. In: Meanings of Design in the Next Era. 4D Conference Proceedings, pp. 162–173 (2019)

11. De Vreede, T., Boughzala, I., De Vreede, G.J., Reiter-Palmon, R.: The team creativity model: an exploratory case study. J. Midwest Assoc. Inf. Syst. **1**, 19 (2017)
12. Boughzala, I., Garmaki, M., Chourabi, O.: Understanding how Digital Intelligence contributes to digital creativity and digital transformation: a Systematic Literature Review. HAL Post-Print (2020)

AI and Learning in the Context of Digital Transformation

Martin Kröll([✉]) and Kristina Burova-Keßler

Institute for Work Science, Ruhr-University Bochum, Universitätsstraße 150, Building NB 1/174, 44801 Bochum, Germany
{martin.kroell,burova.kristina}@rub.de

Abstract. The use of digital solutions and artificial intelligence (AI) in corporate education and training is analyzed in this article using the example of an agile IT-supported learning concept. The focus is on the preliminary findings of the EU project "Career 4.0", including the findings from a usability evaluation of a European learning platform. The central role of constructive feedback processes and user-friendliness for the success of collaborative and self-organized learning and the mentoring process are discussed. At the same time, it is shown which potentials and resistances can occur for the different user groups when using the platform and how these effects can be strengthened or weakened by AI tools in the future.

Keyword: Entrepreneurship education · Constructive feedback · European learning platform · Agile learning · Usability evaluation · Artificial intelligence

1 Starting Point

In order to promote entrepreneurial thinking and action among adolescents as well as their digital skills across Europe's national borders, a learning platform specifically for this purpose has been developed by an international research team within the framework of the EU project "Career 4.0" in the context of the "Erasmus plus - Strategic Partnerships" programme. Also, against the background of the Covid-19 crisis, the best practice-based platform developed in the project offers great value for vocational education and training as well as career guidance at the European level.

With the help of the learning platform and the underlying holistic entrepreneurship education concept, which has been successfully tested in other EU projects, participating adolescents learn how to develop a sustainable professional perspective, create a new marketable service and methods for its marketing. By means of the platform and participating experts and learning guides, they receive constructive, professional and practical feedback.

The learning platform enables learning with the help of agile management methods. Especially against the background of the current challenges in working life, such as digitalization, demographic change, development towards a knowledge society and globalization, the competences of self-organization, self-learning and self-initiative prove to

S. Nazir et al. (Eds.): AHFE 2021, LNNS 269, pp. 36–43, 2021.
https://doi.org/10.1007/978-3-030-80000-0_5

be central. With the help of the platform, adolescents between the age of 16 and 25 are to be enabled to professionalize their personal development plan in co-operation with an internationally active mentor. The primary aim is to match the strengths of the adolescents with the skills expected by the labour market.

In four agile projects that build on each other, participating adolescents can explore their talents in regular feedback loops with the mentors in short learning steps based on clearly defined learning objectives, look for and find new employment opportunities and check these by consulting the concept of an expert hearing. If needed, they are also supported in developing their own business plan.

The learning platform enables direct cross-national communication between mentors and mentees for more effective and efficient feedback processes and better quality management. The adolescents can access the learning offers via the platform flexibly in terms of time and location. An international database establishes contact between adolescents, mentors and experts and enables the exchange and documentation of ideas and experiences. The learning platform has been used in the EU project "Career 4.0" in Germany, Bulgaria, Greece, Italy, Spain and Hungary so far.

Organizers, adolescents (mentees), mentors (learning companions) and experts are involved in the use and development of the platform. They work on a voluntary basis and are usually made up of (former) executives or managers who have many years of practical experience. They are qualified in the project based on a corresponding competence development programme for learning companions on the following topics: Communication management and pathways, design thinking, constructive feedback, agile project management and self-organization management in the age of digitalization.

2 Research Questions

This article deals with the results of the over 2.5 years lasting project "Career 4.0" funded by the EU, which focuses on the development and testing of an agile IT-supported learning concept and addresses the following questions. (1) How do learners arrive at their personal development plan and develop a sustainable career perspective? To achieve this, learners can draw on best practice approaches in the field of Entrepreneurship Education (EE), which have been bundled into a holistic, process- and quality-oriented EE concept according to the European EntreComp Framework. The provision of constructive feedback via this platform is of decisive importance for collaborative and self-directed learning. This leads to the second question: (2) What conditions should be created to ensure the quality of the feedback processes? IT and AI solutions play a crucial role in this context, which leads to the question: (3) To what extent can the process of giving constructive feedback be supported by an IT-supported learning platform in conjunction with AI tools?

A usability evaluation of the learning platform was carried out, the results of which were recorded in an empirical study. Based on scientifically founded questionnaires that deal with the usability (SUS) [1], the design and the structure of learning platforms, a questionnaire adapted to the platform was developed and tested. The focus of this study was the question to what extent the use of selected AI tools improvse the perceived usefulness and the ease of the learning platform.

3 Theoretical Framework

The basis for this project among others forms the Digital Education Plan 2021–2027 of the European Union [2], which refers to the need for further development and establishment of digital education and training. It differentiates between two starting points: (1) the improvement and increasing use of digital technologies, e.g. in the form of apps, platforms or other software tools, (2) the development and expansion of digital competences in order to be more successful in both learning and working and to be able to meet the challenges of the digital transformation.

In this context, the "Career 4.0" project builds on the existing findings on the application of agile approaches in corporate competence development such as "agile learning" [3, 4] and "sprint learning" [5] and develops these further in the form of an agile teaching/learning concept for learning companions and adolescents, taking into account the project needs. In terms of methodology, the learning concept is strongly oriented towards learning in short "learning sprints" and, based on Scrum [6], defines three central roles in the learning process that can be performed by the adolescents and mentors: Learning team, content learning companion, and learning process companions. In the agile learning process, which focuses on self-directed and autonomous group learning processes, great importance is attached to continuous feedback. The retrospectives at the end of the sprints, i.e. after pre-defined work phases, are used for joint reflection and the search for improvement potential [7].

Another foundation is the "Technology-Acceptance-Model" (TAM) developed by Davis, Bagozzi and Warshaw [8], which aims to explain, understand, and predict the acceptance by users of information systems. In the context of the TAM approach, success is measured not least by whether an information system is actually used by users. The TAM is based on the premise that two variables have a decisive influence on the acceptance and usage behaviour of users: (1) the perceived usefulness and (2) the perceived ease associated with the use of the information system. Furthermore, the following four variables influence technology acceptance [9, 10]: (1) experience, (2) voluntariness, (3) age, and (4) gender. In this context, experience can be regarded as the extent of familiarity with the corresponding or similar information systems and is estimated to be one of the central influencing variables for the use of IT and AI systems. For this reason, the two aspects of usefulness and ease were central for the evaluation and further development of the digital learning platform.

4 Usability Evaluation of the Learning Platform

In order to carry out a usability evaluation of the learning platform during the development process, so-called cross-functional teams have been established in each EU country, consisting of selected mentors, adolescents, representatives of the partner organizations and other experts. This approach aimed to include all user roles to involve a wide range of expertise in the evaluation process and to achieve the most comprehensive product evaluation possible. The main task of the teams was to simulate the learning and feedback process with the help of the platform in the sense of a "test learning sprint" to evaluate the usability. Finally, the questionnaire was used, which was developed in the pre-test phase based on the System Usability Scale (SUS) [1] and the results which were evaluated and discussed with the participants.

In the next step, the methodological procedure and the results of the written survey are presented successively, followed by a discussion of the potential and resistance analysis.

4.1 Methodical Approach

In total, the questionnaire for the evaluation of the learning platform comprises 53 questions, including 49 closed and four open questions. The questionnaire is divided into four blocks of questions on the topics: (A) User suitability of the learning platform, (B) Learning content and learning tasks, (C) Personal experience with digital forms of learning and (D) Personal details. The second block of questions focused on the following two sets of questions: (1) How will you assess the learning content and learning tasks? (2) When summarizing your experience with the digital learning platform, how do you evaluate learning with it? The data collection on the topic of "Personal experiences with digital forms of learning" was based on the questionnaire for teachers that was developed in the project "Monitor Digitale Bildung" (Digital Education Monitor) by the mmb Institute for Media and Competence Research on behalf of the Bertelsmann Foundation and used for the evaluation of digital education in universities, vocational training and others [11].

4.2 Results

In sum, twelve learning companions, 25 adolescents and nine organizers, i.e. 46 people in total, from Bulgaria, Hungary, Italy, Spain and Greece took part in the pretest phase. The results of the usability test from Bulgaria and Greece are presented here as examples.

For instance, 90% of the participants in Bulgaria and 87% of the participants in Greece found the learning platform easy to use. 64% of the participants in Bulgaria felt at least very confident using it. Only one adolescent was of the opinion that using the platform was rather difficult, with no participant feeling completely unsafe using it. In Greece, all participants except for one felt at least very confident using the platform, with one person rating the experience of use and feeling of safety as neutral. In Bulgaria, 86% of the participants said they did not have to learn much about using the platform before working with it. In Greece, 75% of the participants were of the same opinion, 25% were of the opposite opinion. When asked if the learning platform was easy to use without help or a manual for other people, two of the participants in Bulgaria tended to disagree, while the rest feeling that the use was easy. In Greece, on the other hand, the evaluation of this question was very diverse, with 25% rather disagreeing, 25% rather agreeing and 50% regarding this to be good to fully possible. In Bulgaria, the participants felt that using the learning platform was not cumbersome and not unnecessarily complex, although one of the mentors was of the opposite opinion here. In Greece, on the other hand, 25–50% of the participants considered the use of the platform to be rather cumbersome and unnecessarily complex, with the rest having a contrary opinion. However, no participant felt that the learning platform showed many inconsistencies, while in Bulgaria two participants were of the opposite opinion. In both countries, the integration of the different functions in the platform was consistently rated positively.

All participants moreover agreed that the learning platform is beneficial for the learning process, well tailored to the requirements of the learning concept and that it offers all necessary functions to make the learning process successful. Overall, 75% of participants in Bulgaria and 50% of participants in Greece fully agree that they would like to use the platform often, and the remaining 25% and 50% would also find it positive to use.

The results of the SUS test illustrate that the learning platform is generally considered as useful and to a certain extent easy to use by the actors surveyed. This indicates that there is a certain acceptance of the platform, including the learning concept behind it, both among the adolescents and the learning guides or mentors. The SUS test, however, offers little indication of which usability problems exist in the use of the learning platform. Therefore, the present study was supplemented by a resistance and potential analysis. Despite the positive responses, it is necessary to clarify how the learning platform and its implementation can be improved. To this end, the focus should be led on the potentials and resistances that can arise when using the platform.

Within the scope of workshops with the partner organizations and the survey of the adolescents, the mentors and the organizers, potentials and resistances were identified that could arise in connection with the use of the agile learning concept with the help of the platform. Among the organizers, the following potentials were pointed out: The possibility has been created to build up a large transnational pool of mentors and experts as well as the reduction of organizational effort to bring mentees and mentors together. In addition, the organizers can draw on a structured learning concept to promote entrepreneurial and digital skills and disseminate best-practice approaches across countries. This would result in the following potentials for the mentors or learning guides: They can give the adolescents timely feedback and immediately correct undesirable developments. In addition, e.g. working experts can also participate in the mentoring process and react flexibly to the adolescents and their ideas and support them. In addition, the mentors can contact other mentors and expand their personal network. For the group of adolescents there is the possibility of having easier personal access to experts. Furthermore, it makes it easier for the adolescents to organize themselves with the effect of eliminating the travel distances for meetings. The group members can acquire common knowledge and then delve individually into the topics that they find interesting. For each individual learner, there is the opportunity to learn in small sprints and to be able to determine the pace of the learning process.

The following challenges arise for the organizers when implementing the agile learning concepts with the help of the learning platform: First, they have to make the concept of agile learning understandable not only for mentors, but also for adolescents (mentees). This proves to be difficult for potential mentors due to their lack of knowledge and qualifications of the concepts of agile learning. Against this background, the selection of suitable mentors is of particular importance. Last but not least, it was considered a challenging to ensure the same technical standard everywhere in order to successfully use the learning platform. The different levels of knowledge as well as a different pace in acquiring new skills proved to be a challenge for mentors and mentees. In virtually conducted learning activities, learning guides sometimes take on the role of a mere entertainer, which can lead to a feeling for inexperienced mentors of being overwhelmed. Furthermore, using the concept of agile learning with the help of the platform highly demands their time management. The technical problems initially faced in the learning groups prevent interaction within the learning group, with an impact on the coordination of the group members' schedules. Difficulties in a common understanding of the task arose due to the lack of non-verbal communication. The learners themselves were distracted or disturbed while learning at home. Furthermore, not all of them had the necessary ability to self-organize and self-motivate. External control in solving the tasks was limited and lacking for some adolescents. Furthermore, the learners were often confronted with new technologies and learning methods in the last months.

5 Further Development of the Learning Platform by Means of AI Tools

Following on from the results of the empirical study, it turns out to be advantageous to improve the usefulness and ease of the learning platform by resorting to selected AI tools in essential points. These include the use of chatbots, speech recognition systems and technical assistance systems as well as the topics of cyber security and big data. The development of quality standards, which form the basis for further development with AI tools, plays an important role here.

Based on the results of a recent benchmarking study on AI in vocational education and training, the use of chatbots and AI-based analysis tools in in-company training seems to be of particular importance [12]. Chatbots, for example, are regarded as having a growth potential of 30% in the next few years. One advantage of these AI tools is seen in the consistent quality of the performance of tasks and services. Scientific studies have investigated the question of whether a higher quality of service can be achieved with the help of chatbots, e.g. compared to a customer hotline. Some empirical studies have come to the conclusion that the positive effects of chatbots on service quality could be determined on the basis of 16 characteristics [13]. Furthermore, empirical studies on the use of speech recognition systems have shown that these systems, especially regarding documentation in the medical field, offer advantages in terms of faster documentation speed, better document content and higher user satisfaction. The speed of documentation could be increased by 26%, the scope of documentation content by 82% and user satisfaction increased by 23% [14]. Having this in mind, the current focus is to use the AI tools addressed to improve the learning platform.

Language barriers are a major challenge for the communication between the mentee groups and the mentors or experts from the different EU countries. In order to improve the communication between the different actors at the EU level, AI tools for speech input and recognition as well as corresponding chatbots could prove beneficial. By using AI based on the principle of machine learning, chatbots could recognize user behaviour patterns and appropriately react in complex situations. This serves an efficient and yet reliable service, which can be provided at any time of the day or at night, independent of human activity. Speaker-independent speech recognition systems are an innovative solution to improve the international exchange and the limitation of productivity by manual transcription of results of the learning platform. At the same time, the use of speech recognition and chatbots based on the principle of AI enables international networking.

When using selected AI tools, the regulations of the European Data Protection Regulation must be kept in mind. It also makes sense to consider the guidelines for the responsible use of AI and other digital technologies in HR work from the HR Tech Ethics Council [15].

The experience from the COVID-19 crisis has shown that the use of digital solutions including AI tools for in-company education and training can no longer be designed and implemented as an isolated solution. The various digital solutions to support teaching and learning should rather be an integral part of in-company education and training. At the same time, the study by Siepmann and Fleig [16] has made clear the great need for orientation among the responsible actors when it comes to the topic of "AI and corporate competence development".

6 Outlook

In order to ensure the basic prerequisites for the implementation of the agile learning concept in the EU countries and to increase the acceptance of using the digital learning platform including AI tools, the next step will be to weight the previous results from the potential and resistance analysis on the basis of a participatory approach. The resistances will be differentiated into those that require little effort, those that require a great deal of effort and those that cannot be overcome. The potentials are differentiated according to whether they are easy or difficult to use. Based on these results, concrete solution approaches will be developed that are particularly well suited for the surmountable resistances and help to use the existing potentials in a targeted manner [10]. With the help of these approaches, resistances and potentials can professionally be dealt with when implementing AI tools for the further development of the learning platform.

The new technological developments, especially related to digitalization and AI, offer special opportunities to create a learning architecture that promotes self-learning [17]. Regarding the various services that are to be provided or supported by AI, it would have to be clarified which quality requirements these services should take into account.

Acknowledgements. A short version of this article has already been published in German: Kröll, M., Burova-Keßler, K.: KI in der betrieblichen Aus- und Weiterbildung. In: Gesellschaft für Arbeitswissenschaft e.V. (ed.) Arbeit HUMAINE gestalten. GfA Frühjahrskongress 2021, Bochum.

References

1. Bauer, A., Jungclaus J., Korge, G.: Agiles Sprintlernen – Eine neue Lernform für die digitalisierte Arbeitswelt. ZWF Zeitschrift für Wirtschaftlichen Fabrikbetrieb, vol. 113, pp. 637–640. Carl Hanser Verlag, München (2018)
2. Bräutigam, S., Höhne, B., Longmuß, J., Schindler, F.: Agiles Lernen am Arbeitsplatz – Eine neue Lernkultur in Zeiten der Digitalisierung. In: Zeitschrift für Arbeitswissenschaft, Ausgabe 2/2017, pp. 110–119. Springer, Heidelberg (2017)

3. Brooke, J.: SUS – a quick and dirty usability scale. In: Jordan, P., Thomas, B., Weerdmeester, B., McClelland, I. (eds.) Usability Evaluation in Industry, European Committee for Standardization, Ergonomics of Human-System Interaction, pp. 189–196. Taylor and Francis, London (1996)
4. Davis, F.D., Bagozzi, R.P., Warshaw, P.R.: User acceptance of computer technology – a comparison of two theoretical models. Manage. Sci. **35**(8), 982–1003 (1989)
5. Dehnbostel, P.: Lernorte, Lernräume und Lernarchitekturen in der digitalen Transformation der Arbeit. In: Richter, G. (eds.) Lernen in der digitalen Transformation – Wie arbeitsintegriertes Lernen in der betrieblichen Praxis gelingt, pp. 19–34. Schäffer-Poeschel-Verlag, Stuttgart (2020)
6. Ethikbeirat HR Tech: Richtlinien für den verantwortungsvollen Einsatz von Künstlicher Intelligenz und weiteren digitalen Technologien in der Personalarbeit. https://www.ethikbeirat-hrtech.de/wp-content/uploads/2020/03/Richtlinien_Download_deutsch_final.pdf
7. European Union: Digital Education plan 2021 - 2027 – Resetting education and training for the digital age (2020)
8. Graf, N., Gramß, D., Edelkraut, F.: Agiles Lernen – Neue Rollen, Kompetenzen und Methoden im Unternehmenskontext, 2. Auflage. Haufe, Freiburg (2019)
9. Kohnke, O.: Anwenderakzeptanz unternehmensweiter Standardsoftware. Springer, Wiesbaden (2015). https://doi.org/10.1007/978-3-658-08206-2
10. Kröll, M.: Innovationsprojekte und organisationalen Wandel professionell gestalten. Theorie der Reflexion und Reflexionskompetenz. Springer, Wiesbaden (2020). https://doi.org/10.1007/978-3-658-29137-2
11. Mmb Institut – Gesellschaft für Medien- und Kompetenzforschung mbH: Monitor Digitale Bildung – Befragung von Lehrerinnen und Lehrern. https://www.bertelsmann-stiftung.de/fileadmin/files/Projekte/Teilhabe_in_einer_digitalisierten_Welt/Fragebogen_Lehrer_restliche Bundeslaender_BST_DigiMonitor.pdf
12. Rossmann, A., Zimmermann, A., Hertweck, D.: The impact of chatbots on customer service performance. In: Spohrer, J., Leitner, C. (eds.) AHFE 2020. AISC, vol. 1208, pp. 237–243. Springer, Cham (2020). https://doi.org/10.1007/978-3-030-51057-2_33
13. Schwaber, K., Sutherland, J.: Der Scrum Guide. Der gültige Leitfaden für Scrum: Die Spielregeln (2013). https://www.scrumguides.org/docs/scrumguide/v1/Scrum-Guide-DE.pdf
14. Siepmann, F. (eds.) E-learning BENCHMARKING Studie 2019 – Teilstudie: Künstliche Intelligenz. eLearning Journal-Verlag, Hagen im Bremischen (2019)
15. Siepmann, F., Fleig, M. (eds.) eLearning BENCHMARKING Studie 2018 – eLearning & Weiterbildung. eLearning Journal-Verlag, Hagen im Bremischen (2018)
16. Vogel, M., Wolfgang Kaisers, W., Wassmuth, R., Mayatepek, E.: Analysis of documentation speed using web-based medical speech recognition technology – randomized controlled trial. J. Med. Internet Res. **17**, e247 (2015)
17. Werther, S.: Feedback in Zeiten der Agilität, 1. Auflage. Haufe, Freiburg (2020)

Impacts on Cognitive Decay and Memory Recall During Long Duration Spaceflight

Terry Rector[1(✉)], Curtis Cripe[2], and James Casler[1]

[1] University of North Dakota, Grand Forks, ND, USA
{terry.rector,James.casler}@und.edu
[2] NTL Group, Scottsdale, AZ, USA

Abstract. Living and working in outer space introduces unique physiological, psychological, and psychosocial stressors to the human body. While most stressors are known and well researched, Long Duration Spaceflight creates additional and more worrisome stressors. This paper describes preparation and research plans for a 30-month repeated-measures, cognitive decay, and memory recall study utilizing ten practical space mission tasks performed by 32 astronaut-like subjects in an Isolation, Confined and Extreme (ICE) analogous environment.

Keywords: Long Duration Spaceflight (LDSF) · Training · Isolation · Confined and Extreme (ICE) · Analogous environment · Human factors · Human-machine interface (HMI)

1 Introduction

As NASA plans for deep space and long-duration missions, it has identified a gap in understanding how the personal relations/interactions (family, friends, and colleagues) affect astronauts' behavioral health and performance during missions, including negative behaviors, and in understanding how such behavior may affect other crewmembers. As discussed, there are other stresses to be considered such as space adaptation syndrome (SAS), micro-zero gravity, ionizing radiation exposure, mission time and location, orientation clues and knowledge, noise types and levels, temperatures, workload and schedule, boredom, cultural differences, and types and length of physical fitness activities [1].

As of the time writing this paper, fewer than 600 humans have traveled to space, with fewer than ten remaining in orbit for more than 300 days. Future spaceflight will entail long-duration missions constrained by technical challenges now being methodically investigated. One of the largest concerns is mitigating any effects from high energy radiation, fluid shift due to microgravity, and any damage either may engender in cognitive decay or memory recall [2].

This paper introduces the preparation and research plans for a 30-month repeated-measures, cognitive decay, and memory recall study utilizing ten practical space mission tasks performed by 32 astronaut-like subjects in an Isolation, Confined and Extreme (ICE) analogous environment. The genesis for this research is the expectation that

S. Nazir et al. (Eds.): AHFE 2021, LNNS 269, pp. 44–51, 2021.
https://doi.org/10.1007/978-3-030-80000-0_6

Long-Duration Spaceflight (LDSF) crews must operate autonomously requiring real-time access to an estimated double to triple current knowledge content. [3] This increased demand for knowledge creates critical learning and forgetting dilemmas for spaceflight crew members already densely cross-trained in two or more disciplines. With adequate pre- and during-mission training systems, many of the other LDSF stressors can be mitigated. However, detecting changes in the "forgetting curve" and identifying when training is most needed to ensure continued fitness for duty is pivotal for sustained and safe spaceflight [4].

Living and working in outer space introduces unique physiological, psychological, and psychosocial stressors to the human body. While most stressors are known and well researched, LDSF crews will travel well beyond Earth's protective lower orbits, which will create additional and more worrisome stressors. More research is needed into how these stressors can be adequately mitigated. Currently, the best solution is to train, practice, and test for all possible outcomes, recognizing the need for responding to the unknown-unknown possibilities [5].

LDSF alters the structure and function of the human brain. Many spaceflight stressors, including altered gravity, sleep loss, radiation, and isolation and confinement, may dysregulate the brain's structure and microenvironment, leading to an imbalance in the neuronal and glial networks' function and the neurovascular unit [6].

2 Background

Training for human spaceflight operations is demanding, costly, and time-consuming, starting with the space agency's use of omnifarious facilities, mock-ups, simulators, and emulators. Before today's lengthy stays aboard the International Space Station (ISS), spaceflight missions remained short enough that the natural forgetting process was not a safety concern. However, NASA's goal of landing humans on Mars by 2030 necessitates two to three times current trained content to prepare astronauts for such missions, increasing the risk of inadequate or latent training potentially inadequate decision making due to delayed or un retrievable memories [7, 8].

The critical component motivating the need for a change in training is anticipated communications latency due to the increased distance between the spacecraft and Earth and the increased speed with which communication signals must travel. Since communications can only travel as fast as the speed of light (approximately 300,000 km or 186,000 mi per second), once the spacecraft proceeds beyond our Moon's orbit (238,900 mi), duplex communications will quickly drop to simplex, increasing to over 40 min (worst case) round-trip from Mars [9, 10].

The communications delay is expected to double or triple the knowledge content needed for the crew to operate autonomously from NASA's Mission Control Center (MCC). The MCC has always monitored and managed the spacecraft and crew's activities but LDSF precludes these MCC functions. The MCC is staffed with decades of perishable experience and knowledge in many different disciplines, including lessons learned from past successes and failures as well as detailed spacecraft and support systems knowledge, some of which cannot be taught. [11] This latency will preclude readily available decision-making support from the MCC. LDSF crews will be the first in space exploration history to operate fully autonomously without MCC oversight [12].

Historically, the MCC has served a pivotal role in monitoring the spacecraft and crew health, anomaly detection, crew tasks scheduling, and decision support 24 h a day, seven days a week. The MCC's role must transfer to LDSF crews and their spacecraft to facilitate future LDSF missions. However, to enable future LDSF crews, most, if not all of the knowledge currently maintained at the MCC must transition to the crew [13, 14].

There is no adequate understanding of how the brain or body will react to being cooped up in a spacecraft with three to five other people traveling to and returning from Mars. Additionally, after arriving at Mars, the crew may spend most of their time together in one or more habitat modules no larger than a large bus or recreational vehicle with minimal opportunity for privacy other than inside a spacesuit exploring the inhospitable Martian landscape. The crews' survival and overall mission success will depend heavily on new technologies, training methods, and refresher training [14].

The commonly accepted and revaluated Ebbinghaus forgetting curve will provide a foundation for test subjects to establish a test baseline (establishing control group) before each mission. [15–17] Test subjects will receive pre-mission training on individual and crew tasks, both simple and complex (estimating ten complete tasks) identified by Stuster [8, 18]. Each subject will be trained using standardized methodologies and allowed to practice acceptable competencies corresponding to the task(s) complexities before testing, to establish a baseline for future comparisons.

As spacecraft crews prepare for LDSF, training must utilize methods and tools that encompass the full training continuum for an exploration mission scenario. Complementing pre-flight training with an adaptive, in-flight training regimen may offer several advantages, including reducing neurocognitive workload burden on the crew and optimizing training content to enhance brain function and provide meaningful work, particularly en route to Mars. [19] A training regimen must keep individual crew and team members motivated and engaged by maintaining brain areas and motor skills areas honed for readiness during post-landing mission requirements [20].

The human brain is a dense, dynamically reconfiguring connectome of multiple network-hub architectures, hosting five commonly accepted frequency bands of oscillating electronic signals. [21–24] As previously learned task performance (critical skills) will undoubtedly diminish over time due to natural forgetting, there is an expectation that adaptive on-board training systems integrated within the spacecraft's systems will be necessary. However, when during the spaceflight journey is the most optimum time for refresher training? How far in advance of a critical mission event such as landing on Mars should refresher training be taken so the crew's performance can be at its peak when most needed? These questions and others both guide and inform this research.

Although the ISS has several different types of on-board training systems, they operate independently ("stovepiped") instead of being collected on a central computer where the combined training effect can be monitored and measured. The stovepiped training systems require space (volume), weight, power, and connectivity independent from other competing on-board systems, using highly limited resources which produce additional noise and heat that must be mitigated. Additionally, each training system has a different human to machine interface (HMI) that requires requiring the trainee to respond differently, thus using additional time and mental resources. For increased effectiveness and efficiency, future training systems should be seamlessly integrated within the spaceflight crew's everyday activities and, where possible and appropriate, should provide some level of release and mental respite from operational activities, thus reducing boredom and monotony [3, 25].

3 Problem Statement

Future spaceflight exploration to Mars and beyond will expose human crews to prolonged microgravity, Galactic Cosmic Radiation, and many other stresses while living and working within an ICE environment. These factors are expected to negatively impact the body's central nervous system, cognition, memory, and behavior. This research goal is to enable new technologies needed to mitigate these effects.

Research is needed to comprehensively measure and model human cognitive decay and memory recall limitations in a spaceflight-realistic environment—the purpose of this research opportunity.

4 Research Questions

This research project seeks responses to the following questions:

- Q1. By employing an electroencephalogram (EEG) 32-Channel device to detect instances of increased brain activations (frequency and signal strength), can skill retention levels be determined with enough accuracy to maintain a predestined cognitive skill level?
- Q2. Given performance degradation concerns related to the broadly recognized "forgetting curve" model research, can an acceptable degree of degradation be measured to determine when training must be conducted to maintain acceptable proficiency?
- Q3. For a given set of mission tasks, can cognitive performance degradation be detected reliably at levels that permit timely application of refresher training protocols in advance of a future mission event?
- Q4. Can factor(s) or variables affecting (accelerating or decelerating) cognitive skill retention and loss rates be determined, prioritized, and mitigated?
- Q5. Can mission planning and execution be designed to accommodate individual training protocols to fine-tune mental or physical performance training a manner similar to athletic training?
- Q6. Can the cognitive decay rate be used to delineate requirements for on-orbit high fidelity training systems and simulator requirements?

5 Methods

This paper introduces a planned 30-month, repeated-measures, ad hoc research study of 32 subjects possessing similar education backgrounds, experiences, skill sets, etc., recruited from the UND's John D. Odegard School of Aerospace Sciences department. The 32 subjects will be evenly divided between an experiment and a control group and then further divided into four, four-person crews (teams), yielding eight teams of four persons. The experiment teams (four teams of four subjects) will live and work for 30 days in the Inflatable Lunar/Mars Analog Habitat (ILMAH) facility simulating LDSF rigors similar to those expected in future Mars exploration missions, including communications delays and no other in-person human contact. The control group (four teams of four subjects) will run concurrently with the isolated (treatment) group) while living everyday lives as UND graduate students and completing the same mission activities outside the habitat environment. The ILMAH serves as an analogous facility mimicking an ICE environment facility, located at the University of North Dakota, Grand Forks, North Dakota. The ILMAH was funded by NASA and is owned and operated by the UND Department of Space Studies Human Spaceflight Laboratory.

Test subjects will receive training in specific tasks, participate in practice sessions, and be tested before each mission starts. Cognitive tasks will have ten levels of difficulty and be designed to imitate spaceflight activities, An average of 320 measurements will be taken during each of the four missions. In addition to individual surveys, questionnaires, observation opportunities, and verbal feedback sessions during testing events, subjects will wear a high-resolution (24-bit) electroencephalogram (EEG) device to record electrical activity produced by the human brain at varying mission times and stress levels. The recorded electrical signals will be post-processed by Matlab®, EEGLab®, eLORETA, NeuroCoach™, and SCCN neuroscience tools to develop a real-time predictive cognitive performance/health profile. Neurophysiological metrics will be used to illustrate neural functional performance changes resulting from changing multiple environmental effects. These metrics will monitor the performance and communication changes from a healthy baseline within and between the brain regions of interest described above.

Researchers will manipulate environmental factors over varying lengths of retention intervals using an operationally realistic LDSF analog to produce each test subject's performance modeling. While many mature training models exist, they generally have not been utilized outside the clinical setting.

Task performance will be observed and measured using EEG recordings while monitoring subjects' knowledge level transitions within the Skill-Rule-Knowledge (SRK) systematic taxonomy. [26, 27] Rasmussen's SRK model will be used to characterize or detect various categories of human behavior, widely applied in many operational environments such as aviation and healthcare.

Objective measurement of task proficiency will focus on the time required for task completion and the number of errors. Test subjects and crews (teams) will also assess their own performance of each task. Individual cognition will be measured by situational awareness, decision-making, and an EEG 32-Channel device as indicated by the time taken to recognize and respond to the emergence of a task or problem and the number of steps (or time) required for a solution.

All EEG measurements will be corrected to individual pre-mission baseline readings seeking variations between individual brains that results in different responses to stimuli as measured by the instrumentation. All research data will undergo a complete statistical analysis, including calculating Pearson's correlation coefficient and subsequent significance testing, *t-tests, F-tests,* and repeated measures analysis of variance (ANOVA) over time with subject and control groups comparisons. Pre- and post-test event surveys relevant to the task will be conducted (referred to as the "ICE" or "Control" group) for later comparison and analysis with post-surveys conducted concurrently with de-briefing after the month is concluded [28].

6 Analysis Approach

After selection and signing of the consent agreement, all questionnaires and surveys will be administrated privately and online using the University-supported QualtricsXM Research Core™ software system accessed by the crew member's Pi-Top-3 laptop computer. Qualtrics is a secure, robust, integrated software suite that interfaces nicely with all essential statistical tools for data reduction and analysis. Should a need arise for hardcopy documents to be added, they will be scanned after completion and stored electronically within the limited access system, then destroyed for privacy.

Research data will be analyzed for significance within the groups, testing differences between mean cognitive psychometric scores over time. This will allow any changes in psychometric responses to be modeled concerning time and predictability. The control group will be the sample of participants for which differences in mean scores can be compared between groups to ascertain whether ICE group participants experience cognitive deficiency due to isolation, confidence, or extreme stressors (as a treatment factor).

7 Research Benefits

This experiment is expected to provide high-level considerations for redesigning preflight and in-flight training regimens. [29, 30] This offers several advantages, such as reducing the neurocognitive workload burden on the crew and optimizing training content to enhance brain function and provide meaningful work, particularly en route to Mars. [5] A well-designed, planned, and executed training regimen is expected to keep crewmembers motivated and engaged by maintaining brain areas and motor skills areas honed for readiness during post-landing mission requirements [20, 31].

Additionally, discoveries from this research may contribute to defining onboard training systems specifications, given the projected limits in available spacecraft weight, volume, and power. This research will help identify content that should be trained before the spaceflight and content that can be delayed for in-flight training, to reduce potential cognitive overload. [7, 33] New training systems must adapt to the trainee's performance [29, 32].

The broad-reaching significance of these findings may lead to efficient and effective training methodology, systems, and programs to cover the entire mission duration. These findings are expected to impact training applications in other fields such as spaceflight mission control, military and commercial flight crews, surgeons, law enforcement personnel, and other professionals trained to deal with high-risk, complex, and uncertain tasks in highly stressful environments. The next step may be to conduct these same or similar experiments aboard the ISS for continued evaluations and refinement.

References

1. Stanley, G., Love, R.P.H.: Crew autonomy for deep space exploration - Lessons from the Antarctic Search for Meteorites (2014)
2. Garrett-Bakelman, F.E., et al.: The NASA twins study: a multidimensional analysis of a year-long human spaceflight. Sci. **364**(6436), eaau8650 (2019)
3. Pieters, M.A., Zaal, P.M.T.: Training for Long-duration space missions: a literature review into skill retention and generalizability. IFAC PapersOnLine **52**(19), 247–252 (2019)
4. Fisher, J.S., Radvansky, G.A.: Patterns of forgetting. J. Mem. Lang. **102**, 130–141 (2018)
5. Dempsey, D.L., Barshi, I.K.P.: Evidence report: risk of performance errors due to training deficiencies. risk statement: Given that existing training methods and paradigms may inadequately prepare long-duration, autonomous crews to execute their mission, there is a risk that increased flight and ground crew errors and inefficiencies, failed mission and program objectives, and increased crew injuries will occur (2016)
6. Roberts, D.R., et al.: Effects of spaceflight on astronaut brain structure as indicated on MRI. N Engl. J. Med. **377**(18), 1746–1753 (2017)
7. Dempsey, D.L., Barshi, I.: Applying Research-Based Training Principles: Towards Crew-Centered, Mission-Oriented Space Flight Training. Training for a Mars Mission (2019)
8. Stuster, J.W., et al.: Human Exploration of Mars: Preliminary Lists of Crew Tasks (2019)
9. Moore, M., et al.: Simulation Based Investigation of High Latency Space Systems Operations (2017)
10. Rader, S.N., et al.: Human-in-the-loop operations over time delay: lessons learned. In: International Conference on Environmental Systems (ICES), NASA Center for AeroSpace Information (CASI), Vail, CO (2013)
11. Gushin, V.I., Yusupova, A.K., Shved, D.M., Shueva, L.V., Vinokhodova, A.G., Bubeev, Y.A.: The evolution of methodological approaches to the psychological analysis of the crew communications with Mission Control Center (2016)
12. Orgel, C., et al.: Scientific results and lessons learned from an integrated crewed Mars exploration simulation at the Rio Tinto Mars analogue site (2014)
13. Gershman, B., et al.: Low-latency teleoperations for human exploration and evolvable mars campaign. In: IEEE Aerospace Conference Proceedings (2017)
14. Lupisella, M.L., Bobskill, M.R.: Human Mars Surface Science Operations (2014)
15. Cowan, N., Rachev, N.R.: Merging with the path not taken: Wilhelm Wundt's work as a precursor to the embedded-processes approach to memory, attention, and consciousness (2018)
16. Murre, J.M., Dros, J.: Replication and analysis of Ebbinghaus' forgetting curve. PLoS One **10**(7), e0120644 (2015)
17. Fuchs, A.H.: Ebbinghaus's Contributions to Psychology after 1885 (1997)
18. Stuster, J., et al.: Generalizable Skills and Knowledge for Exploration Missions - Final Report (2018)
19. Landon, L.B., et al.: Selecting astronauts for long-duration exploration missions: Considerations for team performance and functioning. REACH Rev. Hum. Space Explor. **5**, 33–56 (2017)
20. Fiore, S.M., et al.: Critical Team Cognitive Processes for Long-Duration Exploration Missions, Institute for Simulation and Training: NASA (2015)
21. Kaiser, M.: A tutorial in connectome analysis: topological and spatial features of brain networks. NeuroImage (Orlando, Fla.) **57**(3), 892–907 (2011)
22. Tompson, S.H., et al.: Functional brain network architecture supporting the learning of social networks in humans. NeuroImage (Orlando, Fla.) **210**, 116498 (2020)

23. Zhang, W., et al.: Dynamic reconfiguration of functional topology in human brain networks: from resting to task states. Neural Plast. **2020**, 1–13 (2020)
24. Stam, C.J., van Straaten, E.C.W.: The organization of physiological brain networks. Clin. Neurophysiol. **123**(6), 1067–1087 (2012)
25. Goldberg, J.H.: Training for long duration space missions (1986)
26. Rasmussen, J.: Skills, rules, and knowledge; signals, signs, and symbols, and other distinctions in human performance models. IEEE Trans. Syst. Man Cybernet. **SMC-13**(3), 257–266 (1983)
27. Fleming, E., Pritchett, A.: SRK as a framework for the development of training for effective interaction with multi-level automation. Cogn. Technol. Work **18**(3), 511–528 (2016). https://doi.org/10.1007/s10111-016-0376-0
28. Baguley, T.: Calculating and graphing within-subject confidence intervals for ANOVA. Behav. Res. Methods **44**(1), 158–175 (2012)
29. Kelley, C.R.: What is adaptive training? Hum. Factors **11**(6), 547–556 (1969)
30. O'Keefe, W.: Training Concept for Long Duration Space Mission, NASA (2008)
31. Cuevas, H.M., Schmorrow, D.D.: Exploring Cognitive Readiness in Complex Operational Environments: Advances in Theory and Practice (2012)
32. Gallego-Durán, F.J., Molina-Carmona, R., Llorens-Largo, F.: Measuring the difficulty of activities for adaptive learning. Univ. Access Inf. Soc. **17**(2), 335–348 (2018)
33. Aidman, E.: Cognitive fitness framework: towards assessing, training and augmenting individual-difference factors underpinning high-performance cognition. Front. Hum. Neurosci. **13**(466), 466–466 (2020)

Implementation of Learning Technologies in Education

Spiritual Accompaniment Training in Pediatric Palliative Care: A Case Study from Spain

Florentin Popescu[1](✉), Mario Gabriel Ciccorossi[2], Inma Ferrer Svoboda[2], and Pau Miquel Diego[2]

[1] HAN University of Applied Sciences, Ruitenberglaan 31, 6826 CC Arnhem, Netherlands
Florentin.popescu@han.nl
[2] Hospital Sant Joan de Déu, Passeig de Sant Joan de Déu 2, 08950 Esplugues de Llobregat, Barcelona, Spain
{Mciccorosi,cferrers,Pmiquel}@sjdhospitalbarcelona.org

Abstract. This article examines a case study of Spiritual Accompaniment Training in Pediatric Palliative Care (PCC) at the Sant Joan de Déu (SJD) Hospital in Barcelona, Spain. It describes the current level of training of PCC professionals in Spain and suggests action proposals to improve it. More specifically, the present study explores the challenges, needs, and gaps in the training of the health care professionals who provide pediatric palliative care and grief support in Spain. Furthermore, it investigates the skills and competences needed by pediatric health professionals in spiritual accompaniment care for both patients and their families by trying to specify the degree of spiritual competence. Finally, we propose a structured specialized training in spiritual care for PPC professionals which is based on a standardized Spiritual Accompaniment in Pediatric Palliative Care Training Program Care in Spain.

Keywords: Spiritual care · Spiritual accompaniment training · Pediatric palliative care · Spiritual competences · Palliative care professional

1 Introduction

Healthcare involves a rich mosaic of perspectives, both from a cultural and social point of view. This richness, despite the diversity of approaches to the reality of human pain and health, is a significant axis of the contribution of religions to the social sphere, both in terms of the development of the medical sciences and the significant contribution to the creation of institutions for the reception and treatment of patients, and the development of the principles of public health.

This translates, in the health field, into the need to take into account and respect the spiritual and religious dimensions of the people who are cared for. The deep beliefs or convictions of people and the religious practices they carry out can affect the health field in various ways: for instance, the partial acceptance or rejection of certain diagnostic procedures and treatments, or a specific way of interpreting the hospital admission (also in terms of food, dress, religious calendar and rituals of birth and death) [1].

S. Nazir et al. (Eds.): AHFE 2021, LNNS 269, pp. 55–62, 2021.
https://doi.org/10.1007/978-3-030-80000-0_7

In times of illness and fragility, it is common for individuals to question about the meaning of the world and our existence. Spirituality often acquires an even more significant role in these circumstances. And, while spirituality is not limited to religion, religion often becomes a channel of expression of spirituality. The expression of spirituality is always something personal, which each individual experiences and lives in a unique and particular way [2]. In this sense, it is important that public institutions work to guarantee the exercise of religious freedom rights and that they promote an equidistant position with respect to the different beliefs or ideals in health care [1].

An international consensus on the importance of detecting, evaluating, and accompanying spiritual aspects in health care has been established, for instance, by the Joint Commission for the Accreditation of Health Organizations (JCAHO) which is the largest health accreditation body in the United States [3]. JCAHO now requires the administration of an assessment of patients spiritual needs and of their families. Although most professionals endorse the concept of assessment and support in the spiritual needs of the patient and his family, studies suggest that health professionals have received little training in spiritual and religious care linked to the development of their care work [4].

Despite this evidence and emergence of publications and studies, the approach to the spiritual dimension remains the least developed and therefore the least evaluated in palliative care, both in the adult and pediatric worlds [5]. Countries such as UK via the National Health System (NHS), Germany via the International Society for Health and Spirituality (IGGS) and the United States via the Global Network for Spirituality and Health (GNSAH), have developed evidence-based knowledge in relation to spirituality as well as training and training programs in spiritual care for professionals [5]. However, little is known about the care of the religious, spiritual, or philosophy of life (hereinafter RSLP) needs of patients in Spain [6].

In Europe, most professionals participating in the studies consulted consider spiritual care as part of their professional role [5] which is a trend that has increased in comparison to previous studies [7]. Although 94% of a sample of 191 palliative care professionals in Spain consider the provision of spiritual care as part of their professional role, only 58% of those surveyed consider themselves competent to provide this type of care [8]. These studies conclude that there is a high demand for clinical spirituality training.

The aim of the present study is to address current shortcomings of the training of health professionals in spiritual competence in Spain, and thus to improve the quality and quantity of available training to Spanish health professionals from a comprehensive person-centered care model.

2 Background

Sant Joan de Déu Hospital in Barcelona was founded by the Hospitaller Order of the Brothers of Saint John of God in 1867 and has almost 150 years of experience. It is currently a leading hospital and international reference for pediatric care.

The Hospital aims to encourage and stimulate the training of professionals to improve the health of patients, society at large, and to advance scientific and human knowledge. Its mission as an organization is to provide comprehensive care to patients, combining the more humane side of care with the development of new scientific advances.

The anthropological model that governs the care in the hospital considers the person as a biological, psychological, social and spiritual being, providing an integral attention that regards all the dimensions of human experience and is patient- and family-centered. This transdisciplinary model of evaluation and referral by a direct care professional is aligned with the recommendations derived from the Consensus Conference (2009) of C. Puchalski [9].

Attention to spirituality is a central value of the Hospitaller Order of the Brothers of Saint John of God and the Sant Joan de Déu Hospital. For the last 5 years, the Hospital has aimed to become an international reference hospital attracting patients from other parts of the world, which poses the challenge of adapting spiritual, cultural, and religious care to the needs of an increasingly diverse population of patients and their families.

This research explores the challenges, needs, and gaps in the training of the health care professionals who provide pediatric palliative care and grief support in Spain. Finally, we propose a structured specialized training in spiritual care for Pediatric Palliative Care professionals which is based on a standardized Spiritual Accompaniment in Pediatric Palliative Care Training Program Care in Spain.

3 Literature Review and Benchmarking

A literature review and benchmarking of the design of clinical spirituality training curricula for health professionals were conducted in order to develop our training plan, which follows the recommendations and indications of the main study published to date, summarized in:

- **Interprofessional Spiritual Care Education Curriculum** [10] in which we follow the conclusions of this article for the design of the training program in relation to the needs of the curriculum, including
  - Make a spiritual history
  - Identify the spiritual resources of patients
  - Identify the patients' spiritual distress
  - Practice compassionate presence
  - Differentiate between spiritual and psychological factors
  - Develop a spiritual care plan
  - Identify appropriate spiritual interventions
  - Address ethical issues and dilemmas
  - Examine the impact of spiritual care on the quality of patient care
  - Develop leadership skills in spirituality and health
- The following Consensus objectives were also considered:
  - Recognize the role of spirituality as an essential element of patient-centered care.
  - Review the guidelines for spiritual care and identify the basic concepts for implementing these guidelines.
  - Define spiritual distress in adults and children and distinguish the difference between spiritual distress and other types of distress.

– Recognize the skills needed to practice compassionate presence.
– Identify communication strategies that facilitate the approach to issues of spirituality.
– Develop an evaluation and treatment, or care plan, of the whole person.
– Recognize the spirituality of the clinician as an integral element of professional development.

4 Development of case study methodology

With the goal of improving the quality of spiritual care in our hospital to address the previously described needs (social, professional and institutional), we developed a benchmarking task on spiritual care teams in hospitals specializing in pediatrics. Such benchmarking was essential to transfer the knowledge of good practices and their application to our changing context, with the support of the hospital through its R&D nurse team.

To assess the needs of professionals, a preliminary study on the needs in spirituality training for professionals in Spanish pediatric hospitals was also conducted from February till June 2020 an ad hoc Likert scale questionnaire. Sixty professionals from 20 hospitals answered the questionnaire, with provided a wide geographical representation of the different professionals specializing in pediatric palliative care in Spain. In a similar fashion to the findings of our preliminary study, the full questionnaire revealed that 75% of the pediatric professionals have some form of PPC (pediatric palliative care) training and about 60% have more than 3 years of PPC experience. Also 94% of the respondents believe that spiritual accompaniment is part of their role. These findings confirm similar results found among health professionals working with adult populations in a questionnaire survey conducted by SECPAL (Spanish Medical Society of Palliative Care) [8]. It is worth noting that 67% of our respondents say they know how to identify the spiritual needs of their patients and families, and that and 57% perceive that they do not feel qualified to do spiritual accompaniment despite knowing how to identify their spiritual needs. Also, the vast majority of respondents, 97%, understand the need to deepen their formation in spirituality.

Using their own case methodology, we received training in the development of clinical educational programs at Birmingham Children's Hospital's Spiritual Care Service. We implemented what we had learned with the case methodology designed by Birmingham Children's Hospital's Spiritual Care Service to our care context at Sant Joan de Déu Hospital. This involved the discussion of cases in team meetings with 2 external supervisors during 12 months. These meetings were essential to the dynamic evolution of the process which resulted in the course design and application of the practical modules (see Unit 6 and Unit 9 of the training).

Based on the above mentioned responses, a literature review, and the benchmarking that we conducted, we argue that there is a perceived gap between what professionals see and identify in their daily practice and what they feel they are able to do when their work requires them to engage with patients' and families' spiritual dimension.

We think that this gap can be reduced with training that addresses such a need expressed by professionals. The results obtained in our study confirm the findings of previous studies which lead the authors of this study to conclude "that the professional teams demand specific training programs in the approach of the spirituality in clinic" [10].

We propose that this training is carried out in an interdisciplinary and consensual way through the creation of a network of professionals sensitive to the issue, which will allow to build a common and shared model through the platform provided the Spanish Society of Pediatric Palliative Care (PEDPAL) and their specific team works. We plan to collect these proposals and to share our training programs at a Spanish national level through these networks.

5 Spiritual Accompaniment in a Pediatric Palliative Care Training Program

In what follows, we list the professional competencies of spiritual accompaniment (spiritual competence) to be developed in the Spiritual Accompaniment in Pediatric Palliative Care Training Program:

1. **Cognitive (knowledge and understanding) Learning to LEARN**

   - Knowledge of the basic principles of spirituality, religious denominations and philosophies of life.
   - Ability to promote the healthy interpretation of religious/spiritual belief systems/philosophies of life, identifying and avoiding approaches that limit or hinder people's illness processes or experiences
   - Knowledge and skills for an ethical quality care practice
   - Ability to foster a spiritual dimension of care throughout the health care environment in accordance with appropriate accreditation standards.

2. **Attitudes (attitudes, feelings and values) Learning to BE**

   - Awareness of strengths and weaknesses and personal work style.
   - Respect, openness, and closeness to all people
   - Attitude of active listening and interest to generate a safe, stable and peaceful meeting space that allows to recover the acceptance in situation of crisis.
   - Ability to be a soothing presence in stressful situations.

3. **Behavioral (skills, abilities, and behaviors) Learn to DO**

   - Ability to generate spaces for reflective creation of spiritual experiences according to the different forms of cultural and linguistic construction. (intercultural competence)
   - Dexterity to know how to manage the impact of death and grief
   - Ability to know how to protect the right of people to determine their own spirituality/religion/philosophy of life and to implement special protections against proselytizing in vulnerable populations.

- Abilities to develop cooperative relationships with congregations or faith groups as well as with religious referents in other settings.

We have designed the following Table "Modules and Units of the Spiritual Accompaniment in Pediatric Palliative Care Training Program Course" which incorporates the recommendations of the literature consulted, the available evidence explained in the previous chapters and the practical and clinical experience (Fig. 1).

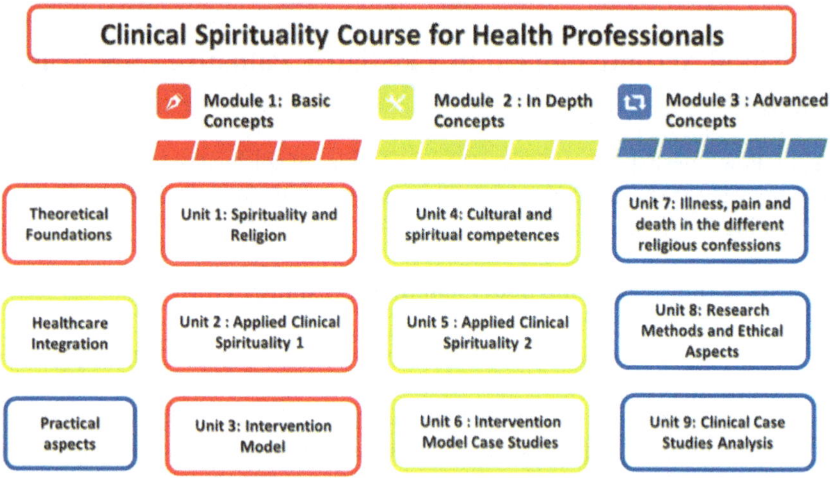

Fig. 1. Clinical spirituality course for health professionals

6 Conclusions and Further Research

The first edition of the 14-h pilot training program "Clinical Spirituality Course for Health Professionals" was held in November 2020 for 52 professionals (physicians, nurses, psychologists, and social workers). The qualitative feedback received from the Pilot "Clinical Spirituality Course for Health Professionals" can be organized according to two distinct areas:

1) Students report an improvement in understanding about spirituality in the following areas:

   - **Conceptual learning:** Concepts of cosmovision, culture, religion, spirituality, scheme of worldviews associated with different religions.
   - **Learning about the clinical implementation of spiritual care:** intervention models used, assessment of patient needs and abilities, qualitative assessment of spirituality.
   - **Practical learning in spiritual accompaniment:** How to begin to broaden one's own worldview by exploring the other's gaze, as well as ability in listening without judgment, help patients ask their own questions.

2) Students demand greater depth in the following areas:

   - Spiritual support for the professional, spirituality and decision-making, spirituality in children in cancer.
   - Deepening of clinical practice: Accompaniment in times of spiritual crisis, practical examples within comprehensive care, spiritual care in Covid times.

The development of a specific training program in pediatric palliative care in Spain responds to the training needs expressed by professionals to improve the competence of professionals to evaluate their patients in this area. The cultural and religious diversity present in Spain at present demands new training approaches based on research in spirituality in pediatric palliative care. Support for spiritual care in the clinic by the public health institutions of our country is a key element to improve the quality of spiritual care at the end of life.

Acknowledgments. The authors are grateful to Birmingham's Children's Hospital for this training. They thank all PEDPAL and Sant Joan de Déu (SJD) Hospital professionals and participants that made this article possible.

References

1. Direcció General d'Afers Religiosos (DGAR): Guia per al respecte a la diversitat de creences als centres sanitaris de Catalunya. Col.lecció "Guies sobre la diversitat Religiosa". Generalitat de Catalunya, Barcelona (2015)
2. Hexem, K., Cynthia, J., Mollen, M.D., Carroll, K., Lanctot, D.: How parents of children receiving pediatric palliative care use religion, spirituality or life philosophy in tough times. J. Palliat. Med. **14**(1), 39–44 (2011)
3. Joint Commision Resources: Advancing effective communication, cultural competence, and patient-and family-centered care: a roadmap for hospitals (2010)
4. Kumar, S.P.: Reporting of pediatric palliative care: a systematic review and quantitative analysis of research publications in palliative care journals. Indian J. Palliat. Care **17**(3), 202–209 (2011)
5. Gijsberts, M.H.E., Liefbroer, A.I., Otten, R., Olsman, E.: Spiritual care in palliative care: a systematic review of the recent European literature. Med. Sci. **7**(2), 25 (2019)
6. Miquel, P.: Sin ruido: un lenguaje para el cuidado espiritual. In: Amo, R. (ed.) Bioética y Cáncer, pp. 169–178. Universidad Pontificia de Comillas, Madrid (2020)
7. Paal, P., Helo, Y., Frick, E.: Spiritual care training provided to healthcare professionals: a sytematic review. J. Pastoral Care Counsel. **69**(1), 19–30 (2015)
8. Sánchez, M.S., Bimbaum, N.C., Gutierrez, J.B., Bofill, C.G., Mora-Figueroa, P.B., Oliver, E.B.: Cómo percibimos los profesionales el acompañamiento espiritual en los equipos de Cuidados Paliativos en España. Medicina Paliativa **23**(2), 63–71 (2016)
9. Puchalski, C., et al.: Improving the quality of spiritual care as a dimension of palliative care: the report of the consensus conference. J. Palliat. Med. **12**(10), 885–904 (2009)
10. Puchalski, C., Jafari, N., Buller, H., Haythorn, T., Jacobs, C., Ferrell, B.: Interprofessional spiritual care education curriculum: a milestone toward the provision of spiritual care. J. Palliat. Med. **23**(6), 777–784 (2020)

11. Partain, D., Ingram, C., Strand, J.: Providing appropriate end-of-life care to religious and ethnic minorities. Mayo Clin. Proced. **92**(1), 147–152 (2017)
12. Direcció General de Planificació en Salut del Departament de Salut: Model organitzatiu d'atenció integral a la población infantil i juvenil amb necessitats pal.liatives i en situación de final de vida. Generalitat de Catalunya, Barcelona (2020)

Evolution of an Approach for Digital Learning and Training in Nursing

Jan A. Neuhöfer(✉), Tim Wöllenstein, and Sabine Hansen

Hamburg University of Applied Sciences, Finkenau 35, 22081 Hamburg, Germany
{Jan.Neuhoefer,Tim.Woellenstein,Sabine.Hansen}@HAW-Hamburg.de

Abstract. Digital learning and training tools have become more and more important across a broad range of professions, which in particular accounts for the healthcare sector, where a shortage of training facilities and trainers is quite common. Thus, alternatives like learning over the Internet has become a necessity, although it might still not replace face-to-face teaching and hands-on training to its full extend, which is especially true when recruiting qualified labor from foreign countries. Regarding this, digital formats can significantly shorten ramp-up times, as prospective employees can start gaining knowledge on required terminology and processes independent from their current location. In our paper, we present details about the design and the architecture of our modular, interactive multilingual learning solution for nursing trainees, developed within a research project funded by the German Ministry of Health. Also, we reveal results of an empirical study on learning effect and user acceptance, conducted with a group of Vietnamese nursing trainees.

Keywords: E-learning · Virtual reality training · Nursing education

1 Introduction

1.1 Demographic Change, Skill Shortage and Migration

Like many European countries, Germany is a facing a demographic change, where the age distribution is shifted towards the elderly. This means that more and more people are entering retirement age with a continuously increasing life expectation while birth rates are stagnating or even regressive [1]. In the medium term, this demographic change implies an increasing number of people in need of care, but at the same time a decreasing number of professional caretakers. This social development is compounded by the already prevailing shortage of nursing staff, which becomes particularly crucial in times of a pandemic, as can be seen since the beginning of the Corona Crisis in 2020 [2]. Solutions must be found to overcome this skilled labor shortage within the healthcare sector. To a certain degree, automation technology and robotics will help to assist nursing staff in hospitals and residential care homes for the elderly [3], but in the foreseeable future, there are still many tasks and processes which require human flexibility and empathy. While long-term strategies (e.g. increase of birth rates, social upgrading of the nursing

S. Nazir et al. (Eds.): AHFE 2021, LNNS 269, pp. 63–70, 2021.
https://doi.org/10.1007/978-3-030-80000-0_8

profession) are important, they need time to be implemented. In the meantime, short-term strategies can help mitigate the effects of democratic change. Using migration as an opportunity to recruit nursing staff is such a short-term strategy. In practice, this means either education of young immigrants who enter the country for diverse reasons and who are willing to go into the profession of nursing, or active recruitment of labor from foreign countries, with willingness for a vocational training in nursing or with proven experience. In both cases, linguistic as well as technical qualifications need to be adjusted to local standards.

1.2 Educational Challenge

Like many officially acknowledged occupations in Germany, the occupational profile of nursing includes an apprenticeship with a usual duration of three years. Alternatively, in case of an existing foreign degree, applicants must pass an equivalency test. While process knowledge gained in class and in training is fundamental, candidates additionally need to develop sufficient language skills as a foundation for oral as well as written communication with colleagues and patients likewise. Moreover, cultural compliance and empathy are key factors for successful exertion, especially in contact with people in need of care. Ideally, these two threads – nursing expertise and language education – are taught in conjunction for easier transfer. Mostly, however, nursing classes presume a certain level of linguistic competence (either B1 or B2), which is usually acquired in language schools. Consequently, the conjunction of language skills and practical know-how comes rather late. Digital learning tools are an effective means to reduce this lag, as these tools easily combine process-related and language-related content into one comprehensive experience.

1.3 Digital Learning Tools in Nursing

As described in a previous paper [4], a range of digital learning solutions for nursing trainees is already available in different countries. Most of these tools are positioned to augment common practice, accompanying traditional classroom instructions and training in skills labs. Still, as there is no international standardization on procedures, equipment, timeframes or examinations for vocational training, learning material always needs to correspond to the guidelines of the country it is used in. Looking available products in Germany, most digital tools cover a defined range of nursing processes, with little or no in-depth information on the actual equipment, and with limited support for foreign languages. A more comprehensive, multi-lingual solution is needed.

While digital learning tools focus on pure knowledge transfer, they rarely include practical training matters. At this point, recent examples for Virtual Reality (short: VR) training of nursing procedures [5] show the potential of this approach for cost-effective hands-on education, intensifying spatial awareness and understanding. However, as a part of research projects, these VR systems are typically too complex and too cumbersome to put into production on a broad scale. Furthermore, ways to integrate them into an existing learning infrastructure are still rare.

2 Hybrid Learning and Training Solution

In 2020, we have started a research project funded by the German Ministry of Health to implement and to evaluate a comprehensive learning solution, covering an expert selection of essential procedures of nursing as a collection of audio-based, interactive learning modules. While German is the primary language, each module includes translations of text and speech into English, Spanish, Portuguese and Vietnamese, to effectively support non-native learners. Additionally, selected modules are augmented by interactive VR training versions for increased spatial awareness and enhanced understanding of movement. As shown in Fig. 1, the solution's production pipeline follows an architecture of four layers. These are described in further detail in the following paragraphs.

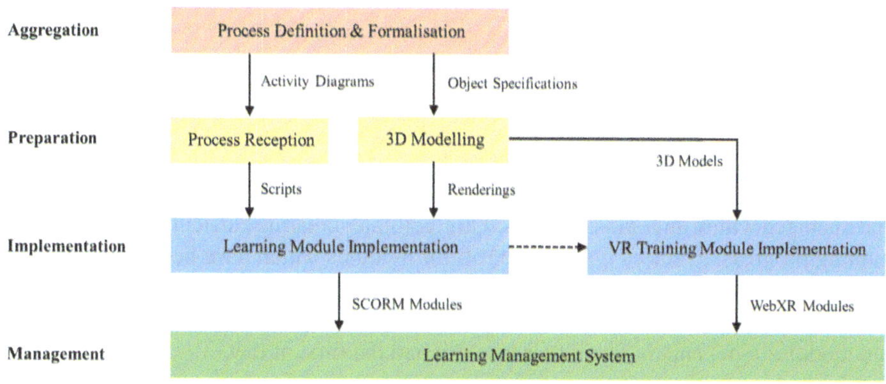

Fig. 1. The solution's production pipeline layers: aggregation, preparation, implementation and management. The preparation layer feeds both the implementation of the learning modules and optionally their VR training extensions. Both the learning modules as well as their VR training extensions are deployed into the same learning management system, to support the concept of a holistic, highly accessible solution. Most transitions (arrows) include multiple review cycles.

Aggregation
At the beginning of the project, we defined an overall amount of 24 essential nursing procedures, together with our project partner, the Knappschaft Kliniken nursing school [6] in Bochum, Germany. For each of these procedures, an activity diagram and a brief step-by-step description is generated, together with a specification of all required equipment and consumables.

Preparation
In this phase, the appropriate instructional methods and media elements are defined on the basis of subject matter expert (short: SME) input. The resulting scripts are the basis for the following implementation. The object specifications are used for creation

of highly detailed 3D models. These models are used for rendering of high-resolution images for the learning modules and, with lower detail, within the VR training modules.

Implementation

The scripts and renderings from the preparation layer are taken as basis for the authoring of the learning modules. Within this development process, three stages are differentiated: beta stage (German text, no audio, with draft renderings), approval stage (German text and audio, final renderings) and final stage (all languages in text and audio). As reactive HTML 5, the final version is packaged as a SCORM module [7] for easy deployment into any standard learning management system (short: LMS) like Moodle [8]. The learning are built to run on PCs, laptops and mobile devices alike.

Once approved, a module can be used as a basis for VR training development. However, only procedures with high demands on spatial awareness are considered for VR training. For easy deployment into the LMS, we chose WebXR [9] as the output format for VR training. Thus, any standard VR headset with handheld controllers, e.g. the Oculus Quest [10], can be used for training.

Management

Both the learning modules as well as the VR training modules can be deployed, stored and managed within any standard LMS. This facilitates creating curricula and monitoring training progress. Trainees typically enroll for modules which have been predefined by one or several instructors who, on their end, can run reports and analyses on particular trainees and/or classes at any time. Provided the LMS is Internet accessible, trainees can run modules independent from their location and the time of day.

3 Learning Module Design and Features

As a general premise of our research project, learning modules shall be usable in the field. In order to maximize both, learning effect and user acceptance, the modules are designed to meet the following core requirements:

1. **Adequacy**: Learning units must be structured and guided, so that learners can use modules individually and on their own, without any assistance. Also, the appearance, especially colors and layout, are in line with the domain of nursing.
2. **Comprehensibility**: As many nursing procedures are critical to the patient's health, instructions need to be unambiguous. Clear wording, appropriate illustrations and professional translations with speech output help to reach this goal.
3. **Attractiveness**: Animations, interactivity and short quizzes can help to increase learners' motivation and support the learning effect. However, gamification elements are used with consideration to match the seriousness of the content.
4. **Adaptability**: The content runs properly on different kinds of devices, e.g. laptops, smartphones and tablets. Also, a module must "memorize" where the user has stopped last time, to ensure a seamless learning experience

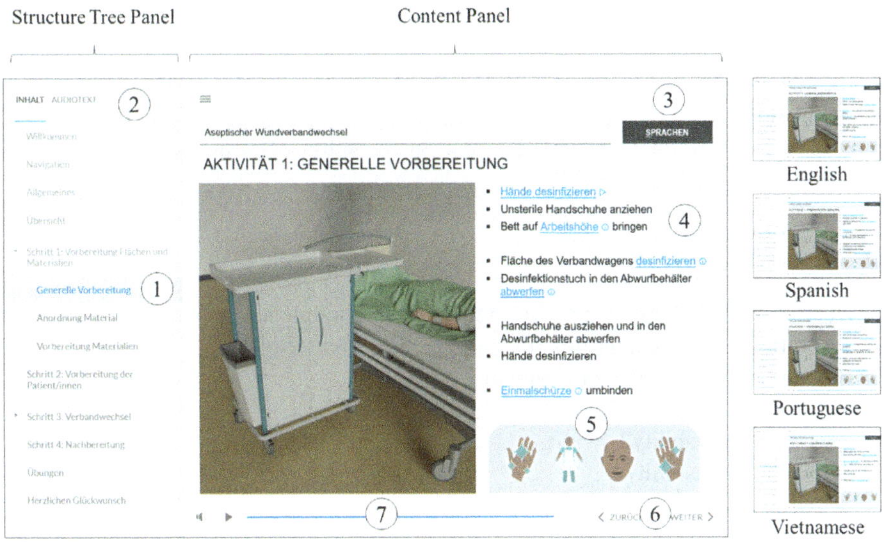

Fig. 2. Module "Aseptic Change of Wound Dressing" in German (left). The user can change to one of the available translations (right) any time. However, only the German content allows to navigate within the process sequence. This specific feature ensures that users do not skip the German learning content.

Figure 2 shows a screenshot of the learning module "Aseptic Change of Wound Dressing", as seen in a browser window on a laptop. The clockwise numbers refer to the paragraph below which outlines how we satisfy the previously mentioned requirements.

Basically, each module is divided into a structure tree panel on the left and a content panel on the right, taking the better part of the screen. On the structure tree panel, the current step or action is always highlighted (1). A step can have subordinate actions, indicated by an expandable subtree structure. Besides the structure tree, the panel holds a secondary tab (2), which contains a transcript of the current audio text. The default/startup language of each module is always German, but learners can change to English, Spanish, Portuguese or French, depending on their language proficiency via drop-down menu (3). Once another language has been selected, all text on the details panel and all audio narration is available in that language. Depending on the step, the content panel can include hyperlinks (4) for additional information and videos. As images on the content panel often present the situation from a first person perspective, an additional, embedded panel (5) indicates the current state of the user's hands (disinfected or not, wearing gloves or not), body (wearing an apron or not) and face (wearing a mask or not, wearing goggles or not). User must navigate through the process sequentially, using the forward and back buttons (6) in the lower right corner of the screen. At the bottom of the details panel, the player controls (7) allow volume adjustment, pausing and rewinding or fast-forwarding to any point on the timeline.

4 Empirical Study

In November 2020, we conducted a preliminary, empirical study to evaluate the general adequacy of the learning module design for education of foreign nursing trainees. The study focuses on a single module's effect on reception of terminology and process knowledge as well as usability and user acceptance.

4.1 Research Questions

1. Does a controlled provision of a single learning module lead to a significant gain in absorption of German terminology and process knowledge?
2. How do learners evaluate the module's

    a. Understandability?
    b. Comprehensibility?
    c. Usability?
    d. Value?

4.2 Test Design

The study took place on November 27, 2020, at the Knappschaft Kliniken nursing school in Bochum, Germany. All participants were located in a regular classroom, equipped with standard PCs. The learning module and all questionnaires were provided via the LMS of the nursing school. As a typical procedure, the aseptic change of wound dressing was chosen. It comprises 12 content pages, equaling approximately 12 PDF document pages, and one interactive quiz. No further aids were provided or permitted.

Participants

The sample comprised 16 nursing students (9 female and 7 male, ages between 18 and 30 years), all with Vietnamese as first language and all in their first year of training in Germany. 14 participants had certified German skills of level B1 and two participants of level B2. Before coming to Germany, all participants had accumulated at least one year of experience in nursing ($M = 1.75$ years, $SD = 0.86$ years).

Procedure

With an overall duration of 4 h, all participants followed the same procedure, comprising a pre-test (1 h), a learning phase (2 h) and a post-test (1 h). Both the pre-test and post-test were held in Vietnamese. The pre-test included an anonymized questionnaire on personal data, followed by a combined test on German terminology (5 questions) and aseptic change of wound dressing process knowledge (11 questions). During the learning phase, participants got access to the learning module. The post-test featured an unannounced repetition of the test on German terminology and process knowledge, followed by an evaluation sheet on understandability, comprehensibility, usability and value. Pausing was allowed at any time.

4.3 Results

Absorption of German Terminology and Process Knowledge
We calculated the participants' success rate (correct answers in percent) for the test on German terminology (5 questions) and process knowledge (11 questions) BEFORE and AFTER the learning phase. Means and standard deviations are reported in Table 1.

Table 1. Means and standard deviations of success rates in percent for questions on German terminology and process knowledge BEFORE and AFTER the learning phase.

	German terminology		Process knowledge	
	Before	After	Before	After
Mean [%]	88.75	95.00	39.97	52.38
SD [%]	17.84	11.55	12.69	18.19

For success rates in learning German terminology, the Kolmogorov-Smirnov Test indicated no normal distribution for neither the BEFORE nor the AFTER sample set, so the (nonparametric) Wilcoxon Signed Rank Test was applied. It reported no significant difference. Focusing on success rates for process knowledge affiliation, the Kolmogorov-Smirnov Test indicated a normal distribution for both the BEFORE and the AFTER sample set, so the Paired Samples t-Test was conducted and showed a significant gain with a medium effect size ($t(15) = -2.47, p < .026, r = .54$).

Evaluation of Comprehensibility, Traceability, Usability and Value
Each participant rated the module's quality regarding comprehensibility, traceability, usability and value for knowledge absorption. The scale was based on the International Telecommunication Union's Absolute Category Rating (ACR) scale [11] which comprises five levels: 1 (bad), 2 (poor), 3 (fair), 4 (good) and 5 (excellent). Means and standard deviations are reported in Table 2. Results show that all categories received an average rating between 4 (good) and 5 (excellent).

Table 2. Means and standard deviations of ratings on understandability, comprehensibility, usability and value. Participants rated on a scale from 1 (bad) to 5 (excellent).

	Understandab.	Comprehensib.	Usability	Value
Mean	4.00	4.06	4.58	4.31
SD	0.82	0.68	0.51	0.70

4.4 Discussion

With over 88% correct answers on average, knowledge on German terminology was already on a high level before access to the learning module. Still, the success rate increased up to 95%. Although the difference is statistically not significant, it indicates that the module helps with learning German terminology, but this hypothesis needs to be verified with a larger sample size and an extended test on vocabulary. The process knowledge success rate gained significantly from below 40% to over 50% on average. This is very promising, considering that the module access time was limited to two hours. As learning strategies can differ between students, additional access time could correlate with an increase in knowledge absorption, but this assumption needs to be investigated with additional studies. Ratings on understandability, comprehensibility, usability and value between 4 (good) and 5 (excellent) indicate that the general design of the learning module is appropriate and should be continued.

5 Outlook

As mentioned in chapter 2, the final solution will comprise an overall of 24 learning modules which we plan to evaluate successively. Test design will include enlarged sample sets, extended questionnaires, and hands-on exercises, as the current results do not allow any assumptions on the effect of the participants' practical capabilities. In this context, the effectivity of VR training will be another focus of investigation. Finally, for the screen-based learning as well as the VR training, professional eye tracking technology will help to evaluate the respective user interface design.

Acknowledgments. This project is funded by the German Ministry of Health.

References

1. Igelmann, A.: Demographic Change in Germany: Reasons, Consequences, Potential Solutions. AV Akademikerverlag. ISBN-10: 978363939996X (2012)
2. Al Mutair, A., Amr, A., Ambani, Z., Al Salman, K., Schwebius, D.: Nursing surge capacity strategies for management of critically Ill adults with COVID-19. Nurs. Rep. **10**(1), 23–32 (2020). https://doi.org/10.3390/nursrep10010004
3. Jiang, J., Huang, Z., Huo, B., Zhang, Y., Song, S.: Research progress and prospect on nursing robot. Recent Pat. Mech. Eng. **2018**(11), 41–57 (2018)
4. Neuhöfer, J.A., Hansen, S.: Digital prequalification for nursing trainees with migration background. In: Nazir, S., Ahram, T., Karwowski, W. (eds.) AHFE 2020. AISC, vol. 1211, pp. 159–163. Springer, Cham (2020). https://doi.org/10.1007/978-3-030-50896-8_24
5. Pfeiffer, T., Hainke, C., Meyer, L., Fruhner, M., Niebling, M.: Virtual SkillsLab – a training application for infusion preparation. In: Proceedings of DeLFI Workshops 2018 at the 16th e-Learning Conference of the German Computer Society, Frankfurt, Germany (2018)
6. Knappschaft Kliniken GmbH: University Medical Center Knappschaftskrankenhaus. Bochum. https://www.kk-bochum.de/en/index.php. Accessed 31 Jan 2021
7. Advanced Distributed Learning Initiative: SCORM – Sharable Content Object Reference Model. https://www.adlnet.gov/resources/scorm-resources/. Accessed 31 Jan 2021
8. Moodle Learning Management System. https://moodle.org/. Accessed 31 Jan 2021
9. World Wide Web Consortium: WebXR Device API Working Draft, 24 July 2020. https://www.w3.org/TR/2020/WD-webxr-20200724/. Accessed 31 Jan 2021
10. Facebook Technologies, LLC: Oculus Quest: The All-in-One-VR-Headset. https://www.oculus.com/quest/. Accessed 31 Jan 2021
11. International Telecommunications Union: Recommendation P.910 (04/08): Subjective Video Quality Assessment Methods for Multimedia Applications. https://www.itu.int/rec/T-REC-P.910-200804-I. Accessed 31 Jan 2021

A Human-Centered Approach for an Autism-Friendly Drawn Space

Anna Lisa Pecora, Erminia Attaianese(✉), and Alessandra Pagliano

University of Naples Federico II, Via Toledo 402, Naples, Italy
{annalisa.pecora,erminia.attaianese,
alessandra.pegliano}@unina.it

Abstract. In recent years, drawing disciplines supported by digital technologies, are playing a fundamental role in the interdisciplinary approach to communication. The ICT may be a versatile instrument for controlling communication at different levels according to the user's personal needs. The autistic clinical frame is mainly connected to perception; then Virtual Environment can become effective support in understanding physical space for people with ASD. However, even if the research is growing in this field, rarely it focuses on relations between human factors and VR. Present paper is aimed at filling in the studies gap about details in VR spatial representation and its perceptual resonance in communication by developing drawing guidelines for the "autism-friendly" design of VR. The study is aimed at shifting the design focus from the technological instrument to the user, in order to enhance the potential of VR as an inclusive education medium.

Keywords: Human factors · Virtual reality · Representation · Inclusion

1 Introduction

In the last twenty-five years, virtual reality (VR) have proven to be very promising for special education, particularly for non-neurotipic people [8, 10]. The trust in this field of studies depends on positives research results in showing the Virtual environment (VE) as both a valid communicative medium and a self space where learning experiences can be approached without the in vivo situations limits.

Particularly in autism, social and communicating problems can complicate many aspects of everyday life. Thus, most researches are aimed at improving life skills, in order to support the autistic people autonomy. Nevertheless, rarely the academic studies focus on human aspects, particularly the visual perception involved into a VR experience. Since VR is a drawn space, the relationship between graphic code and human answer is a primary issue, for understanding how the communication of the space can be enhanced and knowledge improved. On the basis of a literature review, a comparison of a selection of images taken from experimental researches with relative output data have been done, so that some basic guidelines for designing an autism-friendly virtual environment have been deducted. Then, An AVR prototype is developed, in order to test our assumption with the specific target of users.

S. Nazir et al. (Eds.): AHFE 2021, LNNS 269, pp. 71–77, 2021.
https://doi.org/10.1007/978-3-030-80000-0_9

2 Scope of the Study

The paper is aimed at filling in the studies gap about details in VR spatial representation and its perceptual resonance in communication by developing drawing guidelines for the "autism-friendly" design of VR. The study is aimed at shifting the design focus from the technological instrument to the user. The main goal is enhancing the potential of VR as an inclusive education medium, by improving the quality of cultural contents in the architectural heritage communication and learning for all. As Olga Bogdashina affirms: "filtering of an infinite amount of information is necessary to make the processing of information effective and conscious" [2]. For this reason, the graphic design for an autism-friendly VR has to provide essential information about the environment for minimizing sensorial pressure and avoiding distraction. The representation path supports the required process of "synthesis, communication and explicitation" of the physical space, for decoding and subsequent learning its cultural contents. Particularly, complex theoretical and morphological relations among different elements as colors, lights, textures, and pattern are included in the architectural space. So, only through the right graphic choices, the figurative configuration of these features can be deeply understood. Thus, the detail level setting, the chromatic and luminous qualities, the quantity, and the value of graphic signs in the VE, become essential features to define the sense of immersion and presence and, therefore, the communicative power of the VR driven by the perceptual inputs. A right interpretation of spatial codes through a study of the drawing can sustain a human-centered design of a virtual environment, where architectural heritage becomes the medium for communication and inclusion for non-neurotypical people as well as for anybody else.

3 Human Factor, VR and Autism

3.1 ASD Perceptual Aspects

The complex characteristics of autistic people's perception are difficult to summarize in a short description. Frequently, experimentations on ASD perceptual responses reveal contradictory results, leading to opposite assumptions. According to Delacato (1974), we refer ASD to Hyper and Hypo sensitivity [5]. The first condition is associated with the need to process too much information, due to the extreme sensibility of sensorial channels; the second concerns, at the opposite, a low sensorial sensibility, that seems to inhibit the reception of the environmental input. Both are strongly influenced by the environment which can become supportive in transmitting the right spatial information or, at the same time, providing dysfunctional sensations in the case of over or hypo stimulation. In fact, many subjects with ASD, are not always able to effectively and consciously adapt themselves to the context in which they are, probably due to an altered sensory integration.

Despite the visual system is not an impairment in autism, the perceptual incoherence leads to sensorial agnosia that may hinder the ability to fully understanding figurative signs of the environment in semantic terms [2]. This results in the perceptual impairment of the main senses and deficits in executive functions, with alteration of cognitive processes (such as concentration, planning, attention) and general difficulty of interaction and control of the environment.

3.2 The "Autism Friendly" VR Design

On the basis of a literature review, particularly focused on studies reporting ASD users' feedback to different type of images, some basic aspects of an autism friendly VR design have been deducted, in order to define guidelines for application.

Physical Aspects: The navigation modality is a fundamental aspect in the understanding of space and for reduce motion sickness. Continuous navigation is more suitable for autistic people than point and teleport, because it helps the wayfinding, by creating relations between figurative elements of the spatial framework. A natural frame, with an egocentric point of view, is essential for improving comfort and immersion in ASD people. In fact, considering their abstraction impairment, autistics could be stressed and disoriented by an unreal visualization, like a top or bottom view. Objects in a virtual environment play the same role in creating awareness of being there, therefore they should behave in a realistic way, without floating in the space or showing unreal distances. Similarly, observing the arms and legs of the avatar's body helps their sense of presence and their difficulties with proprioception [13].

Perceptual and Emotional Aspects: Some spatial characteristics can provide overstimulation and stress, while others can enhance the comfort and the understanding of the environment. The configuration of clear spatial borders is important to provide a visual reference for displacement and orientation. Figurative guidelines for motion can also work as a tutor for navigation. Often, autistic people have difficulties evaluating dimensions, distances, shapes, deep sense [2], therefore, simple and clear spatial frameworks help users to create their own idea about the virtual world and its position inside it. For this reason, the spatial elements have to be designed in order to give information about depth and distances with specific shade aspects and textures. Breaking space continuity with sensorial rest and deleting the perceptual inputs during the transition between the different scenes, are also useful. Moreover, different functions have to be highlighted by different visual characterizations, like colors or textures, in order to strengthen their role. The comforting sense in ASD people generally increases in recognizable and familiar spaces, while new experiences can be a reason for stress [13].

Cognitive Aspects: Simple frameworks with low sensorial stimulation (few colors and textures only to underlines visual targets) create a comfortable environment for learning reducing pain soliciting the perceptual system [2]. The stimuli can be introduced gradually giving time, to the user, to adapt and take confidence in the previous configuration. Instructions for navigation and training need to be few and clear, preferring images, while targets have to be highlighted by simple or bright colors [14]. Too much light can intensify the perceptual load providing overstimulation, whereas a soft illumination creates a neutral background with low contrast and soft colors communicating a sense of comfort. The main problems are related to abstract concepts since the most common impairment in ASD people is the lack of imagination and abstraction. Referring to these, the use of complex and unusual virtual objects is totally superfluous because they couldn't contribute to understanding the space, worsening the sensory weight.

4 Methodology

By applying the assumed guidelines, an autism-friendly virtual tour is set up and provided to a group of autistic users. The virtual tour overlaps the in-vivo tour with the aim to enhance the understanding of the physical space and to familiarize with the environment. Therefore both tours (virtual and real) are experienced in the same place chosen as a case study for testing the prototype. The virtual experience has the pedagogic role of mediating the heritage cultural content. It is structured for offering different cognitive levels (low, medium, high).

A simple geometry characterizes the whole graphic process with the aim to provide a gradual stimulation, starting from the lowest level until the realistic effect. The visual inputs appear step by step, so the information is separately processed helping the reading of the ambient configuration (Fig. 1 and 2). The route consists of different stages, with rest spaces between them helping the transition towards rooms with different features. In each step, the graphic setting provides a good continuity between the real and the virtual space thanks to a specific frame sequence (Fig. 3). The outdoor tour starts with the realistic image of the real environment because the visitor first experiences the physical space and then he puts on the VR helmet. On the contrary, the indoor tour starts with the lowest stimulation level and ends with the full visual configuration. Therefore, the last virtual image is, as much as possible, similar to the real environment that the user experiences, removing the helmet, step by step. A narrative pattern supports the virtual route enhancing the attention and recording of main topics. Since autistic people, frequently, have wayfinding problems and difficulties in understanding the motion direction, a character, sometimes, appears to catch the attention. The avatar, strongly marked by a colored detail, has many roles: visual target, strain for the narration, help for the familiarization with the scene. Through a recognizable and visible characterization, it becomes a support for the wayfinding and the orientation, driving the sight and pointing on specific details in the scene.

4.1 The Site

The case study, the Real Tenuta di Carditello, in San Tammaro (CE, Italy), has been selected because its characteristics are consistent with our previous assumption. It is a hunting lodge, built for king Ferdinando IV di Borbone in 1787 by the Francesco Collecini architect. The building is characterized by reduced decorations and, at the same time, richness of cultural contents. It contains several frescoes and painted vaults some of them decorated with trompe-l'oeil. The structure is strongly defined by symmetry and reiteration of the same architectonic elements, that configure a simple, schematic pattern. The naturalistic environment is dominated by silence and green spaces with animals, allowing the diversification of several multisensory tours.

4.2 The Autism Friendly VR Tour Prototype

The graphic method for developing the prototype is based on an integrated approach based on different techniques. Most of the tour contents are developed through 3D modeling. A volumetric model of the historical building has been designed for minimizing visual information. The model reproduces only essential architectonic details, without decorations or textures. A global light, with shadow effects, is chosen to improve the naturalistic atmosphere. Another important feature is the sense of presence in order to support immersion and comfort since a good sense of presence enhances the attention and supports the knowledge process. Sometimes, autistic people have difficulties in embodying a virtual subjectivity because of their problems with proprioceptors [2]. According to this, we have set the human point of view for the whole virtual experience, this way enhancing the first-person sensation. In the same way, continuous navigation is used for the motion because it simplifies the development of mental maps and the link between different portions of the space [13]. The animation is obtained by linking different frames composed of panoramic images. The 360 images are deducted by render (Fig. 1), from the 3D model, or from panoramic photos managed to obtain the gradual raising of graphic complexity (Fig. 3). When spatial information increase, both, 360 render and photos, are combined with details extracted by the photogrammetric survey. Thus the target details have a different and deep characterization compared to the neutral background.

Fig. 1. Frame from the prototype: the 360 rendering showing an increase of sensorial stimulation.

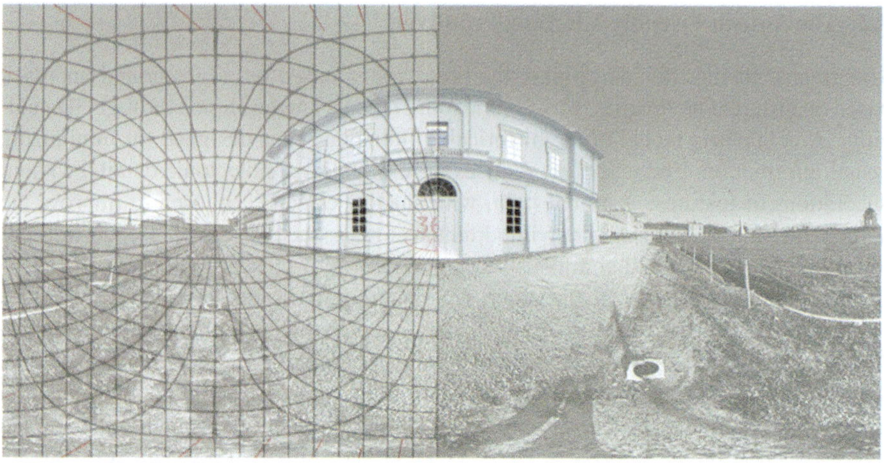

Fig. 2. Frame from the outdoor VR tour: placement of the 3D model in the contest.

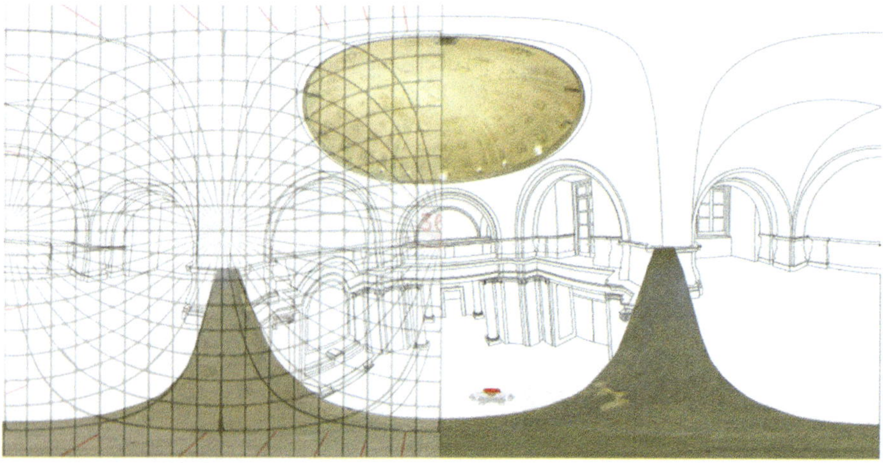

Fig. 3. Frame from the prototype: 360 image from the editing of a panoramic photo.

5 Expected Outcomes and Conclusion

The research focuses on the spatial representation of VR and its perceptual response, filling in the gap of studies about VR and autism. By applying the assumed guideline, the approach with VR systems can turn from the technological aspects to the human factors, conferring it the strategic role of inclusive medium communicating the narrative potential of architectural space. The next step of our research expects to test the VR prototype with a small group of autistic users. The level of severity of ASD will range from moderate to medium. The users will experience the VR and the in vivo tour both in the Real Tenuta di Carditello in alternating sessions. The procedure will provide continuity between the two experiences, in terms of time-frame, narration, and visual information. Some visual details, highlighted in the VR tour, will be used as landmarks to help the wayfinding and the understanding of the spatial characteristics, by linking the virtual and the real experience. These recurring graphic signs/symbols will support the narration conveying the architectural contents in their morphological and conceptual characteristics. The tour could represent a communicative medium not only for the content housed in the building but for the architecture itself as the subject of narration. This way, the relationship with space could enhance human abilities by improving knowledge for non-neurotypical people as well as for anybody else.

References

1. Attaianese, E., Minucci, G.: Progettare Percezione e Piena Fruizione dei Siti di Interesse Culturale da Parte di Persone con Autismo. In: Baratta, A.F.L., Conti, C., Tatano, V. (eds.) Iclusive Living. Il Design for an Autonomous and Independent Living. Anteferma Edizioni S.r.l. Conegliano, TV, pp. 302–311 (2019)
2. Bogdashina, O.: Le Percezioni Sensoriali nell'autismo e nella Sindrome di Asperger, Uovonero, Vignate (MI) (2015)
3. Bradley, R., Newbutt, N:. Autism and Virtual Reality Head-Mounted Displays: a State of the Art Systematic Review, JET (2018)
4. Cobb, S.V.G.: Virtual environments supporting learning and communication in special needs education. TLD No. 3, Vol. 27, pp. 211–225 (2007)
5. Delacato C.H.: The Ultimate Stranger, The Autistic Child. Garden City, NY, Doubleday (1974). ISBN 0385010745
6. Jason, J.: The RV Book: Human Centered Design for Virtual Reality. ACM and Morgan & Claypool Publishers (2016)
7. Mineo, A.B., Ziegler, W., Gill, S., Donna, S.: Engagement with electronic. screen media among students with autism spectrum disorders. J. Autism. Dev. Disord. **39**, 172–187 (2008). https://doi.org/10.1007/s10803-008-0616-0
8. Morganti, F., Riva, G.: Conoscenza, Comunicazione e Tecnologia. Aspetti Cognitivi della Realtà Virtuale. LED On Line (2005)
9. Newbutt, N.: The potential of virtual reality technologies for autistic people: a pilot study. Network Autism (2015)
10. Parsons, S., et al.: Development of social skills amongst adults with Asperger's syndrome using virtual environments: the 'AS interactive' project. In: Sharkey, P., et al. (eds.) Proceedings of the 3rd International Conference on Disability, Virtual Reality & Assoc. Tech., Alghero, Italy, pp 163–170 (2000)
11. Parsons, S., Mitchell, P.: The potential of virtual reality in social skills training for people with autistic spectrum disorders. JIDR **46**(part. 5), 430–443 (2002)
12. Realpe, A., et al.: Co designing a virtual world with people to deliver social cognition therapy in early psychosis. In: Early Intervention in Psychiatry, pp. 1–7 (2019)
13. Rossi, D., Olivieri, A.: First person shot. Convegno UID **2019**, 977–984 (2019)
14. Strickland, D.: Brief report: two case studies using virtual reality as a learning tool for autistic Children. JADD6 **26**, 651–659 (1996)

An Interactive Guide Based on Learning Objects to Train Teachers on the Detection and Support of Children with Attention Deficit Hyperactivity Disorder

Karina Parra-Luzuriaga, Yaroslava Robles-Bykbaev, Vladimir Robles-Bykbaev(✉), and Paúl León-Goméz

GI-IATa, Cátedra UNESCO Tecnologías de apoyo para la Inclusión Educativa, Universidad Politécnica Salesiana, Cuenca, Ecuador
kparral@est.ups.edu.ec, {zrobles,vrobles}@ups.edu.ec, paul.leongomez@mail.utoronto.ca

Abstract. Attention Deficit Hyperactivity Disorder (ADHD) is a neurodevelopmental disorder that interferes with normal children's functioning or development. Children with ADHD show difficulties related to their behavior and school performance. Due to this situation, their teachers must develop assertive communication with the parents and, at the same time, deepen the children's diagnosis. However, in Ecuador, there are no real estimates about ADHD that can help conduct early and appropriate interventions on patients who present this disorder. For these reasons, school teachers do not have therapeutic and pedagogical tools to address students' needs with ADHD. In this line, it is important mentioning that teachers are the second most important educational agents in the child's life, occupying most of their time after the parents. Therefore, it is essential for providing Ecuadorian school teachers with the knowledge and skills to carry out early detection and the consequent management and support of children with ADHD (educational inclusion). Thus, the teachers will develop pedagogical techniques based on psychoeducational guides that facilitate understanding students' special educational needs. In this paper, we present an interactive educational guide for school teachers because of the foregoing. This guide relies on expert systems and quasi-experimental research to train teachers of the "La Salle" Educational Unit (Azogues city) through the application of psychoeducational workshops. The guide address the following topics: recognition, functioning, behavioral control, emotional self-regulation, and social area of children with ADHD. The pilot experiment show encouraging results and high satisfaction levels on the teachers' side.

Keywords: Attention deficit hyperactivity disorder · School · Educational inclusion · Learning objects

1 Introduction

Attention Deficit Hyperactivity Disorder (ADHD) is considered as a "persistent pattern of inattention and/or hyperactivity-impulsivity that interferes with development" [1].

S. Nazir et al. (Eds.): AHFE 2021, LNNS 269, pp. 78–83, 2021.
https://doi.org/10.1007/978-3-030-80000-0_10

The main symptoms of this disorder affect behavior in the social, family, and academic spheres. School performance is particularly affected since teachers only know basic information about this disorder and therefore, they do not how to act with them. The lack of awareness of ADHD, the problems to stabilize behaviors, and negative emotions, are some of the factors that weaken the role and function of teachers within the classroom [2].

In Ecuador, there are no statistics regarding ADHD. Thus, the absence of data and information constitutes one of the main limitations for primary teachers to detect and timely manage this pathology. This is mainly because teachers do not have therapeutic and pedagogical tools that allow them to address the needs of students [3].

The teaching-learning processes in children with ADHD require adequate detection and treatment in cities such as Azogues through pedagogical techniques that facilitate the understanding of school demands [4]. We have achieved a psychoeducational intervention through an interactive guide to improving the knowledge of basic education teachers from La Salle, educational unit. The training took place during workshops focused on recognition, functioning, behavioral control, emotional self-regulation, and social intervention of children with ADHD. The project was executed through a virtual learning object, a tool that allows technological educational reinforcement for the other professionals of the school.

2 State of the Art

According to Bierderman en [5], ADHD is evidenced by the presence of symptoms with a duration of six months of inattention and/or hyperactivity-impulsivity. This type of behavior differs from people of the same level of education. People with ADHD reflect such symptomatology from the age of seven until adulthood. The disorder has a neurobiological origin, as a result of an alteration of the neurotransmitter called dopamine. This leads to behaviors with hyperactivity, impulsivity, and inattention, associated with an important functional impairment in the child's development [5].

The interaction between pre and postnatal genetic and environmental factors hinders the capacity for self-control, emotional self-regulation, inhibition of negative and impulsive behaviors, and activity level in children with ADHD [6]. Therefore, multimodal treatment for this type of pathology is crucial. It consists of a combination of factors: pharmacological, psychotherapeutic, and pedagogical. They vary according to the intensity and frequency of symptoms in each child to exercise inclusive positive progress [4].

The presumptive or definitive diagnosis of ADHD does not depend on the symptom, but actually on the intensity and frequency that occurs in the child. However, there are disputes between research and the reality of the disorder. Thus, the diversity of opinions generates obstacles for the diagnosis of ADHD. Pathognomonic tests and criteria to categorize the disorder are still in development which further exacerbates the issue of detection [7].

Worldwide, the prevalence of ADHD is around 5.3% in children and adolescents, and 2.5% in adults [1]. 76% of ADHD cases are related to a genetic origin, such as premature birth, hypoxic-ischemic encephalopathy, infections of the central nervous system, consumption of psychoactive substances in pregnancy, among others. Yet, the psychosocial factor is not considered part of the diagnosis despite being a fundamental part in the intervention of behavioral control in children with ADHD [8].

The following table reflects the specific prevalence of ADHD, according to its representativeness in some countries (Table 1):

Table 1. Comparision of the quality of life in patients aged 5 to 14 years with a diagnosis of attention deficit hyperactivity disorder who receive treatment at the hospitals of the Ecuadorian Social Security Institute of Quito. Pontifical Catholic University of Ecuador [9]

Country	Prevalence
Spain	5 to 8%, meaning that there are 1 or 2 children with TDAH per class
USA	11%, 6.4 million cases of people with TDAH
México	5.2% 1.5 million cases of people with TDAH
Colombia	3%, 317.665 cases of people with TDAH
Chile	10%
Argentina	4.99%

In Ecuador, the Ministry of Education (2017) registered 7918 students with ADHD in public schools. However, in Quito, it was observed that 246 students (7.3%) between 14 and 18 showed ADHD, 6.5% of inattentive subtype, and 2.8% of hyperactive-impulsive subtype [10].

In Lima, it was observed that after an educational intervention, the level of knowledge about the ability to detect possible cases of ADHD increased by approximately 98% in the first stage of such educational intervention, 49% in the second stage, and 53% in improvement in the last stage.

3 Materials and Methods

A quasi-experimental study was conducted focusing on a psychoeducational intervention in teachers between 27–59 years of age in La Salle school, Azogues, Ecuador. It focused on improving the level of knowledge of teachers regarding the detection and timely management of children with ADHD between 5–12 years old. The intervention lasted a total of 10 days, spread over 5 working weeks (Mondays and Wednesdays). Each educational intervention addressed ADHD issues in 2.5 h (8:00 am–10:30 am), for a total of 25 academic hours.

A virtual course was given to improve knowledge in the detection and management of children with ADHD. It was carried out through the participation of the author and a team of experts in clinical psychology, computer science, medicine, statistics, and anthropology. The course applied an interactive didactic guide, based on a learning object, Exe learning (free interactive educational resources editor, in computer supports). Exe learning is a tool based on free code that creates online educational contents in a rapid and structured way. This tool can also be exported through a link to generate easy access to the knowledge platform. Therefore, psychoeducational training was generated virtually and adapted to the conditions of the COVID-19 pandemic [11].

The experimental group consisted of 33 teachers from La Salle. A questionnaire was applied to everybody before and after the psychoeducational intervention aimed at determining the level of knowledge that they have regarding the detection and management of children with ADHD between 5–12 years of age. The questionnaire was elaborated by experts in the area of clinical and educational psychology, medicine, anthropology, among others. It was later reviewed by a team of experts in education for children with ADHD consisting of clinical and educational psychologists. After careful revision, the questionnaire was applied to the teachers of the institution.

The educational contents from the virtual course were the following:

First section: ADHD, its history, diagnosis, conceptualization, symptomatology, the algorithm in the detection of the disorder, diagnosis, statistics, and multimodal treatment, covering specific and important generalities for the correct detection of students with ADHD. The second section focused on the causes of ADHD, the importance of neuroeducation, and executive functions. This section also dealt with an educational intervention focused on pedagogical techniques and a psychoeducational guide for the emotional and behavioral self-regulation of the student (Fig. 1).

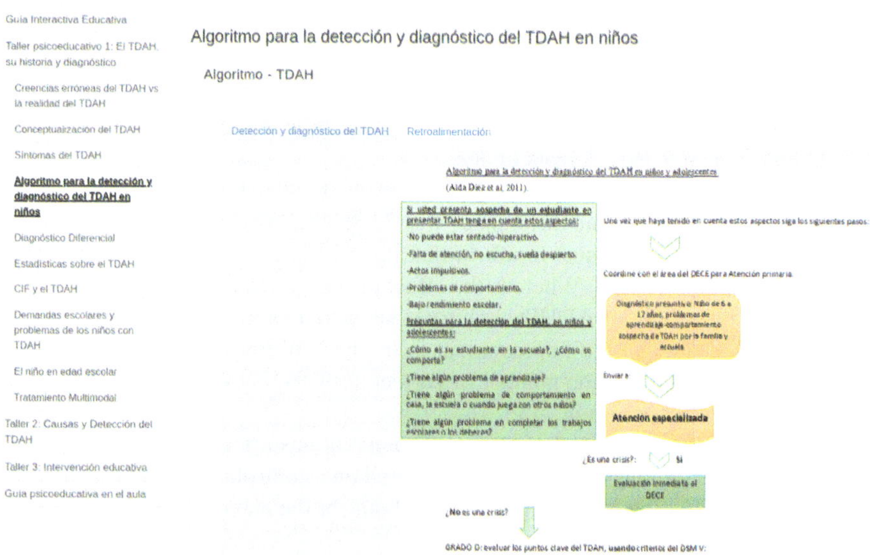

Fig. 1. A screenshot the learning object contents algorithm to perform the ADHD detection and diagnosis.

It is also important to point out that the platform contains annexes of psychoeducational reagents, such as the CONNERS behavioral questionnaire. This questionnaire focuses on the detection of a possible ADHD case. It is aimed at parents and teachers to help them differentiate using an observed behavior questionnaire at school and home. This is to prevent confusion of the disorder.

The educational contents were based on a philosophical and cognitive study. It uses three psychosocial interventions: a psychoeducational guide for teachers, behavioral interventions with peers, behavioral and emotional control in the classroom with ADHD students. They were previously validated and reviewed by a team of experts in the area of clinical and educational psychology with more than 15 years of experience. Experts carefully reviewed the educational contents and made recommendations to modify the project.

After content validation, we proceeded to determine the perception of La Salle Educational Unit teachers regarding the interactive guide for ADHD. This was done through the application of a survey previously validated and elaborated by experts in the area of statistics, clinical psychology, medicine, and anthropology.

The information obtained from the survey and the questionnaire were analyzed using inferential statistics (measures of central tendency and dispersion). Specifically, we used the "Yuend" test, a robust test for paired samples.

4 Results

To the inferential analysis of data, the "Yuend" test was applied. It is a robust t-test for paired samples because the samples corresponding to the pre and post-test were constituted by the same volunteer subjects in the study. It was determined to look for the difference in means between the pretest and posttest results concerning the "wrong" questions of the participating subjects. That is, the questions that reflect lack of knowledge regarding some topics related to the timely detection and management of ADHD in children between 5 and 12 years of age.

After applying the "Yuend" test, we obtained a "T" statistic of 3.8123. P-value of 0.00515. The "Trimmed mean difference" was 7.88889. Therefore, the null hypothesis was rejected. It is possible also to affirm that there is an evident difference in means since the value is far from 0. Whereas the effect of the psychoeducational intervention to improve identification of ADHD in children showed a value of 78% (0.78). In other words, we can argue that the improvement in the level of knowledge of the teachers, reflected by the increase in the number of correct answers, is explained by 78% of the psychoeducational intervention.

We can be concluded that these types of interventions are an added value and are successful. That is, as long as the planning is based on a controlled systematic logic and through the participation of a team of experts, both for the elaboration of contents and for the intervention performance.

5 Discussion and Conclusion

The psychoeducational intervention to teachers of the Educational Unit, La Salle is a pedagogical tool that reinforces their knowledge in the detection and management of ADHD. It also supports teaching in the classroom and/or at home due to its ease of access, organization, and simplicity. All these mentioned advantages correspond to the daily needs that teachers have in the classroom. For instance, during issues at time management activities, or during homework, time control, behavioral and emotional self-regulation in students, as well as during the application of curricular adaptations [8].

In addition, during this project, we evidenced that the methodology worked (virtual psychoeducation, adapted to the conditions of the pandemic). It generated a positive impact on professionals because of the technological educational material and psychoeducational strategies applied. They in turn allow to intervene and correct the needs of students with ADHD. They also enable to work together with parents to strengthen the biopsychosocial sphere of the student.

At the end of the psychoeducational training program, teachers have learned the definition of detection and intervention of students with ADHD. This helped to improve their approach to the education of children with ADHD. The training program has also enabled teachers to improve their ability to detect students with ADHD who require curricular adaptations in their learning. It also helps to generate educational methodological strategies based on a comprehensive approach to teaching-learning.

References

1. American Psychiatric Association, Ed.: Guía de consulta de los criterios diagnósticos del DSM-5. American Psychiatric Publishing, Arlington, VA (2014)
2. Robalino, M.X.: El déficit de atención y las técnicas activas de aprendizaje en los estudiantes del quinto año de educación general básica de la escuela fiscal "Joaquín Sánchez de Orellana" de la Parroquia de Puembo, Cantón Quito, Provincia de Pichincha. Universidad Técnica de Ambato, Ambato, Trabajo de titulación (2017)
3. Paz, C.L.: Formación inicial de docentes para la atención a la diversidad en contextos inclusivos, Paradig. Rev. Investig. Educ., vol. 24, núm. 37, Art. núm 37 (2017)
4. Pacheco, C.M.: Propuesta de diseño de una plataforma digital interactiva que integre y planifique actividades para el seguimiento de los tratamientos a niños con trastorno de déficit de atención e hiperactividad (TDAH) y síntomas relacionados entre 2 a 12 años
5. Biederman, J.: Attention-deficit/hyperactivity disorder: a selective overview. Biol. Psychiatry, **57**(11), 1215–1220 (2005)
6. Zuluaga-Valencia, J.B., Fandiño-Tabares, D.C.: Comorbilidades asociadas al déficit de atención con hiperactividad. Rev. Fac. Med., vol. 65, núm. 1, pp. 61–66 (2017)
7. Rodríguez, C., Gonzalez, M., Arroba, M., Cabello, L.: Prevalencia del trastorno por déficit de atención con hiperactividad en niños de una localidad urbana, Pediatría Aten. Primaria, vol. 19, núm. 76, pp. 311–320 (2017)
8. Tratamiento del trastorno por hiperactividad infantil. http://tdahytu.es/2020/05/29/el-reto-del-tdah-en-el-aula-claves-para-crear-ambiente-de-respeto-confianza/
9. Baños, D., Benavides, E.: Comparación de la calidad de vida en pacientes de 5 a 14 años con diagnóstico de trastorno de déficit de atención con hiperactividad que reciben tratamiento en los hospitales del Instituto Ecuatoriano de Seguridad Social de Quito, p. 95
10. Ramos, C., Bolaños, M., Ramos, D.: Prevalencia del trastorno por déficit de atención con hiperactividad en estudiantes ecuatorianos», Rev. Científica Tecnológica UPSE, vol. 3, núm. 1, Art. núm. 1 (2015)
11. Exe learning, descubriendo la herramienta de autor • Tres punto e-learning, descubrien do la herramienta de autor. https://www.trespuntoelearning.com/exe-learning-herramienta-autor/

Application of Assessment-Analytic Knowledge Building Tools in Collaborative Creative Studies

Amic G. Ho(✉)

School of Arts and Social Sciences, Department of Creative Arts,
The Open University of Hong Kong, Ho Man Tin, Hong Kong
amicgho@ouhk.edu.hk

Abstract. The design process involved complex social, cultural as well as political issues. More skills are required for providing an interactive design experience. The designers organise and structure problems through interactive communication, service, artworks and design. As a result, designers were expected as applied behavioural scientists. The undergraduate design students, those are designers in the future, were disappointingly undereducated for their career. Creative students especially often neglect to recognise their tasks' complexity and connect them with the knowledge they already obtained. They often lack the requisite understanding that their viewpoints must be built on a knowledgeable foundation. Some scholars discussed various teaching and learning approaches, including applying computer-assisted learning, problem-based learning, and inquiry-based learning to explore applicable methods, especially for the design field. However, there were still not yet able to propose practical tools for training design students to interlock the complexity of individual and social behaviours under the influence of technical and social development. Hence, there is a need to explore a collaborative and scientific design education method.

Keywords: Knowledge building tools · Collaborative learning · Design learning process · Knowledge building forum

1 Introduction

The art and design outcomes were primarily concentrated upon tangible artworks or products. In before, industrial designers mainly focused on themselves with forms and functions as well as materials. At present, far more complex and challenging issues are involved in the creative artwork producing process. In the initial stage of creative arts, the outcomes were primarily concentrated upon tangible artworks or products. Industrial designers, in the past period, only need to concern about forms and functions, materials as well as manufacturing approach. Nowadays, far more complex and challenging issues involved in the creative artwork producing process. The creative arts process involved complex social, cultural as well as political issues. More skills are required for providing an interactive design experience. The designers control organising and structuring problems through interactive communication, service, artworks and design. As a result,

S. Nazir et al. (Eds.): AHFE 2021, LNNS 269, pp. 84–90, 2021.
https://doi.org/10.1007/978-3-030-80000-0_11

designers were expected as applied behavioural scientists. However, there were still not yet able to propose practical tools for training Creative arts students to interlock the complexity of individual and social behaviours under the influence of technical and social development. Some of them claimed that they could find novel solutions through different viewpoints. However, some solutions they suggested are seldom or unable to be implemented. The different views of the design students can provide insightful results. Hence, there is a need to investigated the essential elements within the learning and teaching processes of creative art studies. Focusing on the essential elements within the learning and teaching processes, they would develop possible tools for helping design students to be more interactive within their creative processes.

2 The Transformation of Learning Under the Influence of Constructivism

Research conducted on collaboration in creative writing [1], whether face-to-face or over a computer network; necessary for 'exercising and increasing awareness of inert knowledge'. The influence of an arranged procedure for collaborative writing on the learners' quality of creative writing and their thinking and feeling of writing were explored. The research team invited two classes and divided them into "experimental collaborative writing group", and another was set as a control group. The two groups were assigned to write within an eight-week course. The two classes worked on their creative writing tasks individually. In the experimental group, participants collaborated with community members of different abilities. Some of them played particular roles, such as assistants and writers. In the control group, participants worked with members of the community with similar skills, and they represent a supportive role. The research conducted [2] involved training children involved in the project. The performance of personal writing and the quality of cooperative writing during the study were analysed. For cross-ability experiments, the own benefits before and after writing were statistically significant, but not for their control. There was no outstanding differentiation in scores before and after the same ability test and no remarkable differentiation within the control group. However, the experiment with the same ability was improved, but the control had become worse. The gain difference between the experiment and the control was pronounced. Similarly, the collaborative writing score of the same ability pair was much higher than the original writing before its project, although the crossability pair was not. Both pairing writers reported that the method was easy to use, and most people were positive about it. The conclusion was that both designs of group writing might be sufficient. However, certifying the temporary benefits of assistants with more substantial cross-centring capabilities was complicated. More researches were needed, including investigations on cross-capacities of corresponding roles.

The application of computer science in studio-based learning in design-oriented programs was illustrated [3]. In the field of tertiary-level computer science studies, human-computer interaction was regarded to be a combination of user-centric needs analysis, design, implementation, and evaluation in general. Although most courses were taught in the limits of regular undergraduate courses, the design was intricate. The design-oriented programs (such as art, product design, interior design) taught design practices those derived from the design studio culture. This led to an investigation carried out on how did teachers pass on design studio best practices in a traditional program that follows a qualified lecture format? The solution proposed by Greenberg was to set up an atmosphere for creatives in the design studio within the lecture period. Students were introduced to four distinct and new interactive areas throughout the semester, each of which was chosen to minimise the students' expectations of standard design in these areas. They had considerable freedom in designing projects in these fields. They were asked to outline their ideas and present their drawings to other students for criticism. Encouraging the exchange of ideas would help students to apply some of their ideas in each other's work, although it might be regarded as cheating in some traditional courses. Course work, which leaded students to develop their designs during the class, was carried out instead of launching lectures. In this way, students and lecturers were able to observe each other's work during each other's work; they share skills and techniques and participate in constant comments. Students publicly display the final project in a design review environment. Finally, each student had to use a combination of paper and electronic media to create standardised portfolios to illustrate their work.

The proposal, criticism, and iterative process of studio-based learning to provide human-centred classroom management [4]. Based on his research on classroom management, Brocato emphasised that studio-based learning was related to the current teaching method in the school organisation. Brocato described how the method was applied in a course of teacher education program in his study. This model is significant for human-centred management within classrooms because it could position the teacher applicant as the central person in the course or as a professional, and managing the classroom in a more human-centred manner. Survey data was collected and the corresponding analysis provided evidence of how studio-based learning practice to bring more humanism concerning classroom management ideas. There was evidence that studio-based learning's proposals, criticisms, and iterative processes would enable teachers to construct better understand the linkages between human-centred and studio-based learning. Their applications would lead to an ultimate understanding of classroom management.

The implementation of instructional design as well as development in studio-based learning was investigated [5]. They examined the studio curriculum in their study. They took reference from the design, and technology program of a famous research-driven university in the United States represented the intentional implementation of the theories of how learners today learn with complicated information. The Studio course was part of a graduated course, which was implemented in the late 1990s. The objectives of this course were to equipped students to be professionals in designing, developing, evaluating and managing educational multimedia. Understanding theoretical concepts were considered as an essential role in the design of this studio courses. Their outstanding contributions were constructivism, situational cognition learning as well as autonomous learning. Related essential theoretical frameworks included scaffolding, as well as flow theory, were introduced. This study conducted [5] introduced the standard of a qualified learning environment within a studio. They also introduced these theoretical concepts that supported the requirements carried out and discussed the application of theories in the practice of training adult instructional design and development as well.

Based on the previous studies of Studio-based learning, the way of applying Studio-based learning in creative education was explored how [6]. Instructions within studio-based learning generally formulated in various design fields, for example, industrial design, visual communication design, interior design. In this way, it included desk space dedicated to each learner, extended secessions allocated to studio courses, and independent and featured group work on design issues in the classroom. They often carried out public and personal criticism. Although the surface characteristics and teaching methods of the studio had been fully proved, there was relatively limited attention to the participation structure of students and teachers. Through this structure, design knowledge could be jointly generated between teachers and students in the studio. The purpose of the investigation was to study the interactions among teacher and students [6]. Through the interactions, students were able to learn, considering and play the role of a designer. To this end, the research team collected and analysed ethnographic data from chosen studio classrooms in architecture, product design, as well as human-computer interaction. Their findings provided insights about the way conversations are conducted, and how specific social practices during the course supported students to learn how to deal with poorly structured design problems. At the same time, students were guided into practices that reflect professional knowledge. In each studio classroom, teachers have created an environment for students to practice reflection and listen in the form of voluntary participation. They also conveyed design knowledge through modelling and meta-discussions and focused assignments. The criticism added opportunities for individual and group processes, and these opportunities were used to build design knowledge in these studio classrooms jointly.

3 Problem-Based Learning

There was a transformation toward the inductive teaching approach which emphasised learning through consequential duties and open-ended questions [7]. Problem-based learning was a constructivist teaching method of assimilating and developing critical considering skills to solve actuality problems [8]. In the 1980s, problem-based learning originally was developed for medical education. Medical teachers recognised that the effort from Medical students' teamwork for constructing their knowledge lied on making broad generalisations from specific observations as well as expert knowledge from Medical teachers from diverse domains [9]. To identify Problem-based learning, some features of these teaching methods were summarised. The role of the educator was transformed as a facilitator rather than a leader. The learners bore the responsibility of initial directed learning. A general description of a practical procedure of problem-based learning while they conducted a study to explore the pedagogy and practice in the engineering classroom was provided [11]. Students applied a scenario to set their learning objectives in the problem presentation stage, which was the first phases of problem-based learning. Students generally would not have prior knowledge of the designed problem scenario. They had to discuss with the learning community members for planning a direction to explore the scenario according to their existing as well as any updated understanding. The problem-based learning did not only focus on solving a problem but also concerning the appropriate usage of wrong directed problems for leading students' knowledge investigation [11]. According to its nature of cooperative work among the learning community members, problem-based learning was one of the collaborative learning approaches [12]. It combined knowledge construction with innovative higher-order concerning for solving a problem [13].

4 Learning and Teaching of Creative Art Studies

While problem-based learning was developed for years, its application in a rich media-learning environment was still relatively new [7]. Problem-based learning was a teaching method for providing guidelines to students to develop their learning skills initially. Therefore, they explored problem-based learning with digital technology for investigating how it would be extended to developing more persuasive problem-solving skills. Learners construct new understanding according to their current knowledge. They proposed that it enhanced students' abilities in thinking and investigating solutions in creative ways and conducting effective communication [11]. To understand how to extend students' initial learning skills for developing more persuasive problem-solving skills, students' engagement through a classroom observation research study [14]. They took three essential themes including identifying specifications with the learning community members, initial exploration and interactive idea exchanging about their discoveries. The research team found that the teachers took the role of facilitators to guide their students to learn rather than instructors. This was regarded as a maximum effective collaboration among teachers and students within the learning community. Students demonstrated their strong senses of controlling on their projects through carried out meaningful and practical questions. Some of the insights obtained in the interactive communication but there was insufficient evidence to examine how they would lead the insights to transform as effective solutions with the solo digital platform. More scoldfolding support from teachers for foresting students' abilities in examining the potential solutions would be needed. The design of the learning process was outstanding by providing appropriated scaffolding to lead students solving problems, developing initial regulations and cooperation skills. The above-mentioned studies reflected that Problem-based learning students tended to be more motivated and obtained greater potential to become better learners. However, the pedagogy of Problem-based learning requested higher demands on the scaffolding skills of the teachers [7, 15].

5 Eight Pillars for Learning and Teaching of Creative Art Studies

Based on the above literature review, eight pillars of the shifted learning and teaching pattern of education are proposed. The eight pillars identified the essential elements in the learning and teaching of creative art studies (Fig. 1).

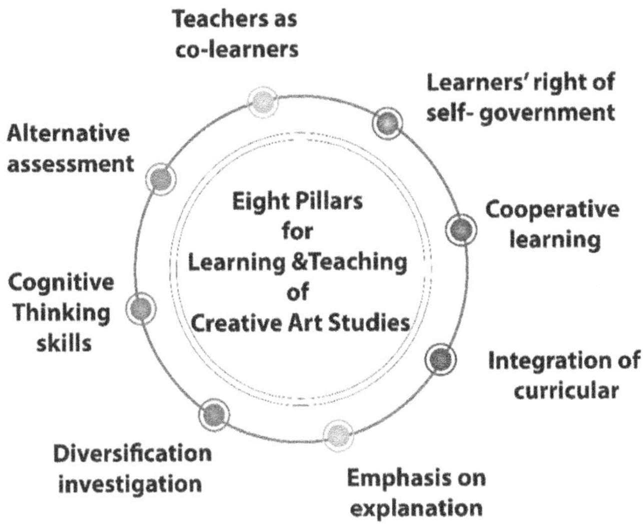

Fig. 1. The figure illustrated eight pillars for learning and teaching creative art studies.

6 Conclusion

In traditional creative education, the creative teaching process mostly followed instructive teaching. Following the change of the creative industry, some scholars applied problem-based learning. However, it was found that students sometimes could not reorganise the real situation in their learning process and so their misunderstanding of the situation led them to process their learning process in the wrong direction. Thus, some students lack motivation and methods for continuing knowledge construction in their design learning. According to the teaching and learning process in other study fields, the application of knowledge building principles was specifically proposed for idea diversity, improvable ideas, democratisation, community knowledge, and seems particularly relevant for low achievers. Knowledge building furnished a moderately pleasant and encouraging environment for learners to build their understanding. It prepared learning opportunities for obtaining experiences through discussions and assigned tasks.

References

1. Sutherland, J.A., Topping, K.J.: Collaborative creative writing in eight-year-olds: comparing cross-ability fixed role and same-ability reciprocal role pairing. J. Res. Read. **22**(2), 154–179 (1999)
2. O'Donnell, A.M., Topping, K.J.: Peers assessing peers: possibilities and problems. In: Topping, K., Ehly, S. (eds.) Peer-Assisted Learning, pp. 255–278. Lawrence Erlbaum, Mahwah (1998)
3. Greenberg, S.: Embedding a design studio course in a conventional computer science program. In: IFIP International Federation for Information Processing of Creativity and HCI: From Experience to Design in Education, vol. 289, pp. 23–41. Springer, Boston (2009)

4. Brocato, K.: Studio based learning: proposing, critiquing, iterating our way to person-centeredness for better classroom management. Theory into Pract. **48**, 138–146 (2009)
5. Clinton, G., Reiber, L.P.: The studio experience at the University of Georgia: an example of constructionist learning for adults. Educ. Tech. Res. Dev. **58**, 755–780 (2010)
6. Cennamo, K., Brandt, C.: The "right kind of telling": knowledge building in the academic design studio. Educ. Tech. Res. Dev. **60**(5), 839–858 (2012)
7. Chan, M.M., Blikstein, P.: Exploring problem-based learning for middle school design and engineering education in digital fabrication laboratories. Interdiscip. J. Probl.-Based Learn. **12**(2), 7 (2018)
8. Ram, P., Ram, A., Spragues, C.: From student learner to professional learner: training for lifelong learning through on-line PBL. In: International Conference Problem-Based Learning (PBL-05) (2007)
9. Barrows, H.S.: Problem-based learning in medicine and beyond: a brief overview. New Dir. Teach. Learn. **1996**(68), 3–12 (1996)
10. Savery, J.R.: Overview of problem-based learning: definitions and distinctions. Essent. Read. Probl.-Based Learn.: Explor. Extending Leg. Howard S. Barrows **9**, 5–15 (2015)
11. Awang, H., Ramly, I.: Through problem-based learning: pedagogy and practice in the engineering classroom. Int. J. Hum. Soc. Sci. **2**(4), 18–23 (2008)
12. Tiwari, A., Lai, P., So, M., Yuen, K.: A comparison of the effects of problem-based learning and lecturing on the development of students' critical thinking. Med. Educ. **40**(6), 547–554 (2006)
13. Hmelo-Silver, C.: Problem-based learning: what and how do students learn? Educ. Psychol. Rev. **16**(3), 235–266 (2004)
14. Blikstein, P.: Maker movement in education: history and prospects. Handbook of Technology Education, pp. 419–437 (2018)
15. Don, Y.: Implementation of problem-based learning to improve students' critical thinking skills. JPhCS **1460**(1), 012058 (2020)

Novel Technology and Methods for Training and Certification

Non-technical Skills Training in Crew Resource Management: Curating YouTube Videos for Educational Purposes

Charlott Sellberg[1](✉), Gesa Praetorius[2,3], and Marin Viktorelius[2]

[1] Department of Applied Information Technology, University of Gothenburg, Forskningsgången 6, 417 56 Gothenburg, Sweden
charlott.sellberg@ait.gu.se

[2] Kalmar Maritime Academy, Linnaeus University, Landgången 4, 392 31 Kalmar, Sweden

[3] Department of Maritime Operations, University of South-Eastern Norway, Campus Vestfold, 3184 Borre, Norway

Abstract. This pilot study serves as a starting point to explore the process of finding and evaluating YouTube videos for non-technical skills training for crew resource management courses. Five videos on situation awareness were identified using a combination of Boolean search strategies, relevance of the content and ratings from YouTube users. In the next step, the videos were evaluated, showing variations in education quality between videos. Moreover, preliminary finding show that user-generated comments and ratings might be of little use for guiding the search for high-quality content on YouTube. A closer analysis of the content shows how tensions between scientific concepts and practical illustrations of situation awareness is seen in the videos. This might create a dilemma for teachers in choosing between theoretically complete and authentic content. Research that further explores the ways discursive transformations of human factors concepts occur through everyday uses in popular media such as YouTube is suggested.

Keywords: Aviation · Crew resource management · Situation awareness · Online teaching · YouTube

1 Introduction

During the COVID-19 pandemic, teachers all over the world struggle with the enforced transition towards online teaching. The use of instructional videos has been a central tenet of online learning for decades, providing students with an effective learning experience [1]. The online video-sharing platform YouTube has since its launch in 2005 become a vast collection of user-generated and corporate videos. While the main function of YouTube might be entertainment, there is also an array of content with educational purposes, for example, different kinds of tutorials, lectures and talks. During the pandemic, YouTube has launched new initiatives for online learning at the platform, for example the Learn@Home hub, but also increased funding for already established educational hubs such as Kahn Academy. The literature on using YouTube videos in education highlights

S. Nazir et al. (Eds.): AHFE 2021, LNNS 269, pp. 93–100, 2021.
https://doi.org/10.1007/978-3-030-80000-0_12

the benefits of being able to both creating and curating selections of educational videos for free, but also the challenges for producing and selecting high-quality content [2]. Connecting to these challenges, this pilot study serves as a starting point to investigate educational videos from YouTube that might be suitable for training non-technical skills (NTS) in Crew Resource Management (CRM).

CRM was developed in aviation in the 1970's after a number of serious accidents, putting focus on developing team members' NTS [3]. NTS can be defined as "*the cognitive, social and personal resources skills that complement technical skills, and contribute to safe and efficient task performance*" [4]. Albeit different taxonomies of NTS have been developed in a number of fields, four primary categories have been identified [5]. These include the interpersonal skills of co-operation and leadership and the cognitive skills of situation awareness and decision making [5]. While CRM courses initially were intended to only include pilots and other staff in the cockpit, the framework has gradually been developed into a multitude of programs including the flight crew, air traffic controllers, and maintenance staff [6]. CRM training is designed to provide systematic means of ensuring that flight crews receive the same type of training in team skills. However, the lack of a systematic approach for assessing training programs makes it difficult to draw conclusions of the effectiveness of training NTS [7]. The courses normally encompass a mixture of classroom-based lectures and group discussions in combination with simulator training. The lectures generally address issues related to human performance in high-risk systems, such as decision making, communication, leadership, teamwork and situation awareness. Further, courses also often make use of a number of exercises, including discussions of incidents and accidents [8]. This article will specifically focus on one key area of NTS; situation awareness (SA). SA is widely used across domains to describe and understand how operators perceive their operational environment and changes within it [9]. Most commonly, SA is defined as a three-step model in accordance with Endsley as "the perception of elements in the environment within a volume of time and space, the comprehension of their meaning, and the projection of their status in the near future" [10]. Another definition has been presented by Moray who describes SA as "keeping track of what is going on around you in a complex, dynamic environment" [11].

In order to explore how YouTube videos can be used in NTS training, this study focuses on curating videos for use in CRM courses. Curating refers to the process of selecting, analyzing, using, organizing and sharing educational videos on YouTube [2]. In this study, the emphasis is on the first steps in this process to explore challenges and opportunities connected to selecting suitable videos for use in CRM courses. The following questions are in focus: 1) what is the educational value of the selection of videos found on NTS training in CRM courses? 2) in what ways can user generated comments and ratings guide selection of materials for educational purposes? In particular the evaluation of videos concern a) the trustworthiness of the educational content, b) the completeness of the educational content c) applicability of the educational content d) user engagement with the educational format and e) the production value of the video. In this way, this study set out to explore educational values connected both to the integrity of the content, as well as motivating factors for using videos in NTS training. Five YouTube videos on the topic of situation awareness were selected for inclusion based on relevance of the content and ratings from YouTube users.

2 Curating YouTube Videos

In the first step, finding YouTube videos that might be suitable for the intended purposes is in focus. In research on curating videos from YouTube, it is highlighted that searching from the vast and unstructured data base is one of the many challenges of curating videos for educational purposes [2]. Familiarity on how search phrases and keywords impact the search results is critical in this step, and in some sense also an understanding on how YouTube algorithms work [12]. This in turn require 21st century skills such as digital literacy, i.e., skills related to the functional use of technology and information literacy, i.e., skills related to locating, evaluating, and synthesizing information [13, 14]. These skills become relevant when searching for content on a specific topic, both to be able to locate videos using Boolean search strategies and for being able to use advanced setting, such as filtering options as recommended in the literature on curating YouTube videos [2]. In particular, information literacy is critical when evaluating the located videos, alongside with subject matter expertise on the topic of interest.

In addition to the challenges of defining a fruitful search strategy, determining how to evaluate the content and assess the reliability of the video sources have been discussed throughout studies on YouTube [2, 15]. While it is common to involve multiple reviewers, it is also usual to determine the qualification of reviewers and to base the evaluation on their mean score [15]. When using videos for educational purposes it is of importance that reviewers have subject matter expertise, in order to determine accuracy and completeness of the information presented [15]. Moreover, we argue that it is important that the theoretical content is communicated in a pedagogical way by connecting theory to practical examples and to present the educational content in an engaging way and in decent audio-visual quality. In the literature on curating videos, strategies for selection often include date range, preferences on language and video length [15]. It is less common that selection is based on number of views and user ratings. However, it is important to point out that these aspects can have a large impact on the search results since YouTube algorithms make use of these kinds of data [12]. It is also possible that teachers using YouTube take into account user ratings when selecting materials. Hence, we will use the pilot study to explore user ratings and comments in relation to the judgements and ratings from the research group.

3 The Pilot Study

The pilot study was carried out between November 2020 to January 2021 in the midst of the COVID-19 pandemic. The search was carried out by the first author, using keywords together with Boolean operators in the following search string "aviation + CRM + training videos + situation awareness". The platform's filtering options to refine the search was used together with the search string, targeting videos based on "relevance" of the content and "ratings" from the YouTube users. The five top videos on SA found through this search were subjected to evaluation in the next step of the process (Table 1). For the pilot study, three researchers with a background in cognitive science, human factors and learning sciences evaluated the videos. The ratings were first performed individually by each member of the group. In the next step a group analysis on the YouTube content took place via an online video conference system, in which the reviewers motivated and discussed their ratings.

Table 1. Overview of included videos

No	Producer	Likes	Dislikes	Comments	Mean score*
1	Commercial	11	0	0	9/20
2	Governmental	62	1	1	16/20
3	Private user	667	12	105	8/20
4	Private user	47	1	3	14/20
5	Governmental	11	0	0	15/20

Video no 1 is about six minutes long and based on talks from human factors experts as well as pilots. The video is mainly focused on planning, following procedures and eliminating distractions in the cockpit. Video no 2 is longer, 28 min, and covers more ground. This video makes a thorough description of SA, but also other NTS such as communication and leadership. In this video, several human factors experts and pilots express their views on SA and connect theory to practice through a practical case. It is interesting that some sequences with aviation experts are the same as the content in video no 1. Video no 3 is showing an approximately six minutes long scenario of two pilots' loss of SA in a flight simulator. Albeit the video gives an insight into maintaining SA in practice, no theoretical framing or explanations of the take-home message is provided. Further, the video also provides examples for several CRM practices in situ, such as challenge-response and closed loop communication, but does not explicitly explains that. Video no 4 is grounded in Endsley's model of SA and connects theory and practice through showing interactions between a novice and an experienced pilot in an instructional situation. It is approximately 16 min long and emphasizes the importance of remaining focus during flight, as well as how SA is connected to decision making. Video no 5 takes the format of a traditional lecture, showing a safety official and his talk in relation to a Power Point presentation. In this approximately 17 min video, theory is connected to practical examples in a systematic way.

During analysis different themes were identified in the video corpus, such as being physically and mentally fit for flight, preparation, planning, following procedures, maintaining SA, thinking ahead, as well as reducing risk, and avoiding disturbances and distractions in the cockpit. On an overall level, a longstanding debate on tensions between situated action and theoretical models of SA became relevant to revisit during the analysis of the video content [17, 18]. In relation to this tension, trustworthiness became quite complex to evaluate. While experienced professionals in the video provided trustworthy narratives and illustrations on flying in practice, the explanations were often theoretically weak. Human factors experts were, on the other hand, able to provide theoretical explanations of SA models, but lacked the practical orientation of demonstrating SA (or the lack of) and engaging the listener. Moreover, issues such as contradictions between statements of the importance between strict following of routines and procedures versus the importance of attention and flexibility towards the moment-to-moment unfolding of the situation was also identified in the corpus. This tension was in one case even noticeable within the same video, showing how different views on SA becomes visible even within this small sample. Hence, a tradeoff between application/authenticity/engagement and theoretical completeness/trustworthiness might become a challenge in curating SA videos (Table 2).

Table 2. Evaluation criteria * and grading system

	Poor (1)	Lacking (2)	Good (3)	Excellent (4)
Trustworthiness: The theoretical content is correctly described				
Completeness: No relevant omissions in theoretical content				
Application: Connecting theory to practical examples				
Engagement: Motivating and fun to watch				
Production value: Technical quality				

Preliminary analyses of YouTube users' likes, dislikes and comments show interesting discrepancies between the number of likes and positive comments in relation to the evaluation of the educational quality of the content. The videos that received few likes and few comments were mostly mentioned in a positive manner. Comments such as *"Great video well done to you both, enjoy the coffee"*, *"Love the video. Wish all pilots could have standards like yours. Very informative"* and *"A BIG THANKYOU for sharing"* shows how YouTube users provide both praise and gratitude towards the producers of educational content. It is also interesting to note that the videos receiving comments were uploaded by private YouTube users.

It is also notable that the video that scored the lowest in the evaluation concerning educational value has by far the most likes and positive comments from the YouTube users. Taking a closer look at the comments, several patters start to emerge: it is mainly pilots within aviation that seem to be commenting. Another pattern is that these YouTube users mainly are positive and recognize the situation presented in the video as authentic, as well as they identify an opportunity to learn from (someone else's) mistakes. The way problems are solved in this clip seem to engage the viewers, and spurs debate amongst professionals on best practices, suggesting alternative solution to the encountered problem and providing assessments on how the scenario is solved.

Even if negative comments are less common, critique of the scenario as non-realistic, outdated and difficult to make sense of also occurred. For example, one user made this comment "*You don't brief the approach and go around as you pick up the glideslope/localizer…this video makes no sense..*", while another user states that "*That's real, but silly. Taking over is matter of live or death, therefore not doing so is suicide.*", relating the simulated scenario to standard work practices in the cockpit. The near disaster seen in the video also engaged viewers emotionally, spurring comments such as "*damn!! that was nerve wracking,, where was the GO AROUND! call? good sim*" and "*Just looking at the video makes my heart pound a bit faster. Great scenario training*". Although the video in itself can be difficult for a novice to learn from, it can provide a good practical example in a CRM course if it is connected to theory on SA and discussed so that the take home message becomes clear for trainees. Hence, a dilemma when evaluating videos is that the quality of the content always is depending on how it is used, i.e., a "bad video" can make a perfectly good starting point for a critical discussion on important issues in CRM training.

4 Discussion and Directions for Future Research

The results of the pilot study, albeit limited, support previous research showing how teachers that intend to make use of YouTube in their teaching need both expertise on their subject matter, as well as skills such as digital literacy, i.e., the ability to find and make use of media from various digital platforms, and information literacy in order to evaluate the uploaded content [2]. These skills are becoming increasingly important, in particular in relation to emerging concerns related to how information is produced and valued in online communities. In terms of liking, disliking and rating educational content on YouTube, the results are supporting earlier research showing how and why using user-generated comments and ratings as a tool for guiding searching and selecting content is problematic [12]. Studies that take an interest in the underlying mechanism of likes on social media show a social practice with its own logic, which is not always easy to interpret [16]. In this respect, the ways that likes, dislikes and comments are used on media with educational content can be a possible direction for future research on YouTube as an educational resource. An issue that stands out in the video corpus, but is outside the scope of this pilot study, is that all SA experts and aviators represented in the materials are male. This can be understood in relation to aviation as a male-dominated professional field, a field that still is struggling to create gender balance [19]. Hence, gendered analyses of aviation pilots and their representation in social media such as YouTube can provide perspectives on problems related to recruitment and retainment of women in the cockpit.

To summarize, preliminary analyses of materials on YouTube show a complex and dynamic platform where variations in educational quality between videos pose a practical problem for teachers. While the evaluation system developed in this study may be quite restricted for producing a scientific valid evaluation of YouTube content for research purposes, it might be useful for guiding teachers' evaluation of YouTube videos. In particular, it highlights some of the considerations in curating videos for educational purposes. The teacher needs, for instance, to reflect on whether the purpose of showing

YouTube videos to students is to *inform* about theory, *demonstrate* an issue relevant in practice or a combination of both. Overall, the complexity of using YouTube in education highlight the need to further explore challenges and opportunities for using social media in NTS training. In particular, it is interesting to continue analyzing the ways discursive transformations of scientific NTS concepts occur through its everyday use in popular media, such as YouTube.

Acknowledgments. This research is funded by the Swedish Mercantile Marine Foundation.

References

1. Kestin, G., Miller, K., McCarty, L.S., Callaghan, K., Deslauriers, L.: Comparing the effectiveness of online versus live lecture demonstrations. Phys. Rev. Phys. Educ. Res. **16**, 1 (2020)
2. Snelson, C.: The benefits and challenges of YouTube as an educational resource. In: Hobbs, R. (ed.) The Routledge Companion to Media Education, Copyright, and Fair Use, vol. 48, pp. 109–126. Routledge (2018)
3. Kanki, B.G.: Communication and crew resource management. In: Kanki, B.G., Anca, J., Chidester, T.R. (eds) Crew Resource Management, pp. 103–137. Academic Press (2019)
4. Flin, R., O'Connor, P., Crichton, M.: Safety at the Sharp End: A Guide to Nontechnical Skills. Ashgate, Farnham (2008)
5. Flin, R., et al.: Development of the NOTECHS (non-technical skills) system for assessing pilots' CRM skills. Hum. Factors Aerosp. Saf. **3**, 97–120 (2003)
6. Thomas, M.J.: Training and Assessing Non-Technical Skills: A Practical Guide. CRC Press, Boca Raton (2017)
7. Bennett, S.A.: The training and practice of crew resource management: recommendations from an inductive in vivo study of the flight deck. Ergonomics **62**(2), 219–232 (2019)
8. Salas, E., Wilson, K.A., Burke, C.S., Wightman, D.C.: Does crew resource management training work? An update, an extension, and some critical needs. Hum. Factors **48**(2), 392–412 (2016)
9. van Westrenen, F., Praetorius, G.: Situation awareness and maritime traffic: having awareness or being in control? Theor. Issues Ergon. Sci. **15**(2), 161–180 (2014)
10. Endsley, M.R.: Toward a theory of situation awareness in dynamic systems. Hum. Factors **37**(1), 32–64 (1995)
11. Moray, N.: Ou sont les neiges d'antan? ("Where are the snows of yesteryear?"). In: Vincenzi, D.A., Mouloua, M., Hancock, P.A. (eds) Human performance, situation awareness and automation: current research and trends, pp. 1–31. Routledge (2004)
12. Rieder, B., Matamoros-Fernández, A., Coromina, Ò.: From ranking algorithms to 'ranking cultures' investigating the modulation of visibility in YouTube search results. Convergence **24**(1), 50–68 (2018)
13. Spante, M., Hashemi, S.S., Lundin, M., Algers, A.: Digital competence and digital literacy in higher education research: systematic review of concept use. Cogent Educ. **5**(1), 1519143 (2018)
14. Lanning, S., Mallek, J.: Factors influencing information literacy competency of college students. J. Acad. Librariansh **43**(5), 443–450 (2017)
15. Drozd, B., Couvillon, E., Suarez, A.: Medical YouTube videos and methods of evaluation: literature review. JMIR Med. Educ. **4**(1), e3 (2018)

16. Jungselius, B.: She liked the picture so I think she liked it": unpacking the social practice of liking. Netcom (2019)
17. Dekker, S., Hollnagel, E.: Human factors and folk models. Cogn. Tech. Work **6**, 79–86 (2004)
18. Suchman, L.: Human-Machine Reconfigurations: Plans and Situated Actions. Cambridge University Press, Cambridge (2007)
19. McCarthy, F., Budd, L., Ison, S.: Gender on the flightdeck: experiences of women commercial airline pilots in the UK. J. Air Transp. Manag. **47**, 32–38 (2015)

Exploration of Team Communication Behaviors from a Live Training Event

Jason D. Saville[1](✉), Randall D. Spain[1], Joan H. Johnston[2], and James C. Lester[1]

[1] Center for Educational Informatics, North Carolina State University, Raleigh, NC, USA
{jdsavill,rdspain,lester}@ncsu.edu
[2] U.S. Army Combat Capabilities Development Command, Soldier Center, Orlando, FL, USA
joan.h.johnston.civ@mail.mil

Abstract. Investigating communication data during team training activities can provide insight into the rich processes underlying team collaboration and coordination. In this paper we explore team communication behaviors collected during a live training capstone involving six squads from the U.S. Army, three of which were experimentally assigned to receive instruction for improving team behaviors. Communication among squads was recorded, transcribed, and labeled for speech acts and team development dimensions. A series of analyses were used to investigate communication pattern differences over time between teams of higher and lower performance. Findings and implications are discussed.

Keywords: Live training · Team communication · Team performance

1 Introduction

Recent advances in the science of teams have provided critical insights into the attitudes, cognition, and behaviors that contribute to effective team performance and how to measure them [1]. Research shows high performing teams share and pass relevant information to the right team members at the right time, seek information from relevant sources, communicate clearly and concisely, and adapt to changing demands and circumstances. Of the many methods team science researchers have utilized to investigate team effectiveness, the analysis of team communication provides one of the most promising approaches for identifying factors that contribute to effective team performance. Team communication is an observable behavior that changes over time, based on task demands, changing priorities, and team goals. Decades of research have shown that communication is critical to effective teamwork [2–4]. It serves as a medium for sharing information about team structure, team roles, team coordination [5], cognition [6], and situational factors [7] that affect performance. Investigating team communication behaviors during team training activities can thus provide insight into the rich processes underlying team collaboration and coordination.

S. Nazir et al. (Eds.): AHFE 2021, LNNS 269, pp. 101–108, 2021.
https://doi.org/10.1007/978-3-030-80000-0_13

1.1 Team Communication Analysis Techniques

Team science researchers have utilized several methods for the analysis of team communication data. One of the most common is to analyze transcripts of team communication and hand-code team communication based on a pre-established coding scheme. The frequencies at which the coded categories emerge from the data, or their patterns, are then correlated with team performance measures. For example, Bowers and colleagues [2] coded utterances collected from flight crews during a simulated flight mission into eight categories that were statements about uncertainty, actions, acknowledgements, responses, planning, facts, and non-task related communications. They analyzed the frequency and sequences of these categories to examine differences in high and low performing flight crews. Results showed the higher performing flight crews had fewer non-task related communications and were more likely to follow communications from air traffic control with planning statements compared to lower performing crews. Fisher et al. [8] used a similar approach to examine communication differences in high and low performing teams who performed a computer simulated search and rescue mission. Utterances were coded into task and response-related categories with analyses showing that successful teams shared information more often.

Another approach team science researchers have used to examine team communication and the underlying team processes is by computing ratios of team communication behaviors. For example, Entin et al. [9], proposed an anticipation ratio index that represents how often information statements or actions are pushed versus how often they are requested from team members. Ratios larger than one indicate that a team pushed or sent information more frequently than they requested information; that is, they anticipated each other's information needs. Ratios less than one suggests that information needed to be requested more often than it was provided. This measure of team anticipation and coordination has been associated with improved team performance in a number of studies [10, 11]. Additional techniques for analyzing team communication have included examining patterns of team communication sequences [12] and examining communication patterns based on task flow [13].

1.2 Goals of Current Effort

Team science researchers have increasingly called for the investigation of team communication behaviors in real world settings to better understand and advance the theory of team development and provide insights for team development training. The goal of this research was to address this need by exploring team communication behaviors collected during a military live training exercise. Spoken utterances were captured among squad members during a live-training capstone involving six U.S. Army squads as part of the Squad Overmatch project [14]. Three of the six squads received team development training to improve teamwork behaviors, team situation awareness, and stress management prior to the training exercise. Spoken utterances were recorded, transcribed, and labeled for speech acts and team development dimensions. Our analyses focused on addressing which speech acts and team dimensions were related to team performance ratings and whether high performing teams demonstrated different communication behaviors compared with low performing teams.

2 Methodology

2.1 Dataset

The dataset used for this study consisted of transcribed audio logs of six U.S. Army squads, ranging from 8 to 10 members with one squad leader and two team leaders subordinate to the squad leader in every squad, who completed a 45-min live training that included a scripted set of training objects and training events designed to elicit team development behaviors. In the scenario, the squad was tasked with entering a village they experienced in prior missions, making contact with key local leaders, and gaining intelligence on local gang movements and activity. During the course of the mission, the squad encounters village characters in need of assistance, an improvised explosive device event, a firefight with local gang leadership, and simulated gunshot wounds to attend to.

Speech Labels. A total of 6,181 utterances were coded using a framework of 27 speech act labels and 18 team dimension labels, where speech act labels represented the basic purpose of a given utterance, such as requesting information or stating an action being taken, and team dimension labels reflected how different forms of information were being transferred up and down the chain of command (CoC). While every utterance was assigned a speech act label, not every utterance constituted a team dimension label ($n = 2743$). The five most frequently occurring speech acts and team dimension labels are presented in Table 1.

Table 1. Top speech act and team dimension label frequencies.

Label	CTL 1	CTL 2	CTL 3	EXP 1	EXP 2	EXP 3	Total
Speech act labels							
Inform	125	139	111	188	235	237	1035
Command	141	143	73	224	186	216	983
Request information	68	123	70	123	113	137	634
Acknowledgement	90	98	59	108	110	133	598
Provide information	62	99	59	83	75	130	508
Team dimension labels							
Provide info up CoC	103	94	59	105	127	149	637
Command team leader	61	60	16	153	71	71	432
Command squad leader	58	52	47	45	85	123	410
Provide info down CoC	24	46	47	61	45	104	327
Request info down CoC	47	64	30	45	71	66	323

Team Performance Ratings. Team performance ratings collected via subject matter experts (SMEs) who followed and observed squad performance in real time during the course of M3 were used as an index of team performance for this study [14]. SMEs completed Targeted Acceptable Responses to Generated Events or Tasks (TARGETs) checklists that broke down the mission into individual events, each with several target behaviors such as providing situation updates up the CoC or providing cover to a squad member completing a task for SMEs to watch for. Ratings for each of 27 behaviors were dichotomous numerical completion ratings (0/1), with a final score for the squad assessed via the total percentage of behaviors completed.

2.2 Data Analysis

A series of correlations and t-tests were conducted to examine the relationship between frequencies of speech act labels and team dimension labels and team performance ratings, as well as to explore the effects of the team development training to which half of the squads were experimentally assigned. Next, a median split of performance ratings was conducted such that squads were designated as either "high performance" or "low performance." Given the limited sample size, the two squads of most average performance were excluded from further analyses so as to generate more variance between performance groups. This resulted in two control condition squads representing the low performance group and two experimental condition squads representing the high-performance group. T-tests were conducted to examine the relationship between frequencies of speech act labels and team dimension labels and a dichotomous team performance rating.

In addition to examining frequencies of team communication behaviors we also calculated two anticipation ratios based on speech act label data to explore how squads shared information. The "information ratio" involved the passage of all general information compared with how often requests for information were made. The "action ratio" involved information being passed relating to how often squad members verbalized what they were doing compared with how often they were asked to do something.

Finally, we examined how squads performed when met with increased situational stressors such as gunfighting and simulated wounds requiring treatment. We coded all events within the mission as either occurring before or after hostile contact occurred. Label values were normalized as a percentage of the total utterances for either before or after hostile contact. MANOVAs and paired t-tests were then used to examine differences between and within performance groups across contact event types.

3 Results

Prior to examining differences in communication behaviors between high and low performing squads' correlations between the experimental condition of team development training, speech act labels, and team performance were examined, where experimental condition was represented by a 0 or 1, with a 1 representing squads who were experimentally assigned to receive team development training. The correlation between condition and team performance was found to approach but not reach significance ($r = .749$, $p = .087$). Condition was significantly positively correlated with the total number of commands given ($r = .867$, $p = .025$), number of statements providing information ($r =$

.936, $p = .006$), number of hail statements gaining the attention of another individual prior to continuing speech ($r = .892$, $p = .017$), and total number of utterances ($r = .869$, $p = .025$), indicating a greater number of these statements by squads who received the team development training.

Speech Act Labels. Next, we aimed to examine differences in counts of speech act labels between high and low performance squads using a median split of team performance score, omitting the two average performing squads. T-tests were conducted on counts of speech act labels and anticipation ratios for squads classified as either high or low performance. Results indicated that high performing squads made inform statements ($t = -14.708$, $p = .005$) at significantly higher rates than low performing squads, while command statements approached significance ($t = -3.925$, $p = .059$). There were no differences between high and low performing groups in the number of information requests ($t = -.664$, $p = .575$), information statements provided in response to a question ($t = -.664$, $p = .575$) or acknowledgement statements ($t = -2.259$, $p = .152$). In relation to anticipation ratios, there was no difference in the ratio of information shared versus requested between performance groups ($t = -1.078$, $p = .394$), but there was a significant difference in the ratio of action statements made versus actions requested ($t = 6.503$, $p = .023$) signifying that members of low performance squads stated the actions they were taking more often than actions were requested at a comparatively greater rate than those in the high performing squads.

Team Dimension Labels. Similar to the examination of speech act labels, t-tests were conducted on counts of team dimension labels for the high and low performing squads. Results indicated that team leaders in high performing squads gave commands significantly more often than team leaders in low performing squads ($t = -21.0$, $p = .002$). No significant differences were found between low and high performing groups for the number of commands provided by a squad leader ($t = -2.547$, $p = .126$), providing information down the chain of command ($t = -1.255$, $p = .336$), or requesting information from down the chain of command ($t = -1.467$, $p = .280$).

Team Communication Behaviors Based on Event Type. Next, a series of multivariate analyses of variance (MANOVAs) were conducted to examine differences in team communication between performance groups for either non-contact or contact categories. Then, to further explore and contextualize changes over time within squads, paired t-tests were conducted on the difference in proportion between speech act and team dimension labels before and after contact took place. Results indicated that, prior to contact, there was a significant difference in the proportion of information statements given by low performing squads ($M = .122$) and high performing squads ($M = .172$) ($F = 35.501$, $p = .027$). Whereas after contact, the relationship between the number of information statements given by low performance squads ($M = .129$) and high performing squads ($M = .185$) approached but did not reach significance ($F = 13.893$, $p = .065$). These findings suggest that compared to low performing teams, members of high performing squads provided more information to each other prior to combat events, but not necessarily after the combat began.

Similarly, we found that, compared with low performing squads ($M = .083$), high performing squads ($M = .140$) had a significantly higher rate of requests for information issued down the chain of command prior to contact (F = 19.891, p = .047). But, after contact, no differences were found between the low (M = .257) and high (M = .161) performing squads (F = 5.08, p = .153). Expanding on these results, paired t-tests revealed seemingly large mean differences between low performance squads' information requests by superiors before contact (M = .083) and after contact began (M = .257), but this only approached significance ($t = -10.046$, p = .063). Furthermore, we found that squad leaders issued a similar number of command statements across groups and events, but team leaders in the low performing groups (M = .209) issued more commands than team leaders in the higher performing groups (M = .144) after contact (F = 40.319, p = .024) (See Table 2).

Table 2. Differences in team communication proportions across events

	Pre-contact			Post-contact		
	Low	High		Low	High	
Labels	M(SD)	M (SD)	*F(p)*	M(SD)	M(SD)	*F(p)*
			Speech act labels			
Inform	.122 (.002)	.172 (.012)	*35.501 (.027)	.129 (.019)	.185 (.009)	13.893 (.065)
Command	.206 (.055)	.229 (.042)	.224 (.682)	.241 (.037)	.287 (.072)	.668 (.500)
Request info	.095 (.056)	.112 (.026)	.148 (.737)	.144 (.005)	0.103 (.04)	1.983 (.294)
Provide info	.018 (.014)	.05 (.010)	7.198 (.115)	.062 (.013)	.044 (.159)	1.563 (.338)
			Team dimension labels			
Request info down CoC	.083 (.016)	.140 (.008)	*19.891 (.047)	.257 (.008)	.161 (.059)	5.080 (.153)
Provide info down CoC	.130 (.083)	.183 (.067)	.473 (.563)	.111 (.017)	.141 (.064)	.398 (.593)
Command team leader	.190 (.045)	.182 (.038)	.038 (.864)	.209 (.005)	.144 (.014)	*40.319 (.024)
Provide info up CoC	.197 (.085)	.157 (.005)	.444 (.574)	.144 (.017)	.102 (.015)	7.381 (.113)

4 Discussion

Squads trained on team development demonstrated greater information sharing and exchange and more closed-loop communication. While the relationship between condition assignment and overall performance score only approached statistical significance, it is worth noting that when squads of the most average performance were omitted, the lower performing squads did not receive the training, while higher performing squads did. Higher performing squads made more "hail" statements to gain someone's attention prior to speaking and shared more information with each other, although the action anticipation ratio based on speech act labels suggests that members of lower performing squads provided information about their own actions at greater rates than higher performance squads when compared to how often actions were requested. Team dimension labels suggest that higher performing squads had team leaders who gave more commands and subordinates who provided information up the chain of the command at greater rates. We then looked at how the communication patterns of teams at different performance levels changed when met with situational stressors such as active combat sequences. We found that high performing squads had leadership requesting information of their subordinates at a consistent rate *before and after* receiving enemy contact, whereas lower performing squads proportionally tripled their requests *after* taking contact.

While this set of preliminary analyses provides some insight into team communication behaviors, future analyses should include stochastic analysis over time to explore team communication patterns and their relationship to team performance. In addition, future research should explore using multi-task learning techniques to determine which sets of labels or features are best combined to understand and predict team performance. A second limitation of this study is the very small sample size. Relatedly, several utterance categorizations such as requesting information from a superior or correcting what someone else said remained largely absent from the squad transcripts, leaving marginal differences in mean scores to have large effects on correlations. Because team knowledge develops over time, a more granular and event-based approach could have future research highlight critical differences in communication behaviors around these key events.

5 Conclusion

In this paper we report on an exploration of team communication behaviors collected during a live military training exercise using a frequency-based approach. Results provided insights into team communication differences between high and low performing teams. In a training context, communication data can provide especially valuable information about what differentiates communication patterns of high and low performing teams and even how patterns of communication within a team shift over time. When used alongside performance data, team communication data has the potential to greatly improve the understanding and assessment of teams in action.

Acknowledgements. The research described herein has been sponsored by the U.S. Army Research Laboratory under cooperative agreement W912CG-19-2-0001. The statements and opinions expressed in this article do not necessarily reflect the position or the policy of the U.S. Government, and no official endorsement should be inferred. The authors would also like to acknowledge Dr. David Traum of the University of California, Institute of Creative Technologies and his research team for providing the labeled transcripts.

References

1. Bell, S.T., Brown, S.G., Colaneri, A., Outland, N.: Team Composition and the ABCs of Teamwork. Am. Psychol. **73**(4), 349–362 (2018)
2. Bowers, C.A., Jentsch, F., Salas, E., Braun, C.C.: Analyzing communication sequences for team training needs assessment. Hum. Factors **40**(4), 672–679 (1998)

3. Marlow, S.L., Lacerenza, C.N., Paoletti, J., Burke, C.S., Salas, E.: Does team communication represent a one-size-fits-all approach?: a meta-analysis of team communication and performance. Organ. Behav. Hum. Decis. Process. **144**, 145–170 (2018)
4. Mesmer-Magnus, J.R., DeChurch, L.A.: Information sharing and team performance: a meta-analysis. J. Appl. Psychol. **94**(2), 535–546 (2009)
5. Jariwala, S., Champion, M., Rajivan, P., Cooke, N.J.: Influence of team communication and coordination on the performance of teams at the iCTF competition. In: Proceedings of the 56th HFES Annual Meeting, vol. 56, no. 1, pp. 458–462. Sage Publications, Los Angeles (2012)
6. Cooke, N.J., Kiekel, P.A., Salas, E., Stout, R., Bowers, C., Cannon-Bowers, J.: Measuring team knowledge: a window to the cognitive underpinnings of team performance. Group Dyn. Theory Res. Pract. **7**(3), 179 (2003)
7. MacMillan, J., Entin, E.E., Serfaty, D.: Communication overhead: the hidden cost of team cognition. In: Salas, E, Fiore, S.M. (eds.), Team Cognition: Understanding the Factors That Drive Process and Performance, pp. 61–82. American Psychological Association (2004)
8. Fischer, U., McDonnell, L., Orasanu, J.: Linguistic correlates of team performance: toward a tool for monitoring team functioning during space missions. Aviat. Space Environ. Med. **78**(5), B86–B95 (2007)
9. Entin, E.E., Entin, E.B.: Measures for evaluation of team processes and performance in experiments and exercises. In: Proceedings of the 6th ICCRTS, pp. 1–14 (2001)
10. Entin, E.E., Serfaty, D.: Adaptive team coordination. Hum. Fact. **41**(2), 312–325 (1999)
11. Sperling, B.K.: Information distribution and team situational awareness: an experimental study. In: Proceedings of the 50th HFES Annual Meeting, pp 477–481. Human Factors and Ergonomics Society, Santa Monica (2006)
12. Gontar, P., Fischer, U., Bengler, K.: Methods to evaluate pilots' cockpit communication: cross-recurrence analyses vs. speech act–based analyses. J. Cogn. Eng. Dec. Making, **11**(4), 337–352 (2017)
13. Nonose, K., Kanno, T., Furuta, K.: An evaluation method of team communication based on a task flow analysis. Cogn. Tech. Work **17**(4), 607–618 (2015)
14. Johnston, J.H., et al.: A team training field research study: extending a theory of team development. Front. Psychol. **10**, 1480 (2019)

Blended Learning. Exploring Contradictory Demands Between Emerging Design Principles and New Learning Practices

Astrid Camilla Wiig(✉)

Department of Educational Science, University of South-Eastern Norway, Raveien 215, 3184 Borre, Norway
camilla.wiig@usn.no

Abstract. The aim of this study is to examine blended learning activities in two Master of Maritime Management courses, preparing students for work with innovation and organisational development. A comparison of the design approaches of digital Canvas rooms and blended learning activities is conducted. The data consists of qualitative interviews, students' midterm evaluation and screen dumps of the learning environment. The findings reveal tensions between different pedagogical approaches and teachers' construction of activities, pointing towards the contradictory demands between new learning practices and emerging design principles. In particular, analyses display the challenges for the students to understand the norms and rules of how to participate and engage in blended learning activities and their expectations of an active teacher role. Hence, the layers of accountability serves as a compelling concept to explain how increased digitalisation of courses change both teaching practices and students ways of participation.

Keywords: Maritime human factors · Blended learning · Teaching · Cognition and learning · Layers of accountability

1 Introduction

In higher education, much emphasis has been put on ways to reinvent the university structures and processes, their curricula and pedagogic practices to meet the capability needs of the knowledge society, as well as the learning preferences of students [1]. Incorporating online teaching resources to the current face-to-face learning environment has been growing in popularity and proved an effective approach to an increasingly diverse student population searching for just-in-time information and answers to questions, rather than weekly attendance at lengthy on-campus lectures [2]. The digital media sector claim that blended learning will be the new traditional model for course delivery in higher education [3]. However, empirical studies show that blended learning is not implemented as one model, but consists of courses designed in many different ways, ranging from building courses from scratch, to adding some extra online activities to a traditional face-to-face course. Thus, researchers argue that selecting the most appropriate design approach to effectively blend, is becoming a major challenge for the majority

S. Nazir et al. (Eds.): AHFE 2021, LNNS 269, pp. 109–121, 2021.
https://doi.org/10.1007/978-3-030-80000-0_14

of teachers in higher education especially because they lack the necessary theoretical preparation and experimental experience with blended learning [4]. Additionally, the lack of a single accepted definition of the term blended learning, forces teachers to designing their courses according to their own experiences and knowledge of the concept [5, 6]. Alammary, Sheard and Carbone [2] classified three different design approaches into low-impact, medium-impact and high impact blends and described the benefits, challenges and recommendations to teachers about when and how each approach should be used. Although academic research has tried to classify the design approaches and compare them to identify the benefits and challenges of applying each of them, less research has tried to investigate how blended design approaches and new learning practices challenge existing norms and rules for how teacher and students engage and interact in learning environment and create contradictory demands.

Against this background, the aim of this study is to explore the pedagogical challenges teachers and students in Maritime Education and Training (MET) face while interacting in blended learning courses. Drawing on situated approach to cognition and learning, which implies analysing interactional aspects by means of qualitative mixed methods of social inquiry [7], gives us the opportunities to explore how teachers and students experience learning and assessment activities during blended learning courses. Specifically, by scrutinizing the pedagogical activities and the designed blended learning environment in two master courses, the aim is to deliver and adequate explication of the contradictions of existing blended learning practices in MET. The study is based on three data sources; interviews, midterm-evaluations and screen dumps of the designed blended learning approaches of two master courses for first and second year master mariner students. The research questions are: 1. how do students experience the design approaches of blended learning courses? 2. How do students experience expectations of participation when engaging in blended learning?

2 Literature Background

A meta-analysis conducted by Means, Toyama, Murphy, Bakia and Jones [8] investigated more than 1100 empirical studies published between 1996 and 2008 and concluded that blended learning seems to be more effective than either online learning or face-to-face instruction. On the other hand, research also showed that blended learning courses are conducted in various ways, ranging from adding extra online activities to a traditional face-to-face course, to building the completely blended learning course as a fundamental reorganization of the teaching and learning dynamics [9, 10]. Consequently, some researchers argue that there is a problem that the field lack one single definition of blended learning, while others suggest it is the term's strength [11]. The classic, most common and broad definition is "The integrated combination of traditional learning with web-based online approaches" (p. 442) by Oliver and Trigwell [12] in [2]. For the purpose of this article, the definition of Alammary et al. [2] serves for describing blended learning as courses focusing on pedagogical aspects; they "1) thoughtfully integrate different instructional methods such as: lectures, discussion group, self-paced activity; and 2) contain both face-to-face and computer-mediated portions" (p. 443). Although extensive academic research has discussed and proposed how to design effective blended learning approaches, less research has classified the design approaches to describe the benefits, challenges and recommendations to teachers about when and how each approach should be used. Through a literature review conducted by Alammary, Sheard and Carbone [2] on studies published between 2004 and 2014, a classification of three different design approaches of blended learning were rendered visible. Low-impact blends means adding extra activities to an existing course, medium-impact blend replace activities in an existing course and high impact blends means to build the blended course from scratch. Although research has tried to classify the design approaches and compare them to identify the benefits and challenges of applying each of them, less research has tried to investigate the pedagogical practices in ongoing courses in higher education. Thus, this study aims to explore how blended design approaches and new learning practices are experienced by students while they interact and engage in these learning practices.

3 Method and Data

The methodological approach in this study is guided by principles for qualitative mixed methods to social inquiry [7], using interviews, midterm questionnaires and screen dumps of digital environments in research. Following these principles, the aim is to explore students' experiences and expectations while reporting from real-time blended learning activities. Additionally, these principles put emphasis on the relationship between temporal, material and social aspects participants' experience [13]. This makes qualitative mixed methods to social inquiry an important source for analysis, since the various data create accounts of the verbal, visual and material aspects under study, enabling dynamic, iterative analysis of the complex experienced practices [14].

The current study is situated within a larger research project funded by the MARKOM 2020 aimed at evaluating how Blended-learning methods in ongoing Master of Maritime Management courses, can enhance students engagement and motivate to better preparations before and after seminars at campus, and on teachers' knowledge and instructional practices. Four teachers in four different courses for first and second year master mariners at a Norwegian University volunteered. In the project, blended learning was not an 'add-on' to an existing teaching approach, but it involved a "fundamental reconceptualization & reorganization of the teaching and learning dynamic" and it was based on "rethinking and redesigning the teaching & learning relationship" [9]. Data were collected during the school years of 2019–2021.

For this study, data from two courses conducted autumn 2019 are analysed. Three data sources have been used to inform the development of another [7]. These are interviews of five students and two teachers, midterm evaluation of two courses and screen dumps of the two online rooms. The first kind of data constitutes the basis for the analysis, while the midterm evaluation have been used to inform the development of topics particularly interesting for the interviews. The topics of the interviews where related to the course content, teaching methods, the design of the online room in Canvas and student engagement. To inform the analysis of similarities and differences in what teachers and students describe as the structure of presenting content, the organisation and design of learning activities and opportunities for student activities, screen dumps of the online rooms in Canvas have been used (Table 1).

Table 1. Data sources

Interviews	5 students and 2 teachers
Midterm evaluation	N = 35 Course 1 and N = 19 Course 2
Screen dumps of the online rooms	

Five students volunteered to participate in semi-structured interviews. The interviews were conducted after exams, at a meeting room in another Faculty by a researcher from another department. The researcher did not know the students. The teachers were interviewed at their offices after grading their students. The interviews were audio-recorded. Responses to questions about the interviewees' experiences with learning activities, digital design and pedagogical approaches to blended learning were transcribed. The data were analysed thematically, using Braun and Clarke's [15] approach, which reduces the transcripts into meaningful groupings, allowing the participants to "speak for themselves" through the data, but not focusing much on the variations among the participants.

Based on previous evaluations of the Master courses, the teachers' focused on developing learning activities that could enhance students preparations before and after seminars at campus and to invite students to a more active role being responsible for their own learning. The teachers aimed to provide students with opportunities to prepare and practice before and after the physical sessions at campus to enhance motivation for studying the curriculum from the beginning of the semester and open for more student active learning (Table 2).

Table 2. The blended learning course in Master of Maritime Management

	Organization and leadership	Innovation projects
Students	First year students	Second year students
Campus sessions	7 face-to-face class seminars	7 face-to-face class seminars
Mandatory assignments	1 mandatory assignment in groups of 3–5, 4000 words Individually written feedback on each assignment, mostly on academic writing	3 mandatory assignments, individually or in groups of 2–3, 4000 words Feedback based on a skill-scale from 1–5 points
Exam	Oral group exam based on a real organizational challenge. Students shall use theories from curriculum and present results in groups. Additionally, read and give feedback on another group's assignment	Written exam based on further elaboration of assignments, individually or in groups of 2–3, 6000 words. Assignments are related to innovation theory and models. Students can chose to create an innovation project for a real company
Aim for Blend	Variation from heavy theory driven instruction course towards more virtual students' preparations and practice before and after campus sessions. Feedback and supervision virtually	Engage and prepare for independent work with Master thesis. Twisting towards more students work with assignments and supervision at campus and self-study of theory individually

Both Master courses were organized with seven face-to-face class seminars and a variation of asynchronous tools and online activities. In addition, both courses were built from scratch and offers asynchronous "on-demand" content such as quizzes, videos, TED-talks, Pod casts and material to practice what you have learnt in a "practice corner". The seven seminar days at campus were organized with synchronous or live lectures, supervision, group work and work with assignments (Table 3).

Table 3. An overview of the design and content of the two blended learning Master courses

In face-to-face classroom sessions Synchronous "Live"	In the online room Canvas Asynchronous "On-demand"
Face-to-face lecturing	Discussion Forums and video-reflections
Peer instruction and feedback	Plain homework such as assignments, tests and quiz
In class activities (exercises, problem based learning, supervision)	Textbook supports and recorded PowerPoint lectures with teachers' voiceover
PowerPoint lecture	Blogs and links to relevant online material
Teacher instructions, explications of concepts, models and theories	Articles and book chapters with links to Universities digital library
Webinars	Podcast and TED-talks
Guest lectures	Online Assignment and feedback
Video/Audio Conferences	Online content (i.e.: videos, newspaper articles)
Group work and work on assignments	Practice corner with quizzes, examples and opportunities for repetitions

Based on students' previous evaluations of the courses and the course plans descriptions of mandatory assignments, exam and learning outcomes, the teachers constructed the blended learning courses independently. The first year organisation and leadership course was created in a rather *"closed manner"* with explicit expectations of students' preparations and delivery of online homework between campus seminars. Similarly, the second year innovation course was organised in an *"open manner"* with clear expectations of student responsibility of their own learning and independency. In particular, while the teacher of the organisation and leadership course structured the online content and activities strictly to themes relevant for preparation and practice between and on the campus seminars, the teacher of the innovation course designed content and activities more as an online content list free of choice.

4 Analytic Results

The analysis of the empirical data display that even though the two courses were designed based on thorough empirical and theoretical studies of blended learning approaches to effectively blend, the design of the content, learning activities, methods and mandatory assignments were experienced differently by the students.

4.1 Students Experiences of the Organization of Blended Learning Courses

The empirical data of the innovation course, display that the focus on independency and freedom to choose what, where and when to study created challenges for the teacher and students. From the midterm evaluation, the class almost divided in two when answering the questions related to the teaching method:

	Totally disagree	.	..	...	Fully agree
I see the benefits in doing the reading and reviewing lectures outside of class *	10,5 %	10,5 %	31,6 %	31,6 %	15,8 %
The in-class learning activities and plenary sessions supports my learning process well *	5,3 %	10,5 %	21,1 %	52,6 %	10,5 %
The teaching method provides me with good opportunities to get feedback *	5,3 %	15,8 %	15,8 %	21,1 %	42,1 %
I would learn more if more lectures/material were presented in class rather than in canvas *	21,1 %	10,5 %	21,1 %	21,1 %	26,3 %
The exam form (portfolio) helps me achieve the course learning objectives *	5,3 %	5,3 %	21,1 %	31,6 %	36,8 %

While 47,4 agreed or fully agreed with the statement that they would learn more if the lectures/material were presented in class rather than in the LMS Canvas, 31,6% of the students disagreed or totally disagreed. Additionally, students' responds deviated when it came to the benefits in reading and reviewing lectures outside of class and in relation to whether blended learning provides god opportunities to get feedback from the teacher. This can be interpreted as differences in expectations and experiences with blended learning approaches in which open for extensive freedom and a reduced amount of teacher support. When suggesting how to improve the course, students wrote for instance:

I believe the class would be even better if we spent a little more focus on theory in class, and a little less focus on the structure in the Canvas room for example

I am not sure about the "radical" approach of removing classes. Even if we are supposed to study outside, I would like to have other complementary activities (cases, group works) rather than "time to work in your assignment"

Also, the interviews displayed that students struggled to understand what was expected and counted as relevant content and learning activities when the teacher interactions disappeared in the blended learning approach. Anna (pseudonym) explicate her challenges:

1 Truly, I really didn't like this way of teaching ehm, because I think that there are a lot of things that get lost in the way.

2 Ehm, I always study at home and I do watch videos to complement my studies, and I read the books, all literature.

3 So in that way, it was not that different.

4 But then, I was lacking the teacher.

5 And [teachers name], ehm I had this help from him and was talking with him on the assignment,

6 and [teachers name] explains very good, he is a very good teacher, very nice and is very easy-going,

7 so I believe it's a shame that we can't have class with him

8 (...) When I talked with [the teacher's name] it was very nice,

9 he gave me a lot of info, very interesting, easy-going, very understandable

10 And I thought oh; you opened my mind

Analysing the interview, documents how the student distinguish between the design of the course when it comes to content and student activities (line 2) and the interactions with the teacher (line 5 & 6). While the student describes that she takes responsibility for reading the course syllabus and watch complementary videos, other things get lost in the way (line 1), such as the interactions with the teacher (line 4). Highlighting the lack of her good teacher as one to talk to, who is nice, easy-going and very good to explain (line 6 & 9), she underlines the shame that these things get lost when they don't have classes with him (line 7). Analysing the interview displays the student's expectations of interacting with their teachers who share time to talk, give interesting, understandable and relevant information and even open the students mind (line 10). The excerpt of the student participating in the Innovation course which was organised in an open matter with expectations of students' independency and responsibility of their own learning, document the students' challenges to understand how to participate and in particular how to handle the lack of interactions with their knowledgeable teacher.

On the contrary, the data from the organisation and leadership course, display that the firm administration of learning activities relevant for preparation and repetition, was based on first year students request of structure, support to read and demand for relevant homework between face-to-face classroom interactions. The midterm evaluation displays that students were satisfied with the course content and teaching methods and in particular the online activities in the Canvas room:

Response distribution (%)

	Strongly Disagree	Somewhat Disagree	Neither Agree nor Disagree	Somewhat Agree	Strongly Agree
It is easy to navigate the canvas room *	8.6%	5.7%	20%	45.7%	20%
The room provides for good opportunities for interaction *	5.7%	5.7%	37.1%	37.1%	14.3%
The room has too much content *	0%	25.7%	31.4%	22.9%	20%
The information about the room and course is easy to understand *	8.6%	2.9%	25.7%	42.9%	20%
The room supports my learning better than most other canvas rooms in this study program *	0%	11.4%	45.7%	31.4%	11.4%
Extensive use of the digital environment helps me learn more compared to traditional classroom instruction *	2.9%	8.6%	28.6%	40%	20%

41,8% of the students experienced that the Canvas room supported their learning better than most other online rooms in their study program, and more than 60% meant the online activities helped them learn more compared to traditional classroom instruction. Still, the organisation of the course created challenges for the teacher doubling the workload for giving feedback on homework, assignments, and activities online and in class, and for students' who needed to do a large amount of activities for preparations and homework to receive their grades. 70% of the students claimed that the Canvas room had too much content and described benefits and challenges such as:

Too much assignment work to complete for a short time

The teacher is very good at inspire to discussion. I also like the mix of cultures in the class.

Other students' knowledge

Clear structure and engaged instructor

Also in this course, the interviews display that the students highlights the interactions with the teacher in order to help them focus on the important things in the curriculum. Liss (pseudonym) explicate her point:

1 Because this interaction with the professor,
2 you have him there; he is telling you what he knows.
3 Ehm, and is like ehm how do you call it,
4 uhm pointing out the important things.
5 Because we have the books, and we can study by ourselves,
6 but we can read all the books and not focus on this part.
7 So we need to know what to focus on, because it is a new topic for us.
8 Because you need to have words to think on.

Examining this interview, documents the student's need for interacting with the teacher to get information on what to focus on in a new topic (line 7). The student explains that they study by themselves and read the books, but need to interact with the teacher to let them point out the important things (line 4), to tell them what they know and give the students words or concepts to think with (line 8). While the investigation of the screen dumps of the online courses display that the teachers have uploaded a variety of relevant online content and learning activities, the analysis of students interviews show that the online content is less important than how to interact with the teachers in order to understand what counts in the course. In particular, students seems to expect that through teacher interactions, they can collect information about the important things and concepts needed to display relevant knowledge when engaging in learning new topics.

4.2 Students Expectations of How to Participate in Blended Learning Courses

Examining the interviews with the students, display their experiences and challenges in negotiating how to participate in the blended learning courses. In particular, investigation of the interviews document the differences the students experience amongst coming to class to interact face-to-face with a teacher contrasted by online interactions. Peter (pseudonym) explicates these differences:

1 You get a full understanding of the curriculum by listening [to the teacher]
2 I think there is a fine line between screen watching and watching someone in person.
3 Actually a very distinct line, not a fine line.
4 The things are very different.
5 Because in YouTube, any topics you search, you have an answer.
6 Already in articles and papers, there is an answer.
7 Everything you thought have an answer.
8 Then what's the point to come to class?

Peter addresses the distinct line between screen watching and interacting with the teacher physically in class (line 2 and 3). While watching videos, for instance searching YouTube for a particular topic, he argue that you will always find an answer. However, by coming to class to watch the teacher in person, his experience is that things become very different through listening to the teacher (line 4) because you get a full understanding of the topics in the curriculum (line 1). Peter further elaborates on the differences and highlights the need for structure in online activities:

1 I think the structure is important when everything is digital
2 Like thousands of hours of videos.
3 I am not expecting to finish watching all thousands hours
4 and I don't know what to see to learn what.
5 But if you come to class, even if we don't read the book,
6 we can persist that this is what is expected of this course,
7 and we find the materials, online or in book and we learn that.
8 I can guarantee; no one reads the whole book or even half of it.
9 Just by getting the point, they look for explanations on different sources and they get their knowledge in that way.
10 And I think in that way the face-to-face conversations in class is important.
11 By watching videos it's like video going and going
12 and you don't exactly pay attention to what is going on
13 and there is no feedback system to video.
14 Like now, I cannot ask anything.
15 Why watch that video? That's the minus.

Analysing the interview, documents how Peter describe the need for structure in the overwhelming mass of digital content, such as thousands of videos that exists in the digital courses. Referring to the problems of paying attention and choosing what is relevant in addition to the lack of feedback system to videos, underlines his relevant question; why should I watch that video? (line 15). By problematizing the whole idea of just adding online content into an existing traditional course, he explicate the vital role of the interactions between teachers and students. By face-to-face conversations in class, students "can persist that this is what is expected of this course" (line 6), and find the sources needed to look for explanations (line 9). His experience of classroom interactions where subjects expected to learn and engage in becomes transparent in teacher – student conversations seems to be in contrast to some courses where extra content is added unconsciously just to blend learning activities.

In another interview, Steve (pseudonym) emphasises the challenge of participating in blended learning courses when it comes to conceptual understanding and need for basic understanding:

```
1 I always think that first we need some basic theory.
2 At the beginning, we need some base.
3 First classes we need these lectures, I believe so.
4 I still believe that we need these interactions with the
  professor.
5 (...) But we need first this baseline on theory,
6 and then, yes we can have days where we will work on the
  project,
7 and maybe even whole days just project work.
8 But at that time we need to have already some ideas,
9 broad ideas of what will be the solution of the problem.
```

The analysis document the students need for lectures presenting basic theories and a baseline at the beginning of a blended learning course (line 1 & 5). Steve highlights the need for interactions with the professor to be able to build a base of theories before they can start working with assignments and projects. When the students have some broad ideas of the field (line 8), they can be more active and independent in their own learning processes and work to find solutions of their problems in the project assignments. Thus, the analysis of the interview describe how a student experience to participate in the blended learning course without the needed baseline, theories and ideas to solve solutions of a project work. While the approach of the course was designed for independency and freedom to choose among content and learning activities available in the digital room, the students experienced the lack of a theoretical baseline provided by the teacher. The empirical data documents the students need for more support, lectures of basic theories and interactions with a teacher who can give ideas to a solution to scrutinize the scope of their project work. Additionally, the students interviews describe how students struggle to participate in the blended learning courses when the norms and rules of how to engage remain vague or invisible.

5 Conclusion and Discussion

This study has explored students' experiences of blended learning activities in two authentic Master of Maritime Management courses. While the blended learning approaches is argued to be effective for fostering students into the maritime management work practice, findings from the current study show that this kind of pedagogical practice also creates challenges. Results displays that the challenges are related to the pedagogical perspectives of how to organise the blend of learning activities to achieve the learning outcomes in the subject matter, the teachers role and how to make explicit expectations of how students should engage to make themselves assessable. The organisation of how to blend content and learning activities found in the data corpus document that courses were designed differently. One blended learning course was designed in a *"closed manner"* with clear demands of preparations and online homework related to face-to-face teaching at campus, and the other was designed in an *"open manner"* with students free to choose what, when and how to make use of the online learning material. In that sense, they differ from previous classifications of different design approaches into low, medium and high impact blends and how it benefits learning preferences of students [7]. However, the students' expectations of interacting with the teachers who help to sort and focus among thousands of videos and other online content, who scaffold, distinguish and lift up what is relevant for meaning making and not, seems to disappear in the focus of designing the degree of effective blended content. This finding suggest that students expect to receive more support, supervision, explications and instructions from teachers in blended learning courses to develop their professional competence. Hence, there are reasons to consider the practice of blended learning activities, focusing more on approaches to blend more effective content, than planning for how teachers can support students meaning making processes online and face-to-face. The findings of this study display that while the design of relevant online content and blend of lectures preoccupied much of teachers focus, students were more or less left alone with their needs for support in the actual learning activities for receiving a professional degree in maritime management.

Even though this analysis is based on a small sample of interviews, midterm evaluations and screen dumps of the online courses, the data corpus offers a complex and interesting array for analysing the complexity of organising the pedagogical issues in existing blended learning approaches in MET. In this data, findings show that the organisation of students' ways to participate in learning activities differs and that students' experiences with new norms and rules of how to engage to make themselves assessable are blurred. In regards to this, findings in the empirical data raises critical and important questions in regards to what it means to be accountable in blended learning activities [16] and [17]. While the teachers organise content and learning activities to support students' learning processes during the course, the overall aim of the blended learning approaches seems to be the selection of the most appropriate design to effectively blend [4]. Developing different design approaches of blended learning to identify the benefits and challenges of applying each of them, is one way to go [2]. Another part of the solution is to develop the teachers' knowledge on how to conduct relevant and meaningful interactions with the students during their learning processes in the blended learning courses. In regards to this challenge, there are reasons to be careful before defining a method and one design approach for organising blended learning activities for enhancing professional development across management courses with different goals for students learning outcome. As pointed out in the introduction, results from empirical research on implementing blended learning courses display a variety in how to blend, ranging from building courses from scratch to adding some extra online activities to a traditional face-to-face course. Hence, there is a need for future studies that analyse the current blended learning practices to identify areas of improvement, and develop a practice where the students need and expectations of interacting with teachers who support, make prompts and engage to explain what is relevant and not among the thousands of possible digital content deliveries are the core subject of analysis.

Acknowledgments. This research was funded by MARitim KOMpetanse mot 2020 (MARKOM 2020), by the Norwegian Ministry of Education, project no. T136. Undervisningsmetodikk basert på Blended Learning. I would like to thank the teachers, students, informants and project members for their support and engagement. In particular, I would like to thank Veronica Jaramillo Jiminez and Marius Imset for their wise comments and contributions on earlier drafts.

References

1. Bridgstock, R.: Educating for digital futures: what the learning strategies of digital media professionals can teach higher education. Innov. Educ. Teach. Int. **53**(3), 306–315 (2016)
2. Alammary, A., Sheard, J., Carbone, A.: Blended learning in higher education: three different design approaches. Australas. J. Educ. Technol. **30**(4), 440–454 (2014)
3. Norberg, A., Dziuban, C.D, Moskal, P.D.: A time-based blended learning model. Horizon **19**, 207–216 (2011). http://www.emeraldinsight.com/loi/oth
4. Huang, R., Zhou, Y.: Designing blended learning focused on knowledge category and learning activities. In: Bonk, C.J., Graham, C.R. (eds.) Handbook of Blended Learning: Global Perspectives, Local Designs, pp. 296–310. Pfeiffer Publishing, San Francisco (2005)
5. Deperlioglu, O., Kose, U.: The effectiveness and experiences of blended learning approaches to computer programming education. Comput. Appl. Eng. Educ. **21**, 328–342 (2013). https://doi.org/10.1002/cae.20476
6. Lee, G., Fong, W.W., Gordon, J.: Blended learning: the view is different from student, teacher, or institution perspective. In: Cheung, S.K.S., Fong, J., Fong, W., Wang, F.L., Kwok, L.F. (eds.) ICHL 2013. LNCS, vol. 8038, pp. 356–363. Springer, Heidelberg (2013). https://doi.org/10.1007/978-3-642-39750-9_33
7. Greene, J.: Is mixed methods social inquiry a distinctive methodology? J. Mixed Methods Res. **2**(1), 7–22 (2008). Sage Publications. https://doi.org/10.1177/1558689807309969
8. Means, B., Toyama, Y., Murphy, R., Bakia, M., Jones, K.: Evaluation of evidence-based practices in online learning: A meta-analysis and review of online learning studies. U.S. Department of Education, Washington, DC (2009)
9. Garrison, D.R., Kanuka, H.: Blended learning: uncovering its transformative potential in higher education. Internet High. Educ. **7**, 95–105 (2004). https://doi.org/10.1016/j.iheduc.2004.02.001
10. Garrison, D.R., Vaughan, N.D.: Blended Learning in Higher Education: Framework, Principles, and Guidelines. Wiley, San Francisco (2011)
11. Driscoll, M.: Blended learning: Let's get beyond the hype (2002). http://www07.ibm.com/services/pdf/blended_learning.pdf
12. Oliver, M., Trigwell, K.: Can 'blended learning' be redeemed? E-learn. Digital Media **2**, 17–26 (2005)
13. Sellberg, C., Lindmark, O., Lundin, M.: Certifying navigational skills: a video-based study on assessments in simulated environments. TransNav Int. J. Mar. Navig. Saf. Sea Transp. **13** (2019)
14. Denzin, N., Lincoln, Y.S.: The SAGE Handbook of Qualitative Research. SAGE Publications, Inc. (2017)

15. Clarke, V., Braun, V.: Teaching thematic analysis: overcoming challenges and developing strategies for effective learning. Psychologist **26**(2), 120–123 (2013)
16. Michaels, S., O'Connor, C., Resnick, L.B.: Deliberative discourse idealized and realized: accountable talk in the classroom and in civic life. Stud. Philos. Educ. **27**(4), 283–297 (2008)
17. Wiig, C., Wittek, A.L., Erstad, O.: Teachers, tools and accountable practices. Engaging with a wiki blog as a learning resource. Learn. Cult. Soc. Interact. **22** (2019). https://doi.org/10.1016/j.lcsi.2018.07.001

Kaizen Method Applied in Higher Education: Case Study of Autonomous University of Baja California

Susana Zambrano-López(✉), María Marcela Solís-Quinteros, Carolina Zayas-Márquez, and Luis Alfredo Ávila-López

Universidad Autónoma de Baja California, Calzada Universidad, número 14418, Parque Industrial Internacional Tijuana, 22390 Tijuana, BC, Mexico
{susana.zambrano,marcela.solis,carolina.zayas, avila.luis}@uabc.edu.mx

Abstract. The origin of the Kaizen methodology is seen in the business sector specifically in the manufacturing industry, its main objective is to achieve the optimal performance of employees in the process of continuous improvement, generating that the transfer of this method to the educational field is analyzed. This is the case study of the Faculty of Accounting and Administration (FCA) of the Autonomous University of Baja California (UABC) in Tijuana, Mexico, where the Kaizen method was applied through an applied research of an experimental and descriptive scope, where The purpose is to identify and improve the educational quality processes involved in the area of Professional Social Service (SSP), given that a series of inconsistencies have been detected in the processes and on those involved, this being an essential requirement for the qualification process FCA and influences the efficiency levels of the terminal.

Keywords: Higher education · Professional social service · Quality · Continuous improvement · Kaizen

1 Introduction

Currently, technological and scientific changes seem to have greater visibility, everything seems to go faster and the production processes undergo constant changes that force us to rethink the concept of quality that they have immersed. In the case of educational systems, exactly the same thing happens worldwide, in addition, there is a marked trend in the incorporation of business tools and methodologies.

Largely, this approach between schools - companies, and particularly universities - companies, responds to the linking processes, which require a certain standardization between both institutions, with the aim of mitigating the naturally existing bureaucratic barriers. In addition, an aspect of special importance is globalization, especially economic and productive, which leads to a convergence between institutions from different countries that come together with the aim of responding to the problems that arise in

S. Nazir et al. (Eds.): AHFE 2021, LNNS 269, pp. 122–130, 2021.
https://doi.org/10.1007/978-3-030-80000-0_15

the productive processes of companies and that are approached by universities with the purpose of providing a solution [2].

The main objective of this research work is to establish the method of continuous improvement in the Faculty of Accounting and Administration (FCA) of the Autonomous University of Baja California (UABC), specifically in the coordination of Professional Social Service (SSP). We have detected delays and failures in the processes, which has therefore a negative impact on terminal efficiency, since the SSP is a partial requirement for the degree of students at the undergraduate level.

In response to this problem, the Kaizen methodology was applied, which is one of the main elements of the Toyota house, in which the production processes and administrative strategies that allow optimal performance are combined, analyzed from the point of view of continuous improvement. [5].

This tool was applied to the coordination of SSP, hoping to find delays in the process, given that the experiences expressed by users denote dissatisfaction with the service provided. It is extremely important to implement improvement plans, since it is an essential requirement to finalize the graduation process of graduates, who sometimes suffer delay effects attributed to the SSP for about six months and one year to obtain the compliance certificate.

2 Theoretical Framework

Kaizen comes from the Japanese words "Kai" and "Zen", which refer to the action of change and continuous improvement. That is why the implementation of this methodology supposes the elimination of the waste generated in the production systems, and as can be seen, it is a process focused on continuous growth, so it never stops innovating and using [1].

The Kaizen methodology is characterized by using a culture of involvement at the global level of the entire company, in order to favor the humanization of the work environment through the empowerment of employees. It is necessary to bear in mind that it is not about generating big changes abruptly, but rather focusing all attention on making small improvements. Small improvement will be continuously and covering all the activities generated within the company, all with the aim to become part of the market by improving processes within the organization rather than pushing products to market and in turn improving the quality of work.

It is designed to reduce waste and eliminate errors, but takes a different approach by focusing on small improvements and addressing problems across the organization that are not necessarily specific, as the main goal is simply improvement. In addition to relying heavily on employee feedback, rather than customer feedback, intended to be integrated into the corporate of organizations as a way of operating on a day-to-day basis [6].

It is a key tool for continuous improvement; it makes possible to identify those situations, processes or procedures that are not adding value to the company. Through the gradual elimination of this waste, the company becomes increasingly efficient, improving quality, processes and products. According to Rodríguez, A. (2018), this tool is based heavily on the analysis of the current situation and allows constructive criticism in order to improve each of the processes.

The Kaizen methodology follows the DMAIC process (Define, Measure, Analyze, Improve and Control). It is an improvement process that is based on the Six Sigma methodology, a model that follows a series of structured and disciplined steps, consisting of five phases that are interrelated in a coherent way.

In practice, the Kaizen method works as follows: 1. Setting clear and realistic goals, well documented, 2. Review of the current situation and development of an optimization plan, 3. Implementation of improvements, 4. Review and application of the necessary corrections, 5. Preparation of a report of results and determination of the monitoring elements. This type of cycle is commonly known as PDCA (Plan, Do, Check and Act). From the development of a hypothesis, an execution experiment is applied, the results of which are evaluated to gain alignment and, after the necessary adjustments, a new cycle begins. That is why the Kaizen method is known as the practice of continuous improvement [3].

3 Kaizen Methodology Within the Educational System

Kaizen philosophy can also be effectively implemented in education to greatly benefit the overall development of the student and the school. The strategy revolves around continuous improvement measures.

Education is a continuous process where information is processed and the resulting knowledge will be applied for the betterment of both the individual and society. In this context, Kaizen greatly helps the very purpose of education. By absorbing the concepts of Kaizen from infancy, students will strive for their best as they grow into adults [7].

4 Justification

At the UABC, the SSP is a transversal activity that undergraduate students must carry out, and it is aimed at developing in them the spirit of service towards society and the environment that surrounds it. For the university it is important that graduates distinguish themselves in the labor market due to this noble characteristic. For this reason, the functions of the SSP coordination were analyzed, and a survey was generated in which the students were asked about the service provided.

Among the main findings is that the workshops offered to publicize the process involved in the SSP do not have the required effectiveness according to the surveyed students. In addition, there is a lack of information on the various processes carried out to register the Receiving Units (UR), which are those institutions where the SSP is carried out. Another important aspect is that the waiting time for assignment to an SSP program is long, in addition to the fact that students consider that the URs available have little relation to the careers offered.

Communication with the SSP coordination is another element that makes the care provided inefficient. The students mentioned in the survey that they lack support from the coordination to search for an UR and register in it, and that there is no monitoring and evaluation of the activities carried out in the service. Finally, more than 60% of the respondents mentioned that they were not notified about the documents that they had to present to process the SSP termination letter.

5 Problem Statement

KThe objective of the SSP is to generate an awareness of social responsibility in students, and to act as a link to bring the benefits of science, technology and culture to the marginalized sectors of society and thereby promote development sociocultural. In the case of the FCA, which is part of the UABC, the service is public and mandatory. The program does not imply economic remuneration in favor of the person who provides it. Finally, there is no employment relationship between the student and the receiving unit. Another feature is that the SSP is a necessary condition for the student to obtain his bachelor's degree.

As it is a requirement for qualification, all FCA students have the obligation to eventually go through the coordination of SSP, hence the importance of establishing continuous improvement strategies that allow better functioning. To achieve this, the functions of this coordination were analyzed in order to identify the problem and thus be able to propose improvement proposals. According to the information collected through the survey, there is an evident problem in the SSP process. Those problems are in the attention it provides to students. 1) General information about the SSP, 2) lack of relevant receiving units with the careers offered, 3) lack of monitoring of the activities carried out by the student, 4) lack of evaluation of the activities carried out by the students and 5) lack of guidance on the process of the completion certificate.

6 General Objective

Apply the Kaizen methodology in the SSP of the FCA to improve the attention to students.

6.1 Specific Objectives

- Analyze the SSP process to identify activities that need improvement.
- Prepare a kaizen methodology proposal to improve established procedures.
- Propose an action plan to establish standardized procedures.

7 DMAIC Process (Define, Measure, Analyze, Improve and Control)

7.1 Define

Next, in Fig. 1 is the process of assignment to professional social service is presented through a flow chart, in order to identify the existing problems in coordination.

The problem detected in the coordination of SSP lies in the access to the information, and the understanding of it by the students.

7.2 Measure

In the survey carried out on a sample of 74 students, it was calculated using a statistical equation for population proportions whose confidence level was 95%, with the universe of 313 participants. It was detected that according to their perception; the published information is not clear, that is; After analyzing it, doubts remain; in addition, there is a duplication in the access web pages. Another essential aspect is that the coordination processes are long and tedious, from the registration of a receiving unit, to the student's own assignment to the service. On the other hand, it should be noted that there are intermediate processes, such as reports of activities carried out, that is; the quarterly report and the final report, where complaints are also presented.

However, there is also the perspective of coordination that indicates that students have a lack of analysis and criteria to read the information and interpret it and thus make the necessary records to carry out this activity. In addition, there have been cases in which students do not verify their records, perform the service and upon completion, the record is presented as incomplete and this causes them to have to resort to it.

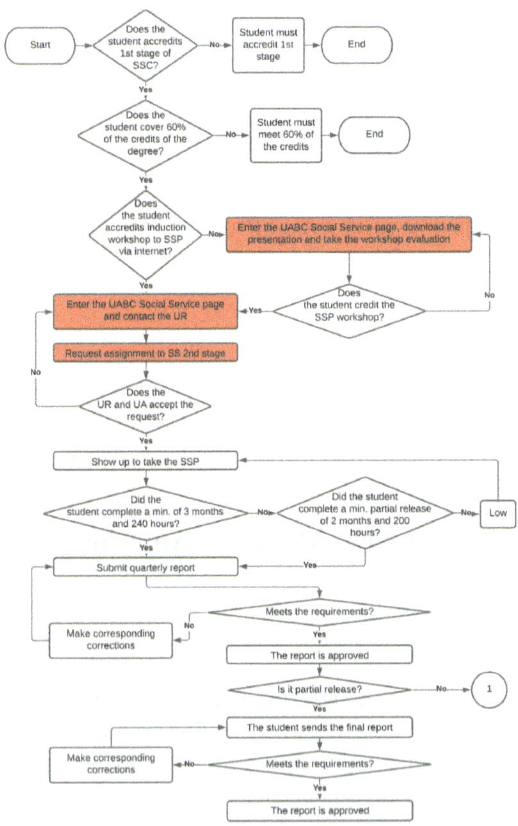

Fig. 1. Flowchart. Own elaboration with data from the ACBSP accreditation.

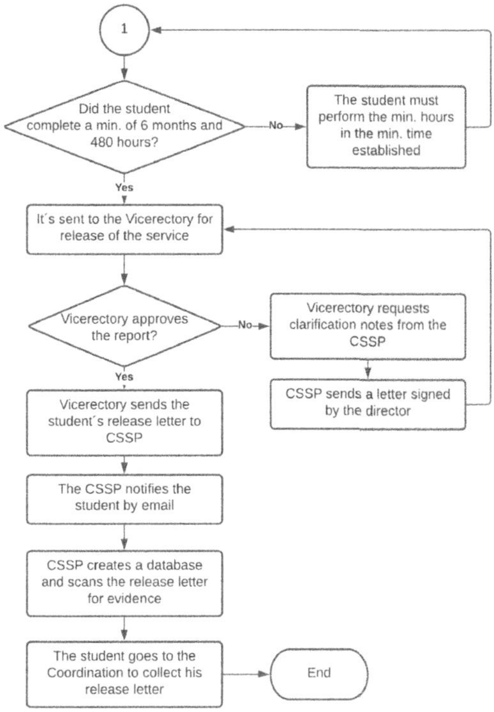

Fig. 1. (*continued*)

7.3 Analyze

In order to clearly identify the real causes that have an effect on student complaints in professional social service care, the following Cause-Effect Diagram was elaborated, finding that the main causes are related to the environment (Fig. 2).

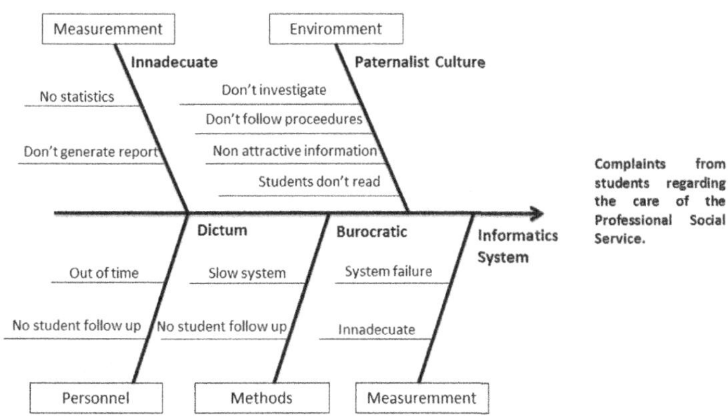

Fig. 2. Cause and effect diagram. Own elaboration.

7.4 Improve

After the analysis and review of the flow process, cause-effect diagram, student surveys, as well as the interview with the SSP coordinator, three problems that can be solved within the FCA were identified.

Below is an improvement plan whose proposals are intended to streamline the processes and thereby improve the care provided by this coordination.

7.4.1 Continuous Improvement Plan

In the measures proposed for SSP coordination, the first (internal) aspect to improve is the lack of audiovisual material. Currently the information is in written form, but taking into account the advancement of generations, you are more likely to receive audiovisual information through short and specific videos.

The second aspect to improve is related to the causes found where the student does not have the culture of reading the information related to the social service procedure. We found that the students does not follow procedures, nor does he investigate on his own.

The design of the page is not attractive and the visibility of the program on the internet page (http://fca.tij.uabc.mx/) is currently within the "extension and linking" tab. In other faculties of the UABC itself, the SSP information is in student affairs tab and that facilitates access to information; therefore, it is proposed to make this change on the login page.

The third aspect to improve is the lack of dissemination of this activity, both by teachers, the coordination of the SSP, social networks, and the figure of tutors whose activity is oriented towards students. Therefore, it is proposed to offer training to tutors to publicize the existing processes and changes in coordination in a timely manner, so that they in turn disseminate it among the students.

Finally, an (external) improvement action was detected related to the lack of an effective and efficient institutional communication channel among all the members of the SSP coordination, with the objective of externalizing system failures, improvements, etc.

7.5 Control

Below is a table with an action plan that responds to each of the proposed improvements and where the criteria are defined so that the control of the proposed actions are not lost (Table 1).

Table 1. Proposal for an improvement plan.

Plan	Do	Check	Act	Schedule
Audiovisual material	Make videos and tutorials on each of the processes	Present materials for review with a sample of students	Conduct a student satisfaction survey to measure effectiveness	First week of the 2021-2 period
Information location	Place the SSP information in the students section of the FCA website	Test a sample of students to verify easy access to information	Conduct a student satisfaction survey to measure effectiveness	First week of the 2021-2 period
Tutor training	Organize trainings to publicize updates in the SSP	Generate the contents of the training courses	Carry out a satisfaction survey to tutors to measure effectiveness	Last week of the period 2021-1
Institutional communication channel	Propose an intern channel for the SSP coordinators of the different faculties	Establish thematic axes to place updates of SSP systems	Conduct satisfaction survey to SSP coordinators to measure effectiveness	Last week of the period 2021-1

8 Conclusion

With the development of this work, it becomes clear that engineering tools can be perfectly transferred to the educational plane. In this case, the Kaizen methodology was implemented with the aim of incorporating a quality culture and with it continuous improvement in the coordination of SSP in the FCA of the UABC. A series of problems were detected, and a kaizen event was developed with the purpose of generating an improvement proposal, but above all, to generate an action plan that allows establishing the necessary guidelines to correct delays in the SSP processes.

References

1. Antevenio: Whats is kaizen methodology? Kaizen methodology, what it is, how it is applied and how it affects your business (2018). https://www.antevenio.com/usa/what-is-kaizen-methodology/
2. Larson, J.: Cooperation between universities and companies, globalization of education in the field of management: European perspective. Univ. Manage. Pract. Anal. **22**(2), 17–24 (2018)

3. OBS: Método Kaizen: aplicación y beneficios. Project Management: Business School (2020). https://obsbusiness.school/es/blog-project-management/temas-actuales-de-project-management/metodo-kaizen-aplicacion-y-beneficios
4. Rodríguez, A.: Metodología de implementación de Kaizen y 7 desperdicios para Tablemac S.A.-Plnata de Yarumal. In: Universidad EAFIT (2015). https://repository.eafit.edu.co/bitstream/handle/10784/8300/CarlosAlberto_RodriguezAlvarez_2015.pdf?sequence=2
5. Sarria, M., Fonseca, G., Bocanegra-Herrera, C.: Modelo metodológico de implementación de lean manufacturing. Revista EAN **83**, 51–71 (2017)
6. White, S.: What is Kaizen? A business strategy focused on improvement. CIO from IDG Communications (2019). https://www.cio.com/article/3408780/what-is-kaizen-a-business-strategy-focused-on-improvement.html
7. Wilson, M.: How to introduce Kaizen philosophy in education? Creative Safety Supply (2012). https://blog.creativesafetysupply.com/how-to-introduce-kaizen-philosophy-in-education/

Interactive Multimedia and Gamification

Gamification Strategies to Teach Algorithmic Thinking to First Graders

Roberto Araya(✉)

Institute of Education, University of Chile, Periodista Carrasco 75, Santiago, Chile

Abstract. Computational thinking (CT) is at the base of the development of today's society. One of the pillars of CT is Algorithmic Thinking (AT). This means be able to follow and write precise instructions, understand how algorithms operate and know their different component blocks. In this paper, we describe coloring activities addressed to first graders to teach them to include logical conditions on the instructions, both conjunctions and disjunctions, use universal and existential quantifiers, use implications and negations, and connect different instructions to lead to conclusions. Teachers assess students' activities in the coloring book with an app and track the progress in a territorial ecosystem with gamification features. In phase one, 144 first-grade classes used the app to assess reading and executing coloring instructions. We found that 87% of students performed correctly the instructions. In phase two in one school, students wrote coloring instructions. We found that 54% prefer coloring to other learning tasks, 95% executed correctly simple coloring instructions, 73% wrote correct instructions for simple coloring tasks, 83% executed correctly complex coloring instructions, and 64% were able to write instructions with quantifiers for complex tasks with arrays. We conclude that coloring books enhanced with smartphones have big potential to teach AT to first graders and track their learning.

Keywords: Gamification · Algorithmic thinking · Computational thinking · Coloring books · Mobile tracking

1 Introduction

Computational thinking (CT) is at the base of the development of today's society. It has an enormous impact that ranges from smartphones with systems that aid navigation to models of the structure of COVID19. Smartphones are making a big impact in our lives. Soon they will take control of our vehicles, changing completely the safety of the passengers and liberating enormous amount of time wasted on driving. They will also negotiate with other people's devices and decide for us to undergo a surgical operation, and thereafter ask the surgical robot to do the surgery. They will choose our mate and couch us in our relationship. Therefore, it is critical that the citizens understand what is going on, how the machines decide, their errors and biases, and how they learn. This will be a fundamental knowledge, similar to be able to read and write. Therefore, it is urgent to add CT as a core knowledge together with literacy and numeracy. However,

S. Nazir et al. (Eds.): AHFE 2021, LNNS 269, pp. 133–141, 2021.
https://doi.org/10.1007/978-3-030-80000-0_16

there is still no curriculum for CT that is universally accepted. A particularly confusing issue is whether students should begin by learning a programming language or rather begin by understanding the computing process [1–3]. Nor CT is a widespread practice in elementary school math classes [4, 5]. Thus, it is important to design and test curriculum and activities that help teachers on what and how to teach CT. It should be at the center of the educational efforts that prepare students for the Super Smart Society [6]. It is learning to think very precisely as a machine does, step by step in a mechanical way, so that students can collaborate with machines and understand their recommendations and behavior. [7] proposes three basic pillars of computational thinking: algorithmic thinking (AT), computational modeling and Machine Learning. AT is the process of decomposing a problem in a sequence of simpler sub problems that can be solved by machines [8]. Computational Modeling is a way to understand the world, recognizing its main characteristics, discovering its hidden mechanisms and articulating them explicitly with rules [9]. Machine Learning is training the computer with cases instead of giving instructions [10, 11]. However, it is critical to understand how the machine detects those patterns and learn from those examples. The focus is in understanding the conceptual foundations of CT.

Given the enormous and increasing impact of these three basic pillars on society and the new nature of work, all citizens should have the right to obtain a basic but deep enough knowledge in order to understand CT. In this paper, we describe the implementation of unplugged lessons [12] to develop AT in first graders.

2 Why Coloring Books for AT?

A coloring book is a type of book with line art to which readers add color [13]. Coloring books are very attractive to adults and children. Nine of the 20 books on Amazon's 2015 bestseller were coloring books [14]. During the COVID19 pandemic, publishers reported a boost in coloring book sales [15]. People associate coloring books with art and therefore with a recreational and playful activity. Coloring books have also effect on mood. On adults, they have short-term effects on decreasing stress and increase relaxation [16, 17]. Moreover, [18] found that brain activation via functional near-infrared spectroscopy during coloring resulted in significant activation of the medial prefrontal cortex compared to the rest conditions, and that participants improved in their self-perceptions of problem solving and in having good ideas. Coloring books are also used for non-art education, such as anatomy-coloring books for adults and for children [19]. There are coloring books for learning to read [20]. For children learning to read and write, [21] claims that coloring enhance fine motor development, improves hand-eye co-ordination, develops spatial awareness, builds concentration and focus, promotes creativity and self-expression, teaches color and shape recognition, and relaxes and releases emotions. There are also coloring books for other disciplines, such as to explain computer programming [22, 23].

Thus, the strategy is to take advantage of the recreational environment that coloring books generate in the students. On the other hand, the strategy is to use them to perform core AT activities. Firstly, the coloring book activities should ask students to read instructions, called code, of what to paint and what color. That is, the students have to imagine that they are printers, and that they have to read the code, interpret it, and then print accordingly. In a more complex set of instructions students have to connect sentences and using conjunctions, negations and implications, they have to be able to infer the location of the hidden object and then color its location. Secondly, students should also have to design instructions for others to paint. That is, they have to write very precise code. Classmates should be able to interpret the instructions and color accordingly. In this way, students have to develop thought processes involved in formulating a problem and expressing its solutions in a way that a computer-human or machine- can effectively carry out [8].

A second strategy is the use of glove puppets or imaginary companions. Children perceive them to be agents that are worthy of social interaction [24]. 67% of children have imaginary friends and having an imaginary friend confers a developmental advantage in a number of important socio-cognitive areas [25]. Children with imaginary friends produced a range of more complex sentence types in a narrative task than children in a control group [26]. Using glove puppets as a tool for learning could be more effective if children use their own favorite glove puppet, since they are less likely to accept an identical replacement [27]. Play and glove puppets facilitate the development of powerful cognitive strategies in children, such as role-reversal imitation and transmission chains, in which children learn something and then teach another child [28]. However, in this case the child discuss the situation with her glove puppet and instructs it to color as shown in Fig. 1, left and center. This in turn facilitates the internalization process, where a child not only follows instructions but also self regulates their own problem-solving activities and instructs themselves.

Fig. 1. Left and center: students coloring the AT activities in the coloring book and discussing the instructions with their hand puppets. Right: teacher using the ConectaIdeas app to assess students´ activities and upload the information

3 App for Teaching Reading and Writing

Smartphones are already making a huge impact on several areas of the economy [29]. In this work, we adapted an App to the ConectaIdeas teaching platform [30–32]. The App allow teacher to register students' activities in their smartphones as shown in Fig. 1, right. Later on, the teacher can track the progress according to specified learning objectives in the curriculum. Teacher and parent then can access each class performance in a territorial ecosystem that display curriculum coverage by schools, districts, and regions as shown in Fig. 2 [30, 33]. The territorial learning ecosystem is as a geographic information system, with maps showing the activity by region, district and classroom. A territorial learning ecosystem for the whole country shows an interactive map with statistics on curriculum coverage per class, district and region. In each region, teachers and parents can see the activity indicators. Darker zones correspond to districts where students have completed more tickets per student.

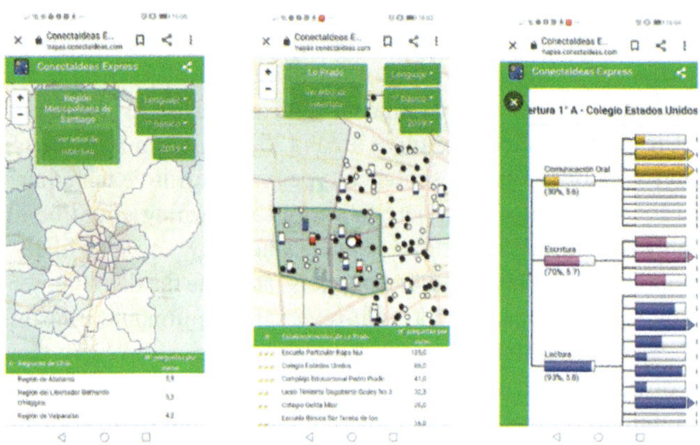

Fig. 2. Screenshots on a smartphone of real time maps of the territorial ecosystem. Left: map of a region with performance by districts. Center: map of a district with performance by schools. Right: curriculum coverage of a school.

The territorial learning ecosystem has gamification features that promotes learning with territorial scores to encourage participation of all districts and regions.

4 Methods

We designed a pilot composed of two phases. First phase was at the end of the 2019 school year, just when schools were leaving for summer vacation. Phase two, was planned for March, just after returning from summer vacation. The first phase was the National Reading Day. We designed this day to test schools' and students' responses to the coloring book strategy. On that day, schools downloaded coloring activities for a 90 min session, and use the ConectaIdeas App to assess students work. It was a voluntary activity. Schools could make the activity on a day of their choice between November 25 and December 6, 2019. The activity consisted of four coloring pages. In each page, students had to color some images according to the written instructions. All of them belong to the Learning Objective (LO) *Read independently and understand simple non-literary texts*, of the National Curriculum. The National Reading Day was a very successful project according to teachers, principals and Ministry of Education authorities. It was a pilot of national reading comprehension test, but the great appeal of coloring activities generated an environment very different of the typical stressing environment of a standardized test.

Moreover, the coloring instructions tested the idea of including coloring to introduce very basic AT in first grade.

For the second phase, we designed the ColorealIdeas coloring book [34] and implemented a pilot in some schools. ColorealIdeas coloring book contains a sequence of activities of increasing complexity in AT. Each student had to imagine she was a printer that colored according to the written instructions. These instructions were the code. On the other hand, in the next page the student have to create instructions (code) for a classmate to execute. We report the implementation in one first grade class with 25 students. The pilot lasted two hours and contained a survey and four pages of the book: pages 1A, 1B, 10A and 10B. The survey ask students their preferences about learning with coloring activities, and their use of hand puppets or other imaginary or non-imaginary companions. The first two pages 1A and 1B are very basic activities, and serve as training. The goal is for students to learn to respond by following very precise instructions. They are basic AT activities. Page 1A requires that the student, in addition to writing her name, carry out the instructions written in 4 boxes. In each box, the student must first read and rewrite it with handwriting, and then respond by coloring following the instructions. That is, she must do as if she were a printer that colors according to the instructions. In page 1B, for each of the 4 boxes, she must make up her own instructions and write whom she is directing the activity to.

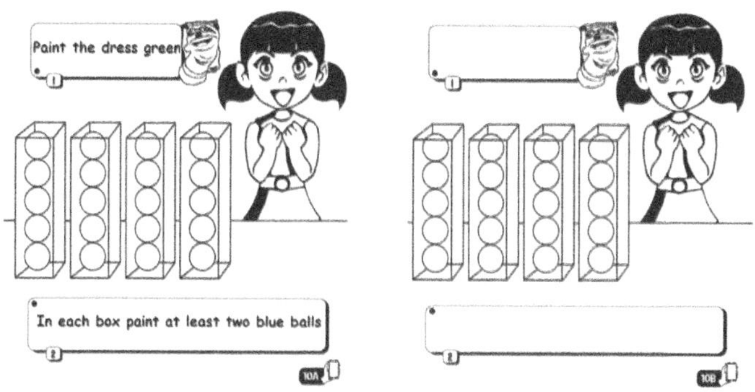

Fig. 3. Left: ColorealIdeas page 10A where the student has to read the instructions and then color accordingly. Right: Page 10B where the student has to write coloring instructions. Both pages show a hand puppet above. The hand puppet accompanies the student and reminds her to discuss the solutions with him, and thus fostering the development of metacognition.

Then comes activities of pages 10A and 10B. They require deeper AT. In page 10A the student has to consider a list if boxes each one with a list of balls. This means, the student has to work with an array. List and arrays are core data structures, widely used in computer science. In page 10B, the student has to write her own instructions as shown in Fig. 3. A trained rater assessed the correctness of instructions. Additionally, the rater assessed whether the written instruction was novel, and not just a mere copy of the instructions in page 10A.

5 Results

In phase one, 144 schools participated on the National Reading Day. We received on real time the assessments of 9,808 coloring tasks from these schools. 87.34% of students participated. This is a high participation rate, since on those weeks student leave for summer vacations. On some schools, parents brought back students just to participate on the activity. 87.33% of the 9,808 answers were correct. This first phase proved that the strategy was attractive to students, and that first grade students at the end of their schoolyear were able to read, interpret, and follow coloring instructions.

Phase 2 started on the first week of March. However, due to the COVID19 pandemic, on the second week all schools closed. For this reason, we implemented phase 2 in very few schools and only 1 school completed with full student attendance. Here we report the results at that school with a total of 25 students. In the initial survey of phase 2, 54% of students preferred coloring to writing, 21% preferred to write, and 25% were indifferent. 74% find funny to write instructions to a classmate. On the other hand, 79% of the students talk to himself, 43% talk everyday with their teddy bear, 48% does it sometimes. Only 9% never does it. Moreover, when the teddy bear is not present, 81% of the students imagine the teddy bear is present and talk with it. Activity 1A required writing the student's name and had 4 instructions. All the students put their name on the sheet, almost nobody rewrote the instruction, and all of them executed the instructions. Activity 1B required the student to put his name and the name of who would have to read and execute the instructions. Then, the student had to design and write instructions in the empty boxes as indicated in Fig. 4. All students wrote new instruction, different to the first instruction in 1A. Activity 10A had 1 simple instruction and one complex instruction with quantifiers. Activity 10B had 2 writing activities, one that could be simple and one complex with quantifiers. The results are in Table 1.

Table 1. Summary of students' performance on the activities of phase 2

Activity	Instruction	% of correct responses
1A	Executing 4 simple instructions	95%
1B	Writing 4 simple instructions	77%
10A	Executing 1simple instruction	96%
10A	Executing 1 complex instruction	83%
10B	Writing 1 simple instruction	72%
10B	Writing 1 complex instruction	64%

Fig. 4. Left: Colorealdeas page 10A shows the coloring response of a student. Right: Page 10B shows the instructions written by the student.

6 Conclusion

One of the pillars of CT is Algorithmic Thinking (AT). This means to follow and write precise instructions, understand the conceptual bases on how algorithms operate and know their different building blocks. In this paper, we describe a coloring book addressed to first graders that teach them to include logical conditions on the instructions. Its focus is to teach a deep understanding of the algorithmic nature of computing processes. In a first phase with 144 first grade classes 87.33% of the 9,808 answers were correct. In the second phase, in one school, 83% of the students were able to color according to a complex instruction using existential and universal quantifier. Moreover, 64% of the students were able to write instruction with these two quantifiers. Therefore, we conclude that coloring books enhanced with smartphones and gamification have a big potential to teach AT to first graders and track their learning.

Acknowledgments. Support from ANID/ PIA/ Basal Funds for Centers of Excellence FB0003 is gratefully acknowledged.

References

1. Yadav, A., Mayfield, C., Zhou, N., Hambrusch, S., Korb, J.T.: Computational thinking in elementary and secondary teacher education. ACM Trans. Comput. Educ. **14**, 1, Article 5 (2014)
2. Lu, J., Fletcher, G.: Thinking about computational thinking. In: Proceedings of the 40th ACM Technical Symposium on Computer Science Education (2009)
3. Bocconi, S., Chioccariello, A., Dettori, G., Ferrari, A., Engelhardt, K.: Developing computational thinking in compulsory education – implications for policy and practice; EUR 28295 EN (2016)

4. Lockwood, J., Mooney, A.: Computational Thinking in Education: Where does it fit? A systematic literary review. Department of Computer Science. Maynooth University (2017)
5. Araya, R.; Dartnell, P.: Video study of mathematics teaching in Chile. In: Proceedings 11th International Conference on Mathematics Education Conference. ICME 11, Monterrey, Mexico (2008)
6. Japan Science and Technology Policy Council for Science, Technology, and Innovation: Realizing Society 5.0 (2018). https://www.japan.go.jp/abenomics/_userdata/abenomics/pdf/society_5.0.pdf
7. Araya, R., Isoda, M., González, O., Inprasitha, M.: A framework for computational thinking for the transition to a super smart society. J. Southeast Asian Educ. **1**, 1–15 (2019)
8. Wing, J.M.: Computational Thinking Benefits Society (2014). http://socialissues.cs.toronto.edu/index.html%3Fp=279.html
9. Tedre, M., Denning, P.: The Long Quest for Computational Thinking (2016). http://denninginstitute.com/pjd/PUBS/long-quest-ct.pdf
10. Villani, C.: Donner un sens a l'Intelligence Artificielle (2018)
11. Araya, R., Gigon, P.: Segmentation trees: a new help for building expert systems and neural networks. In: Computational Statistics, COMPSTAT 1992, vol. 1, pp. 119–124. Physica-Verlag HD (1992)
12. Computer Science Unplugged. https://csunplugged.org/en/
13. Wikipedia. https://en.wikipedia.org/wiki/Coloring_book
14. NYPost. https://nypost.com/2015/12/13/hottest-trend-in-publishing-is-adult-coloring-books
15. Publishersweekly. https://www.publishersweekly.com/pw/by-topic/industry-news/publisher-ews/article/83172-coloring-and-crosswords-books-for-adults-get-boost.html
16. Ashdown, B., Bodenlos, J., Arroyo, K., Patterson, M,; Parkins, E.: how does coloring influence mood, stress, and mindfulness? J.Integr. Soc. Sci. **8**(1), 1–21 (2018). www.JISS.org
17. Mantzios, M., Giannou, K.: When did coloring books become mindful? Exploring the effectiveness of a novel method of mindfulness-guided instructions for coloring books to increase mindfulness and decrease anxiety. Front. Psychol. **9** (2018)
18. Kaimal, G., et al.: Functional near-infrared spectroscopy assessment of reward perception based on visual self-expression: coloring, doodling, and free drawing. Arts Psychother. **55**, 85–92 (2017)
19. Anatomy Coloring Pages for Kids. https://coloringhome.com/anatomy-coloring-pages-for-kids
20. Inharjanto, A., Lisnani: Developing Coloring Books to Enhance Reading Comprehension Competence and Creativity Advances in Social Science, Education and Humanities Research, vol. 394 (2019)
21. https://ripplekindness.org/benefits-of-mindfulness-coloring-for-children-and-adults/
22. Kaufmann, R.: A FORTRAN Coloring Book. MIT Press, Boston (1978)
23. Sandor, G.: A Fortran coloring book. Comput. Struct. **10**(6), 931–932 (1979)
24. Tahiroglu, D., Taylor, M.: Anthropomorphism, social understanding, and imaginary companions. Br. J. Dev. Psychol. **37**, 284–299 (2019)
25. Roby, A., Evan Kidd, E.: The referential communication skills of children with imaginary companions. Dev. Sci. **11**(4), 531–540 (2008)
26. Bouldin, P., Bavin, E.L., Pratt, C.: An investigation of the verbal abilities of children with imaginary companions. First Lang. **22**, 249–264 (2002)
27. Hood, B., Bloom, P.: Children prefer certain individuals over perfect duplicates. Cognition **106**, 455–462 (2008)
28. Tomasello, M.: Becoming Human. A Theory of Ontogeny. Harvard University Press, Boston (2019)
29. World Bank Report. The Changing Nature of Work. http://documents.worldbank.org/curated/en/816281518818814423/pdf/2019-WDR-Report.pdf. Accessed 04 Feb 2020

30. Araya, R.: Supporting teachers and parents with mobile technology to teach reading. IADIS Int. J. WWW/Internet **18**(1), 62–75 (2020). ISSN 1645-7641
31. Araya, R., Diaz, K.: Implementing government elementary math exercises online: positive effects found in RCT under social turmoil in Chile. Educ. Sci. **10**, 244 (2020)
32. Araya, R., Arias Ortiz, E., Bottan, N., Cristia, J.: Does Gamification in Education Work? Experimental Evidence from Chile. IDB WORKING PAPER SERIES N° IDB-WP-982 Inter-American Development Bank (2019)
33. Araya, R.: A territorial learning ecosystem for parents' participation and cooperation. In: Mealha, Ó., Rehm, M., Rebedea, T. (eds.) Ludic, Co-design and Tools Supporting Smart Learning Ecosystems and Smart Education. SIST, vol. 197, pp. 77–89. Springer, Singapore (2021). https://doi.org/10.1007/978-981-15-7383-5_7
34. Araya, R.: ColoreaIdeas. AutoMind, Santiago (2019). https://www.conectastem.cl/colorea-ideas/

Changing Children's Behavior Based on Persuasive Game: Design for Children's Safety Education

Yanan Yang[1], Duoduo Zhang[1(✉)], Xiaoqian Mou[2], and Yuanyuan Yang[1]

[1] Hunan University, Hunan, China
zhang_duoduo@hnu.edu.com
[2] Kingston University, London, England

Abstract. This study uses persuasive game design to improve the safety education of left-behind children in rural China. Persuasive game is a design method that is used to change or shape the user's behavior or attitude in recent years. It has been practiced and applied in many fields such as medical health [1–4], social interaction [5], advertising [6]. By exploring how the Persuasive Technology and Behavior Change Model affect user behavior change, this paper optimizes the design model of persuasive games in the existing research and clarifies the four stages of the persuasive game on user behavior change. Finally, we verify the feasibility of persuasive games in the field of children's safety education through a design practice case named "Class is Over".

Keywords: Persuasive game design · Behavior change · Rural left-behind children · Game mechanics · Experience design

1 Introduction

According to statistics, nearly 50000 children die of accidental injuries every year in China, of which nearly 80% are rural children. After a year-long field investigation in Baishuidong village, Longhui County, Hunan Province, the research team found that the cause of the tragedy was the lack of safety education for left-behind children in rural areas. Due to the existence of unavoidable objective factors such as poor living environment, insufficient teaching resources, and lack of parental guidance and protection, children often do not listen to their elders' advice due to their subjective curiosity and desire to explore, and they often wander in dangerous places such as waterfront and mountain path.

Persuasive game is a branch of serious game, which has been used to guide users to change their behavior or attitude in recent years. Many studies have shown that it is easier to promote children's learning if games are combined with education [7, 8]. As the most important recreational activity of children, play is an important part of their daily life and an excellent carrier of information transmission. By exploring the relationship between persuasive games and behavior change, this paper clarifies the

S. Nazir et al. (Eds.): AHFE 2021, LNNS 269, pp. 142–148, 2021.
https://doi.org/10.1007/978-3-030-80000-0_17

process of persuasive games' effect on behavior change and discusses the influence of four elements of game design on players' behavior change. This conclusion can provide a reference for designers to apply persuasive games to other fields. Finally, we verify the conclusion through a specific project of safety education for left-behind children in rural areas.

2 Current Situation of Safety Education for Left-Behind Children in Rural Areas

According to the data of the Ministry of Civil Affairs of the People's Republic of China, in 2018, the number of left-behind children in rural China exceeded 6.97 million, of which 96% were under the custody of grandparents [9]. After comparing and analyzing the differences between urban children and rural left-behind children in the aspects of traffic mode, school time, road risk, natural disasters, and so on, we find that, compared with the urban children, the safety education of rural left-behind children should be paid more attention to.

1. In rural areas, the terrain is complex and the road conditions are difficult, and there are often wild animals;
2. Children's commuting time is longer on the way to school, and their commuting way is mostly walking;
3. Most of the left-behind children in rural areas are supervised by their grandparents. They lack the supervision and guidance of their parents, their safety awareness is weak and their safety education is insufficient.

Data show that rural residents generally believe that the best way of safety education for children is school teaching (92.1%) [10]. However, there is a general shortage of teachers in rural areas. Sometimes a teacher needs to teach several courses at the same time, so they pay less attention to children's safety education. Teachers often only emphasize superficial safety knowledge such as "don't play by the water" and "don't walk in the middle of the road", which can't let children form a complete safety knowledge system. At the same time, this "preaching" teaching method is easy to cause children's psychological resistance, and passive learning is difficult to really arouse their active safety awareness. To improve the safety education of left-behind children in rural areas can not only rely on rigid preaching, but should start from two aspects of "consciousness" and "behavior", arouse children's safety awareness, let them change their behavior spontaneously and actively, and avoid the occurrence of danger.

3 Gamification and Behavior Change

As an interesting knowledge carrier, games can effectively transfer information and let users learn in entertainment. For children, play is one of the main ways of social interaction. Children interact with others and get information from game situations.

In children's education products, gamification has three necessities: first, in line with children's nature of love to play. Second, game teaching can reduce children's psychological conflict with education. Thirdly, games can create a relaxed entertainment situation for children, which is conducive to the absorption and transformation of knowledge.

Dr. BJ Fogg of Stanford University coined the term "persuasive technology" and proposed a behavior change model (Fig. 1). Fogg believes that behavior is the product of three factors: motivation, ability, and trigger, and only when these three factors occur at the same time can it lead to the occurrence of behavior [14]. From this critical line, motivation and ability are mutually compensatory. Users do not need to have strong motivation and a high ability to induce behavior change. For example, when users do not have strong purchasing motivation, the low price may also promote the occurrence of user purchase behavior [15]. For children, in the case of their own weak ability, we can focus on stimulating their motivation to guide children to cross the critical line of behavior change and promote them to change their own behavior, to achieve the purpose of safety education.

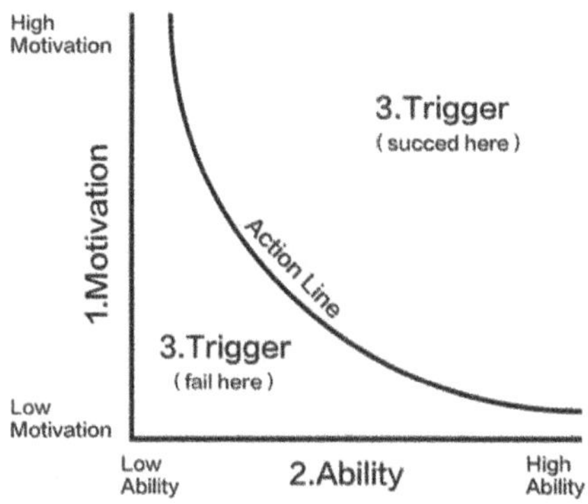

Fig. 1. Fogg behavior model

Persuasive games are often used to persuade users to change their habits, attitudes, or behaviors. Valentijn [16] and others pointed out that designers can form, change or strengthen users' consciousness, behavior, or attitude in the real world when they experience the game world by turning some elements of the real world into game elements.

Persuasive game can create an immersive game environment for users. The game with persuasive purpose is not just a virtual entertainment situation, but a carrier of real life. Users can always feel the existence of real elements in the process of playing the game. By blurring the boundary between the game world and the real world, children can reduce their vigilance and resistance to "education", and absorb the knowledge of the information transmitted in the process of the game imperceptibly, so that they can be more easily convinced.

4 Mechanism of the Persuasive Game

The game mechanism is the core element of game design, which determines the operation mode of the whole game system. McGonigal proposed four characteristics of the game mechanism: goal, rule, feedback, and voluntary participation [12]. Goals are the specific results that players strive to achieve. Rules set limits on how players can achieve their goals. Feedback tells players how far away they are from achieving their goals. Voluntary participation requires all players to understand and be willing to accept target rules and feedback [13]. The result of the interaction among goal, rule, and feedback is to arouse the players' consciousness of voluntary participation and let them use the game continuously.

Based on Fogg's behavior change model, persuasive design, and game mechanism, this paper puts forward the design model of persuasive game action mechanism, as shown in Fig. 2. In the game design based on persuasion technology, designers integrate and reorganize the information in the real world through goal, rule and feedback, to create a game world based on reality for users. The game world and the real world are interrelated and influence each other. The process of behavior change includes four stages: precontemplation stage, contemplation stage, action stage and change stage. Behavior change takes place at the boundary between the game world and the real world and is completed in the real world. The process of behavior change is equivalent to the process of transforming the game world into the real world. We will explain this model in detail in the following design case.

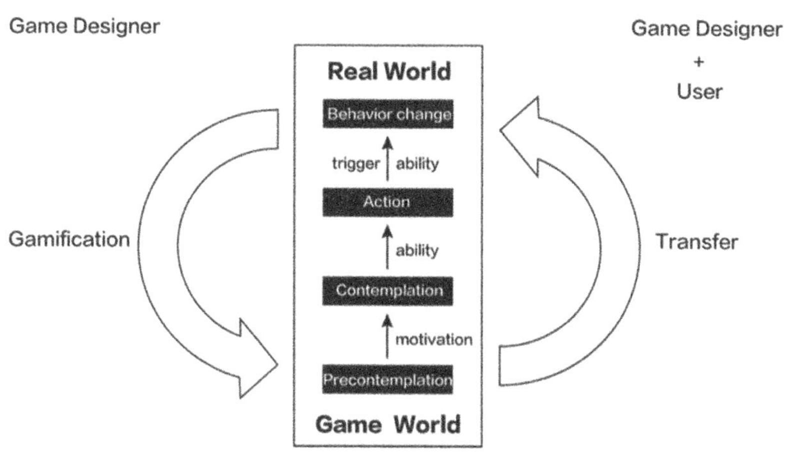

Fig. 2. Mechanism of the persuasive game

5 Design and Practice of Persuasive Games for Children's Safety Education

Based on the mechanism design model of the persuasive game proposed above, the research team designed a "Class is Over" (Fig. 3) safety education persuasive game after a year's field survey in Baishuidong village, Longhui County, Hunan Province.

Fig. 3. "Class is over" persuasive game

"Class is Over" is a three-dimensional card game in which many people participate. Its modeling inspiration comes from the terraces in the mountain area. The three-dimensional board game is closer to reality, blurs the boundary between the game world and the real world, and creates an immersive situational experience for users. At the same time, modular design can enhance children's sense of experience and arouse their interest. We integrate the potential danger and safety education knowledge into the game elements, so that children can arouse the awareness of behavior change and improve the ability of behavior change in the process of participating in the game.

The game mechanism of "Class is Over" is as follows:

1) Objective: on the map, the chess pieces advance to the "home" position;
2) Rules: through the turntable to determine the number of forwarding and backward steps, when the chess pieces reach the designated position, they need to draw cards to answer safety education questions;
3) Feedback: correct or wrong answers will be rewarded (forward) or punished (backward).
4) Voluntary participation: the competitive, challenging, and successful sense of multiplayer games arouse players' willingness to participate voluntarily.

Before children use "Class is Over", they don't realize that some behaviors they often do in their daily life may cause dangerous consequences. At this time, they are in the "precontemplation stage" of behavior change. The goal of winning in the game will initially stimulate children's interest in playing the game. In the process of game experience, children will realize the overlap of game scene and real life, recognize the potential dangerous consequences of behavior and arouse the motivation of behavior change. At this time, children enter the "contemplation stage" of behavior change. The "reward and punishment" feedback brought by the game will stimulate children's desire to win, and enhance the motivation of behavior change. The children enter the "action stage". Through the card Q & A, the knowledge of safety education is transferred to the children to guide them to improve their ability of behavior change. Finally, the competitive characteristics of multiplayer games, the sense of interaction and the sense of achievement may be the trigger factors of children's behavior change, and children complete behavior change under the influence of triggers.

After the completion of the design, all the fifth-grade students of Baishuidong primary school used this game as a practical test of the design. The research team made statistics on what safety problems each child encountered in the game. At the end of the game, the research team designed the corresponding safety knowledge test questions and conducted a separate test for each participant. After data statistics, it was found that 78% of the children can use the safety knowledge obtained in the game to solve their dangerous situations in the game. The research team left the game products in Baishuidong primary school for children to use. A month later, we made a follow-up visit and conducted a safety knowledge test and a semi-structured interview with all fifth graders. After statistics, it was found that 85% of the children had initially mastered the basic safety knowledge and were aware of the serious consequences of bad behavior, which reduced the risk in real life. From the teacher's semi-structured interview, we know that children show a strong interest in "Class is Over", which has become one of the most popular extracurricular activities for children in their spare time.

6 Conclusion

This study explores how persuasion technology and game mechanism influence user behavior change, clarifies the four stages of persuasion game on behavior change, puts forward the design model of persuasion game mechanism, and preliminarily verifies the feasibility of this model through a design case. The persuasive game is a design method to guide users to change a certain behavior or attitude. It integrates the information that designers want to convey to children into the game that children are most interested in, which is more conducive to be absorbed and accepted by children, to achieve the purpose of persuasion.

This study verified that persuasive games can have a positive impact on children's safety education. In the future, the feasibility and applicability of persuasive games can still be explored in more fields.

References

1. Kato, P.M.: Video games in health care: closing the gap. Rev. Gen. Psychol. **14**(2), 113–121 (2010)
2. Miloff, A., Lindner, P., Hamilton, W., Reuterskiöld, L., Andersson, G., Carlbring, P.: Single-session gamified virtual reality exposure therapy for spider phobia vs. traditional exposure therapy: study protocol for a randomized controlled non-inferiority trial. Trials **17**(1), 1–8 (2016)
3. Reid, D.T.: Benefits of a virtual play rehabilitation environment for children with cerebral palsy on perceptions of self-efficacy: a pilot study. Pediatr. Rehab. **5**(3), 141–148 (2002)
4. Hardy, K., Bonner, M., Willard, V.: Computerized cognitive training for survivors of pediatric cancer. Pediatr. Blood Cancer **55**(5), 776–777 (2010)
5. Ruggiero, D.: The effect of a persuasive game on attitude towards the homeless (Unpublished thesis). Purdue University, West Lafayette, USA (2013)
6. Deal, D.: The ability of branded online games to build brand equity: an exploratory study. In: DiGRA 2005 Conference: Changing Views-Worlds in Play. DiGRA, Vancouver (2005)
7. Garris, R., Ahlers, R., Driskell, J.E.: Games, motivation, and learning: a research and practice model. Simul. Gaming **33**(4), 441–467 (2002)
8. Peters, V., Vissert, G.: A simple classification model for debriefing simulation games. Simul. Gaming **35**(1), 70–84 (2004)
9. Yu-ting, G.: The study of left behind children in rural areas in henan province from the perspective of rural revitalization. Rural Econ. Sci. Technol. **031**(001), 248–250 (2020)
10. Chen, Y.: Investigation on the importance of children's safety education by residents of Tounian Township, Panzhihua City, Sichuan Province. Economist, vol. 04 (2020)
11. Huizinga, J.: Nature and significance of paly as a cultural phenomenon. In: Salen, K., Zimmerman, E. (eds.) The game design reader: A rules of play anthology, pp. 96–120. MIT Press, Cambridge (1955)
12. McGonigal, J.: Reality is Broken: Why Games Make us Better and How they can Change the World. Penguin Press, New York, NY (2011)
13. Xu, J.: Identity, power and capital: emotion research in online games of teenagers. Zhejiang University, Zhejiang (2015)
14. Fogg, B.J.: A Behavior Model for Persuasive Design, vol. 1 (2009)
15. Zhang, J., Qiu, T.: Research of user behavior persuasion design. Design **9**, 188–189 (2014)
16. Visch, V., Vegt, N., Anderiesen, H., et al.: Persuasive Game Design: A model and Its Definitions CHI (2013)

Augmented Reality Teaching Resources and Its Implementation in the Teaching-Learning Process

Omar Cóndor-Herrera[1], Pamela Acosta-Rodas[2], and Carlos Ramos-Galarza[1,2](✉)

[1] Centro de Investigación en Mecatrónica y Sistemas Interactivos MIST/Carrera de Psicología/Maestría en Educación Mención Innovación y Liderazgo Educativo, Universidad Tecnológica Indoamérica, Av. Machala y Sabanilla, Quito, Ecuador
omarcondor@uti.edu.ec

[2] Facultad de Psicología, Pontificia Universersdad Católica del Ecuador, Av. 12 de Octubre y Roca, Quito, Ecuador
caramos@puce.edu.ec

Abstract. This work presents a revision of possible applications of the augmented reality in the scholar context, such as the utilization of augmented reality to improve reading comprehension or its application in geometry learning and many others. At the same time, as a contribution to the research field, there is a proposal about the design and development of didactical resources that count with augmented reality objects, images, or illustrative graphs, as well as multimedia videos to improve the teaching-learning process of various academic topics.

Keywords: Augmented reality · Education · Virtual learning environment · Educative applications · Mobile learning

1 Introduction

Nowadays, technological development increases rapidly in every context, this contemporaneous phenomenon, has allowed the innovation of educative practices and the implementation and adaptation of new technologies to the learning process. In the present, the emergence of Augmented Reality applications, from now on (AR), has taken the lead into the educational context, since the AR has the singularity of combining the virtual and physical world in the sense that, virtual content experiences by the user of AR technology will depend on his/her geographical location. The appearance of games such as Pokémon GO and technologies as HoloLens has caught an incrementing number of people that uses the AR [1].

The usage of AR technology could facilitate the comprehension of scientific concepts because this complements the sensorial perception that the user has from the real world through the addition of content generated by computer to the environment, this offers a new way of interactivity between virtual and real worlds [2, 3], and, for educative context it represents an innovative way, making possible for the professor to present to his/her students the objects for being studied.

S. Nazir et al. (Eds.): AHFE 2021, LNNS 269, pp. 149–154, 2021.
https://doi.org/10.1007/978-3-030-80000-0_18

Following, there are presented some applications of the augmented reality (AR) in the educational context and, a proposal of didactical resources designs that combine RA objects with multimedia resources to improve the experience of the students when learning.

2 Augmented Reality

At first instance, there must be differentiated the virtual reality (VR) from the augmented reality, because this last one combines digital information with the real environment [4], about the virtual reality where the individual accedes the information through an immersive and simulated environment.

Augmented Reality (AR) could be defined as the interaction of audio, graphs, text plus other virtual elements overlapped in a reality where objects are visualized in virtual time, some applications of AR are supported in M-Learning using mobile devices such as smartphones and tablets that allow users to interact with digital information.

It is possible to distinguish two ways of using AR technology, these are related to the location and vision. The first one, once identified the location of the user the digital information will be shown to students while they are moving throughout the physical area with a smartphone that has integrated GPS, as it is shown in Fig. 1.

Fig. 1. AR portal of Van Gogh's painting

The second way to use the AR technology is based on the vision, presenting digital means to students after pointing the camera to the object that serves as a shooter [6].

2.1 Augmented Reality Applied to Education

The study of augmented reality applied in the educative field shows that these technologies present many benefits for students, improving their attitude in front of learning processes and, at the same time, influence positively in motivation, interaction and collaboration, these abilities are considered as crucial benefits when using tools that allow overlapping virtual information to real objects [7, 8].

AR has allowed converting books, as in the case of fairytales in books with augmented reality, which have supported in the process of reading, improving comprehension [9], as well as it has been used augmented reality to work on student´s cursory reading [10].

Another application of AR could be in math and geometry areas, in this sense, students may use augmented reality to practice the basic principles of geometry [2], which gives students the possibility to have an interactive experience with concrete objects to assimilate the topics proposed to study and learn.

On the other hand, it is inherent to explore how the augmented reality learning activities could be part of teaching methodologies combined, such as the flipped classroom [11], or through games, since are potent generators of positive emotions in children increasing their intrinsic satisfaction [12].

This explains why the application of gamification would be one of the pedagogical strategies to optimize the application of AR, improving the compromise, motivation, and engagement of the participants in any activity that generally is accomplished without playing [13], adding by this, to the learning activity some emotions like joy and challenge.

3 Implementation of AR in the Educative Process

3.1 Implementation of AR in the Educative Process

A study about the usage of AR with students, suggests that the utilization of AR in educative applications for children of seven years and older improve their learning experiences [14]. It explains the reason that in the present article it is proposed to develop teaching cards with AR using the Unite.AR platform, these teaching cards will combine AR objects with multimedia resources such as videos and illustrative images, as it is shown in Fig. 2, that will enable students to interact with these resources in an interactive learning environment. Once the teaching cards are elaborated, we aim to measure,

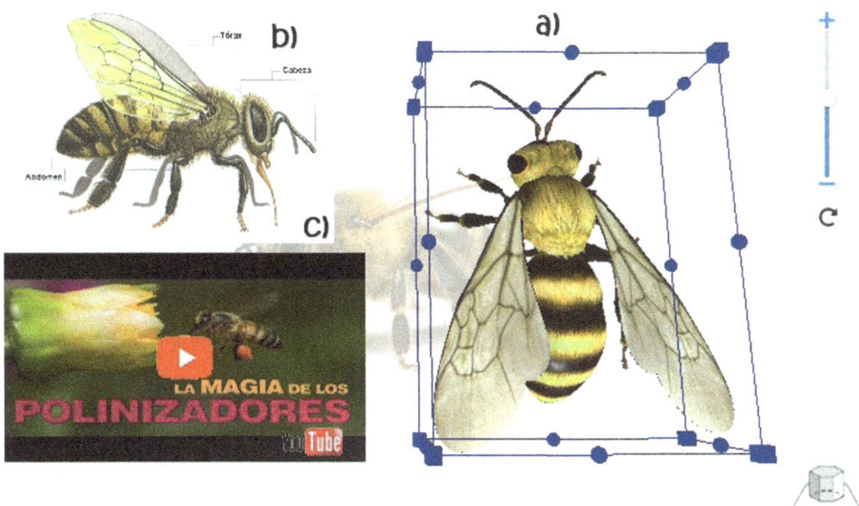

Fig. 2. Proposal of a teaching card with AR. Elements: a) AR Object, b) Illustrative image and c) Multimedia video.

in primary school students, the impact of the implementation of these resources in the teaching-learning process.

4 AR Applied in the Educative Process - Previous Studies

To enhance the present proposal, as following, there is presented a research compilation about recent studies, and at the same time, it shows a short description of their evidenced findings. This information deep about the diversity of options that AR technology offers and which could be applied to educative processes, favoring the improvement of educative abilities (Table 1).

Table 1. Previous studies

Title	Authors	Investigation	Findings
Comparing reading comprehension between children reading augmented reality and print storybooks	Delneshin,Hamid,Yazdan and Hassan [9]	To measure reading comprehension in children that use an AR fairytale book, in comparison to those children that read the same fairytale in the traditional pressed version	Children who experienced the AR fairytale showed better comprehension when they were asked questions about the book and to repeat the story
Impact of augmented reality technology on academic achievement and motivation of students from public and private Mexican schools. A case study in a middle-school geometry course	Ibañez, Portillo, Cabada and Barrón [11]	Design of an AR application for students to practice basic geometry principles, to be contrasted with an application implemented in a web learning environment	Students who used virtual learning environments based in augmented reality obtained higher punctuation in the test applied after the experience, than those who used the application based on the web
MantarayAR: Leveraging augmented reality to teach probability and sampling	Conley, Atkinson, Nguyen and Nelson [15]	The aim of this study is to explore if the experiences of AR could support university students learning	Assigned participants to an AR experience informed a statistically significant greater perception of compromise
EmoFindAR: Evaluation of a mobile multiplayer augmented reality game for primary school children	López and Jaen [12]	The usage of mobile augmented reality without markers to improve socialization, communication abilities, and emotional intelligence in primary school children	The AR version of collaborative games showed a greater impact on emotional affect, social interaction, and participants' interest

(*continued*)

Table 1. (*continued*)

Title	Authors	Investigation	Findings
Augmented songbook: an augmented reality educational application for raising music awareness	Rusiñol, Chazalon, Diaz [7]	Development of a mobile AR application that has the aim to sensitize little children to abstract musical concepts	The application allows the overlapping of augmented contents over the pages of a songbook which allowed obtaining participants' positive results

5 Conclusions

Thanks to the bibliographic revision it is possible to evidence that the application of the AR to work on different academic abilities in the teaching-learning process results in benefit for the students.

The contribution to the educative field from the technological scope is proposed at first designing augmented reality teaching cards complemented by multimedia resources to be implemented in the teaching-learning process.

Also, as the future investigation, it is of great interest for the research team to implement the AR teaching cards as educative resources in the teaching-learning process, and according to prove the effects of these types of technological resources it is proposed to conduct experimental studies.

References

1. Neely, E.L.: Augmented reality, augmented ethics: who has the right to augment a particular physical space? Ethics Inf. Technol. **21**(1), 11–18 (2018). https://doi.org/10.1007/s10676-018-9484-2
2. Ibañez, M., Portillo, A., Cabada, R., Barrón, M.: Impact of augmented reality technology on academic achievement and motivation of students from public and private Mexican schools. A case study in a middle-school geometry course. Comput. Educ. **145**, 1–20 (2020)
3. Kaur, N., Pathan, R., Khwaja, U., Sarkar, P., Rathod, B., Murthy, S.: GeoSolvAR: augmented reality based application for mental rotation. In: 2018 IEEE Tenth International Conference on Technology for Education, vol. T4E, pp. 45–52 (2018)
4. Liberati, N.: Augmented reality and ubiquitous computing: the hidden potentialities of augmented reality. AI Soc. **31**(1), 17–28 (2014). https://doi.org/10.1007/s00146-014-0543-x
5. Dundeleavy, M., Dede, C.: Augmented reality teaching and learning. In: Handbook of Research on Educational Communications and Technology, pp. 735–745 (2014)
6. Green, J., Green, T., Abbie, B.: Augmented reality in the K-12 classroom. TechTrends **61**, 603–605 (2017)
7. Rusiñol, M., Chazalon, J., Diaz, K.: Augmented songbook: an augmented reality educational application for raising music awareness. Multimedia Tools Appl. **77**, 13773–13798 (2018)
8. Bacca, J., Baldiris, S., Fabregat, R., Sabine, G., Kinshuk, R.: Augmented reality trends in education: a systematic review of research and applications. J. Educ. Technol. Soc. **4**(17), 133–149 (2014)

9. Delneshin, D., Jamali, H., Mansourian, Y., Rastegarpour, H.: Comparing reading comprehension between children reading augmented reality and print storybooks. Comput. Educ. **153**, 1–24 (2020)
10. Pei-Luen, P., Jian, Z., Zhi, G., Jiaqi, L.: Speed reading on virtual reality and augmented reality. Speed Read. Virtual Reality Augmented Reality **125**, 1240–1245 (2018)
11. Ibañez, M., Delgado, C.: Augmented reality for STEM learning: a systematic review. Comput. Educ. **123**, 109–123 (2018)
12. López, L., Jaen, J.: EmoFindAR: evaluation of a mobile multiplayer augmented reality game for primary school children. Comput. Educ. **149**, 1–42 (2020)
13. Nah, F.-H., Zeng, Q., Telaprolu, V.R., Ayyappa, A.P., Eschenbrenner, B.: Gamification of education: a review of literature. In: Nah, F.-H. (ed.) HCIB 2014. LNCS, vol. 8527, pp. 401–409. Springer, Cham (2014). https://doi.org/10.1007/978-3-319-07293-7_39
14. Savita, Y., Pinaki, C., Kochar, G., Ansari, D.: Interaction of children with an augmented reality smartphone app. Int. J. Inf. Technol. **12**, 1–6 (2020)
15. Conley, Q., Atkinson, R., Nguyen, F., Nelson, Y.B.: MantarayAR: leveraging augmented reality to teach probability and sampling. Comput. Educ. **153**, 1–22 (2020)

Student Training and Environment Interaction

Design Approach of Digital Numeration Training for Students in a Primary School

Adrian Roussel-Fayard[1], Nadine Vigouroux[1], Frédéric Vella[1](✉), Jean-François Camps[2], and Charlotte Tabarant[3]

[1] IRIT, CNRS, UPS, 118 Route de Narbonne, 31062 Toulouse Cedex 9, France
{Nadine.Vigouroux,Frederic.Vella}@irit.fr
[2] URI Octogone-Lordat, UT2J, 5, allée Antonio Machado, 31058 Toulouse Cedex 9, France
jcamps@univ-tlse2.fr
[3] Academic's Direction, French National's Education, Lourdes, France
charlotte.artus@ac-toulouse.fr

Abstract. This article report the principles and the design of HMK-Learning, a digital tool for students at primary schools in France. The design is based on Mounier's numeracy work in didactics. Then, we describe the questionnaire design method to study the usability of HMK-Learning. Next, we present the questionnaire results answered by four teachers of primary schools. Finally, we discuss them from the point of view of the design of HMK-Learning and the pedagogical questions they raised. This paper also highlights the highly important role of didactics in the process of designing a digital tool for numeracy learning for primary school children that we have identified in the answers of the usability questionnaires.

Keywords: Centered design · Usability · Digital numeration · Primary school

1 Introduction

According to the results of the Trends in International Mathematics and Science Study survey, carried out in May 2019 on writing in figures a sample of 4,186 (9 years old) children (primary school), France ranks last in the European Union countries for the level of knowledge in mathematics. In France, the Villani-Torossian [1] report "21 measures for the teaching of mathematics" stated learning difficulties from primary school onwards (6 to 10 years old). These relate to acquiring the sense of numbers, counting, memorizing the names of numbers, writing in figures and the transcoding between oral numeration (one, two, etc.) and written numeration (1, 2, etc.). One of the recommendations in the report concerns the meaning of the four operations (addition, subtraction, multiplication and division) and the teaching of the quantities and measures in primary school that should support number and operation sense. According to E. Mounie, (2016) [2] "*The problem then arises, with 6 years old pupils, of showing that numbers refer to units of numeration, tens and simple units for the acquisition of skills to be built by students in*

S. Nazir et al. (Eds.): AHFE 2021, LNNS 269, pp. 157–164, 2021.
https://doi.org/10.1007/978-3-030-80000-0_19

primary school." To overcome these difficulties, we wish to approach the learning of numeracy through a digital application.

Several recent studies have examined the effects of digital applications on students' skills. Drijvers [4] wrote that "the success of digital technology in mathematics education include the design of the digital tool and corresponding tasks exploiting the tool's pedagogical potential, the role of the teacher and the educational context". Geiger et al. [6] presented a model of numeracy based on the use of digital technologies. They also demonstrated the good influence of digital tools on student's development of skills and mathematical knowledge. Hodaňová et al. [5] also reported that the efficient use of digital technologies is undoubted a prerequisite for both quality education. These same authors suggested digital technologies in mathematics teaching at primary schools belong to innovative methods and activity forms.

Our hypothesis is that the design of interactive digital tools integrating Mounier's [2] numeracy work could be more efficient for students at primary schools in France. The aim of this paper is to describe the methodology used to design the application HMK-Learning for pupils at primary school. Then, we will describe the survey design method to study the usability of HMK-Learning. Finally, we will present the results of the usability study and discuss them from the point of view of the design of HMK-Learning and the pedagogical questions they raised.

2 Related Work

Evaluating the usability of interactive systems is no longer a new scientific issue but remains a scientific challenge [13] for educational software for children in primary school. According to Nielsen [11] usability is about learnability, efficiency, memorability, errors, and satisfaction. However, the definition of usability from ISO 9241–11 [14] is "*the extent to which a product can be used by specified users to achieve specified goals with effectiveness, efficiency and satisfaction in a specified context of use*". Several methods are available to measure the usability. One of them is Nielsen's heuristics [11] which allows to identify usability problems in the process of designing an interactive system. Another is Heuristic Evaluation for Child E-learning (HECE) [9] based on Nielsen's heuristics adapted to children and e-learning applications. In parallel to these heuristic-centered approaches, other classically methods are used during user tests: interviews, observations by experimenters in use situations, surveys, activity logs. The SUS (System Usability Scale) [7] is widely used because it easily provides an interpretable score [12]. However, children cannot use this survey in the reading learning phase. Child-friendly tools for assessing the learning environment are relatively rare. Khanum and Trivedi [10] analyzed different assessment techniques (**Think Aloud** (His reasoning, thread of thought.), **Constructive Interaction** (interaction between two testers who interact), etc.) to determine which are the most suitable for children according to their age. These authors also suggest crossing several methods in order to have more results that are relevant and recommend a series of recommendations for the conduct of the experiments. Evaluation involving end-users during the co-design phases are highly recommended to identify usability problems precisely when using the application when it is possible.

Within the framework of the design of the HMK-Learning software, we have chosen to involve the teachers of the primary schools in the iteration of HMK-Learning design and the usability of designs.

3 Design of HandyMathKey-Learning

In collaboration with a group of teachers eventually called Villani Torossian, we implemented a participatory design method for the HMK-L application. The constraints of use were that primary school children did not have a sufficient reading level, and that the interface should be simple to use and fun. The particularity of our approach concerns the involvement of the ecosystem in the sense of Guffroy *et al.* [3].

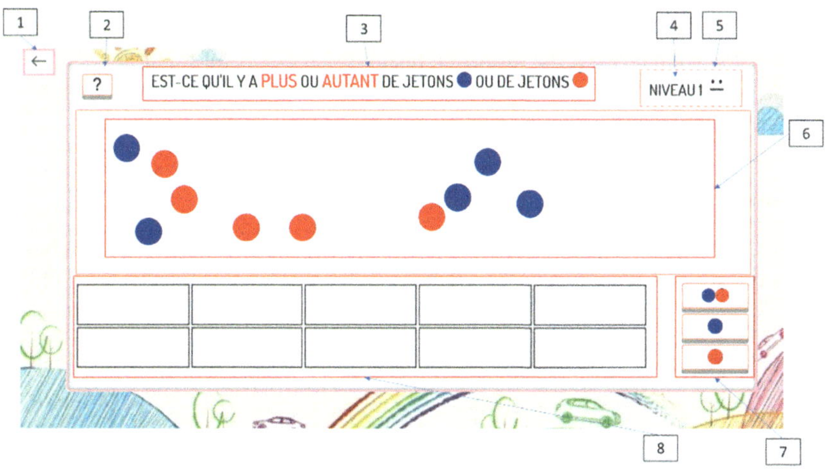

Fig. 1. The number comparison exercise has several interactive components: 1) the button to return to the home page; 2) the help button for handling the tokens; 3) the textual instruction; 4) the level of the exercise; 5) the pictogram representing the child's progress in the exercise; 6) the 2 classes of tokens to be compared; 7) the button to be selected for the choice of the correct answer; an animation and audio feedback are given according the answer 8) the digital tape to help the pupils visualise the quantity of tokens.

The ecosystem is a group of people who express needs in the place of end users when they are unable to formulate the demand: lack of knowledge, difficulties in expressing themselves, etc. The ecosystem is composed of mathematic's didactical expert (five teachers in primary school and one expert mathematical referent). A first activity, the comparison of numbers, has been prototyped. The ecosystem has defined the specification based on Mounier's work [2] on learning to number, gave feedback on the proposals for interactive models during two focus groups, discussed the choices on how to formulate the instructions, the representations of the buttons and the overall structure of the HMK-L interface (Fig. 1) represents the first iteration of the number comparison interface.

4 Design of the Usability Method

4.1 Approach to Designing the Usability Survey

There are also many standardized surveys for assessing usability, notably the System Usability Scale (SUS) [7] or the Heuristic Evaluation for Child E-learning (HECE) [9]. User tests are not limited to usability surveys [10]. However, in the health context of COVID in March 2020, the survey is the most suitable tool for remote evaluation. This makes it possible to overcome the heavy constraints associated with face-to-face work (social distancing, cleaning of experimental equipment, etc.). The Primary teachers carried out usability tests. They will use HMK-L as a teaching tool in their ordinary classroom.

A multidisciplinary team (1 cognitive psychology researcher, 2 researchers and a student in human-computer interaction) has designed the survey. An expert mathematical referent and two primary school teachers of her working group have checked the comprehension (objective of the question, meaning of words) of the first version of the survey. We corrected ambiguous questions so that the evaluation could be carried out without the presence of the experimenter.

4.2 Content of the Survey

To ensure that we addressed all aspects of usability, the design of the survey was based on the heuristic set, HECE adapted to the evaluation of e-learning applications for children.

Table 1. Links between Nielsen's heuristics, Alsumait's criteria and questions of the survey.

Tasks	Nielsen's Heuristics	Alsumait's criteria	Question
Discover the exercise layout	Match between system and the real world	All learning objects and images should be recognizable and understandable to the child, and speak to their function	Do you understand the meaning of all the elements on the screen? If not, which ones?
		The e-learning program interface employs simple words, phrases and concepts familiar to the child and makes information appear in a natural and logical order	Is the page layout understandable? If not, why not? Is the structure of the screen layout logical and understandable?
View text instruction and answer buttons	Visibility of System Status	The child understands all terminology used in the program	What do you think of the shape of the answer buttons in terms of readability and comprehension?

We then classified the 21 Alsumait's heuristics by grouping them according to Nielsen's 10 heuristics. Then we distributed the 47 criteria (decomposition of those heuristics) of HECE in regard of Nielsen's ten reference heuristics.

To complete the survey, users had to perform a scenario. This scenario is structured into 8 tasks divided into sub-tasks, covering all the functionalities we want to evaluate. These tasks are: Choice of interaction technique, Discovery of the interface, Exercise using the help button, Unassisted exercise, Exercise with error, Go to next level, Exit the application, Free use. A question section follows the guidelines of the subtask to be performed. The survey consists of semi-open questions and questions with a 6 item Likert scale (absolutely not good, not good, rather good, good, and absolutely good). The Table 1 presents how questions are linked to Nielsen's heuristics and Alsumait's criteria.

The survey contains 65 questions using 76.6% of the criteria. Table 2 gives the details of the rate of use of the criteria.

Table 2. Percentage of Nielsen's Heuristics used in the survey.

Nielsen's Heuristics	Number of criteria used	Number of criteria by category	Percentage use of criteria
Visibility of System Status	6	8	75%
Match between system and the real world	4	5	80%
User Control freedom	3	4	75%
Consistency and standards	4	6	66,67%
Error prevention	3	4	75%
Recognition rather than recall	3	4	75%
Flexibility and efficiency of use	7	8	87,5%
Aesthetic and minimalist desi,	3	3	100%
Help users recognize, diagnose and recover from errors	0	1	0%
Help and documentation	3	4	75%

5 Analysis of Survey

The usability' survey were send to 4 teachers during the containment (October November 2020 in France). Teachers were invited to explore and use the HMK-Learning application and then follow the scenario associated with the survey before answering it. The Table 3 summarizes relevant results from the analysis of the survey for the second iterative design of HMK-Learning.

Table 3. Some relevant results of the usability surveys by 4 teachers.

Interface elements and pedagogic strategy	Evaluation report
Visual aspect of the interface	Pleasant, airy, understandable, playful and well-structured;
Textual instructions	Understandable, presentation in sound modality required
Spatial layout of tokens	Good distribution, token overlapping to be avoided, no need of audio feedback
Feedback	Progress scale for exercises is suitable, feedback of success or failure is also appropriate
Strategy to compare the token collection	Need to add different strategies: clustering by sets (5 or 10 tokens) or pair to pair comparison
Selection mode of tokens Help	Allow several strategies (one by one, by set) Need to be redesigned from a pedagogical point of view

The following section gives some more verbal results from the questionnaires that complete the analysis.

According to Strategy to Compare the Token Collection: Respondents reported that pupils using HMK-L seemed not to develop an optimized strategy involving clustering of tokens to compare bigger collections. We also ask the **question of simultaneous token selection** or not to ease the comparison on higher numbers than 19. 2 teachers are for selection per pack of 5 or 10 faster; the other two are against it because it is too complex but less adapted to the term-to-term comparison.

The Progress scale for exercises is suitable for three of the four teachers. The last suggests improving the smiley representation (number 5 on Fig. 1) to better distinguish the levels of exercise. The animations displayed according to the student's results are very playful and fun for 3 of respondents. The fourth teacher suggests a more encouraging animation when a wrong answer to the exercise is given. The usefulness of the sound associated with the animated image is also mentioned.

The arrows for navigation through the exercise (go to the next exercise, repeat the exercise, go to the next level of difficulty) are affordant. The role of the digital tape is also understandable for the pupils according to the 4 teachers. According to the 4 teachers respondents, the HMK-L application is adapted for children at this educational level as shown in this verbatim by one of the fourth teachers: "*Yes: possibility to repeat the levels if there is a lack of confidence and possibility to pass to the next level quickly enough for those who are more at ease*".

The help device is an essential function: the analysis of the results shows that it need to be context-specific, to depend on the level of the exercise, to be adapted to the pupil' skills and to take into account the various pedagogical strategies.

Analyses of this usability survey had an unexpected and interesting result. More than ergonomics data, results led to didactic and pedagogical debates particularly on instructions'rewording (number 3 on Fig. 1) on and the strategical use of the digital tape (number 8 on Fig. 1).

6 Conclusion and Perspective

Based on the didactic work of Mounier's numeration we co-design the HMK-Learning with the teachers of primary school. Then, we implemented a method for designing a usability survey based on Alsumait's and Nielsen's heuristics. Four primary school teachers answered this survey. The analysis was very rich and usable from two angles: 1) redesign of interaction components (instruction, progress feedback, etc.) to improve ergonomics and accessibility of HMK-Learning; 2) implementation of pedagogical strategies (token comparison strategies, formulation of instruction, help, etc.).

Based on this first ergonomic study and to become a more inclusive and adapted tool at school, HMK-L need further implementations. It is of interest to note that ergonomic data participated to didactic reflection on strategies, procedure.... This unexpected result could involve evolutions in usability studies by integrating some didactic 'like heuristics in future studies. This project will be ongoing by analyzing utility and usefulness in context with pupil's interactions, learning performances.

Acknowledgments. This project was partly funded by the MSHS-T in Toulouse. The authors would like to thank the teachers from the primary schools who participated in the HMK-Learning specifications and who answered the usability survey.

References

1. Villani, C., Torossian, C.: 21 mesures pour l'enseignement des mathématiques. Ministère de l'Éducation nationale, Paris (2018). http://cache.media.education.gouv.fr/file/Fevrier/19/0/Rapport_Villani_Torossian_21_mesures_pour_enseignement_des_mathematiques_896190.pdf
2. Mounier, E.: Nouveaux outils d'analyse des procédures de dénombrement pour explorer leur lien avec la numération écrite chiffrée et la numération parlée. Recherches en didactique des mathématiques **36**(3), 347–396 (2017)
3. Guffroy, M., Vigouroux, N., Kolski, C., Vella, F., Teutsch, P.H.: From human-centered design to disabled user and ecosystem centered design in case of assistive interactive systems. Int. J. Sociotechnology Knowl. Dev. (IJSKD) **9**(4), 28–42 (2017)
4. Drijvers, P.: Digital technology in mathematics education: why it works. In: Selected Regular Lectures from the 12th International Congress on Mathematical Education, pp. 135–151 (2015). https://doi.org/10.1007/978-3-319-17187-6_8

5. Hodaňová, J., Škultéty, M., Hyksová, H.: Digital technologies in mathematics teaching at primary schools. In: Conference: 11th Annual International Conference on Education and New Learning Technologies,Palma de Mallorca, Spain (2019). https://doi.org/10.21125/edulearn.2019.1057
6. Geiger, V., Goos, M., Dole, S.: The role of digital technologies in numeracy teaching and learning. Int. J. Sci. Math. Educ. **13**(5), 1115–1137 (2014). https://doi.org/10.1007/s10763-014-9530-4
7. Brooke, J.: SUS: a "quick and dirty" usability scale. In: Jordan, P.W., Thomas, B., Weerdmeester, B.A., McClelland, A.L. (eds.) Usability Evaluation in Industry. Taylor and Francis, London (1986)
8. Quiñones, D., Rusu, C.: How to develop usability heuristics: a systematic literature review. Comput. Stand. Interfaces **53**, 89–122 (2017)
9. Alsumait, A., Al-Osaimi, A.: Usability heuristics evaluation for child e-learning applications. In: Proceedings of the 11th International Conference on Information Integration and Web-Based Applications and Services, pp. 425–430 (2009)
10. Khanum, M.A., Trivedi, M.C.: Take care: a study on usability evaluation methods for children. arXiv preprint arXiv:1212.0647 (2012)
11. Nielsen, J.: 10 usability heuristics for user interface design. Nielsen Norman Group, vol. 1, no. 1 (1995)
12. Lewis, J.R., Sauro, J.: The factor structure of the system usability scale (PDF). In: International Conference (HCII 2009), San Diego CA, USA (2009)
13. Baraudon, C.: Principes de conception et de validation d'un outil d'évaluation de l'utilisabilité perçue adapté à des enfants, Septièmes Rencontres Jeunes Chercheurs en EIAH (RJC EIAH 2018), Besançon, France, hal-01769535 (2018)
14. ISO/IEC. 9241–11 Ergonomic Requirements for OfTice Work with Visual Display Terminals (VDT)s-Part II Guidance on Usability, ISO/IEC 9241–11 (E) (1998)

Risks to Student Achievement in Higher Education

Angela Atwell[1,2](✉), Bettina Mrusek[1,2], Jim Solti[1,2], and Keith Wilson[1,2]

[1] Worldwide College of Aeronautics, Embry-Riddle Aeronautical University, 1 Aerospace Blvd, Daytona Beach, FL 32119, USA
{Atwella2,Mrusekb,Soltij,WilsoJ15}@researchlocation.edu
[2] Rothwell Center for Teaching and Learning Excellence/Online Education, Daytona Beach, USA

Abstract. A university system sets out to deliver educational experiences that meet set goals such as the achievement of learning outcomes for individual courses and program outcomes for degree programs. There are many factors that impact the successful achievement of student learning outcomes and therefore successful program design and implementation. If courses are not effectively designed with assessments properly aligned to learning outcomes, student achievement is challenging to measure. If faculty do not consistently adhere to college and/or university policies regarding submission of assignments, student behavior and perceptions of expectations in future courses may be skewed. In addition, students may, for various reasons make choices that result in failure to submit assignments that serve as measures of achievement for learning objectives. All of these factors could lead to a system breakdown and subsequent research location failure to meet the established goals, i.e. student learning outcomes. In this case study, an introduction to aeronautics course used to determine if the failure to submit assignments significantly impacted the achievement of stated program outcomes using a systems engineering approach. Data from core courses required for degree completion were used in the study. The results indicated that the lack of assignment submission presents a flaw in the system design and that the risk of not meeting learning objectives and program outcomes is very high when students fail to submit assignments.

Keywords: Higher Education · Systems engineering · Student Achievement

1 Introduction: Case Analysis

Balanced educational experiences, whether online or in traditional classrooms [1] require the use of varied approaches. This includes assessments that are both written in nature, such as research papers and case analyses as well as oral presentation assessments where students are practicing and demonstrating general education competencies such as public speaking and presentation development skills. This varied approach is not only warranted from an educational perspective, but it also mirrors skills and abilities students will need beyond the classroom, in the workplace. According to a 2015 Employer

S. Nazir et al. (Eds.): AHFE 2021, LNNS 269, pp. 165–172, 2021.
https://doi.org/10.1007/978-3-030-80000-0_20

Survey conducted by Hart Research Associates [2], employers place the highest value on demonstrated proficiency in interdisciplinary skills such as written and oral communication when hiring recent college graduates. Specifically, the report found that oral communication rated an 85% on the employer priorities for most important learning outcomes. Written communication rated 82% in the same report [2].

At the research location, learning objectives arc more specific than program outcomes to allow students to explore concepts on a more granular level during each individual course [3]. The cumulative impact of learning is thus measured by program outcomes that demonstrate a student's mastery of all program content. However, failure to complete the more specific assessments and effectively demonstrate mastery of a learning objective, calls into question, a student's ability to demonstrate mastery of an overall programmatic learning outcome.

At the research location, courses are built using the backward design method, where learning objectives are developed to ensure achievement of learning outcomes. Then, assessments are aligned with learning objectives and created so students can demonstrate mastery of these learning outcomes [4]. Students are asked to demonstrate mastery through a variety of educational tasks or assignments throughout individual courses to demonstrate mastery of these learning objectives which cumulatively demonstrate mastery of program outcomes. In some cases, though, to streamline the course and program, student learning outcomes are assessed by a singular activity.

The degree programs at the research location is designed with this process in mind, emphasizing the achievement of program outcomes via learning outcomes assessed in individual courses and activities. For example, one program outcomes states, "upon completion of this program, students will be able to communicate effectively using both written and oral communication skills" (Table 1).

Table 1. Factors contributing to learning objective/Program outcome achievement failure.

Program	Program outcome	Learning objective	Assessment	Measure of achievement
Undergraduate Degree in Aeronautics	Upon completion of the program, students will be able to communicate effectively, using both written and oral communication skills	Upon completion of this course, students will be able to demonstrate professional communication and oral presentation skills using appropriate media	Writing and Presentation activities	70% or above on all writing and presentation activities

2 Review of the Literature

Complex systems, such as online education often appear as wicked problems, where incomplete, contradictory and changing requirements make it extremely difficult to not only identify all of the essential components, but to also link the connections and draw meaningful conclusions to improve the overall system. Student preferences, for example, may influence risk assessment in academic decision making. Studies have found preferential differences regarding assignment formats between the genders. Males have been found to prefer multiple choice formats over essay type assessments [5]. In contrast, females have preferred essay formats [6]. A recent study sought to uncover more details regarding differences in opinion regarding various assignment types. For this study, assessment preference was defined as "imagined choice between alternatives in assessment and the possibility of the rank order of these alternatives" [7, p. 647]. Students, regardless of gender, were shown to have preferences for written assignments, like research papers. This research demonstrates that if students have preferences for certain assignments, the assumption is, they are more likely to complete them.

Additionally, risk tolerance and assessment are highly individualized and personal. However, these individualizations must be considered during systems engineering processes to allow for successful goal achievement. Specific student situations, while varied in nature can contribute to the decision-making process. At the research location students are typically non-traditional students. According to the National Center for Educational Statistics [8], non-traditional students are defined as a diverse population of adult (over the age of 24) students with work and family responsibilities along with other life circumstances that may interfere with educational experiences. Fitting a degree program into an already busy schedule can be stressful and anxiety provoking. This additional work load may lead students to prioritize and make decisions about what gets done and what doesn't. Limited resources, like experience and knowledge can lead to poor decisions. To make the most of these limited resources, heuristics are utilized. Heuristics, or rules of thumb can be misleading. For example, the availability heuristic may lead an individual to believe that a certain decision or action is the most appropriate simply because it is the first one that comes to mind [9]. Individuals "satisfice" by seemingly considering all available options and selecting the one that seems to best meet a predetermined minimum level of acceptability [10]. Non-traditional students may be looking for the best use of their time. If an option, where they do not need to submit an assignment seems to appear, some students may take the chance. This is especially true if students can still earn a preferred grade. All of this information, accurate or otherwise contributes to the decision-making process. Furthermore, how individuals approach risks and make assessments partly depends on their understanding of the issue at hand as well as the available options [10]. For students to adequately assess their risk, definitions must be clear to them.

At the research location have the discretion to fail a student should they choose not to submit all assignments, as outlined in the syllabus. However, if students have had an experience contrary to this statement, in that a faculty member allowed them to earn a zero on an assignment and still pass the course, this information would skew the student's definition and therefore impact their risk assessment. This reality aligns with Risk Homeostasis Theory where behavior and decisions are made with the intent of remaining within a pre-determined level of acceptable risk [11]. For instance, students who desire an honor distinction at graduation may not risk earning a low score on an assignment because a low score could take them beyond their comfortable threshold and risk the achievement of a lower grade. In contrast, students may not wish to spend any more time or effort on assignments than is absolutely necessary because they have identified a level they are willing to commit to this endeavor. For example, students may choose not to submit an assignment that is only worth 10% of their final grade because they have already determined they are comfortable with a lower final course grade. The variability of student threshold and risk determination is highly individual, making it difficult to calculate and almost impossible for an instructor and/or course designer to predict. Furthermore, given the variability in faculty expectations and behaviors, this calculation, done by students could be flawed. Where one faculty member may be flexible in allowing students to miss one or two assignments, another may not. In order to support students' ability to adequately assess their risk, definitions, such as all assignments must be submitted to pass the class, must be clearly communicated, as they are in the syllabus and uniformly adhered to by faculty.

In an attempt to tackle this wicked problem, systems engineering models and themes can be directly applied. Attempts were made to illuminate the shortcomings within the system which justify the need for further exploration. Systems engineering concepts can then be further applied to make adequate and effective adjustments to the system to ensure goals, in this case, student achievement of learning and program outcomes, are met throughout the system.

3 Methodology

This case study is an applied, descriptive research project. The techniques and methods of this project set out to inform a body of knowledge about a situation or potential problem with student learning objective and program outcome achievement to impact further understanding about the situation and potentially impact future policy [12]. A case study methodology is utilized by which an in-depth analysis of a particularly concerning condition will be explored utilizing existing data sources. Failure Mode and Effects Analysis works to identify and address the most critical concerns in processes, products or within a system [13]. As such, it was also utilized in the analysis.

Data was gathered utilizing existing online databases from the research institution; Campus Solutions and Canvas. To begin, graduate courses were removed from the sample of all courses. Then, non-relevant activities and assignments such as discussions were filtered out. Then, the sample was further limited to the academic terms of interest. The resulting data set included information regarding final grades and grades for specific assignments including high-stakes written and presentation assignments for undergraduate students during the terms identified. Furthermore, demographic data (age and gender) was collected on students from the research location. Campus Solutions system and aligned with the Canvas data. All collected data was deidentified using a seven-digit integer. Collected data was then conditionally formatted for use with Excel and SPSS, a statistical software platform. To generalize the data in this study, the Power Analysis Equation was utilized to determine adequate sample size [14].

Sample Size = {z^2 * p * (1 − p)/e^2}/{1 + {z^2 * p * (1 − p)}/{e^2N}}.

Using this formula, the original 16,040 individual data points from the ASCI 202 course for the given time period was decreased to 580 individual data points. A random sampling of 580 individual data points proved statistical sufficient for the analysis. The criteria below were students who passed (>70% overall) with at least one non-submission. It does indicate however that the 580 results are consistent with the 16,000, that no gender bias is apparent, no assignment type (written vs presentation) bias is apparent, and that overall fewer than 5% of students chose not to submit for this sample. See Table 2 (Table 3).

Table 2. Sample size criteria.

	ASCI 202			
# Activities Reviewed	**16,040**	**100.0%**	**580**	**100.0%**
Female	2180	13.59%	82	14.14%
Male	13155	82.01%	473	81.55%
Written (Total)	**7140**	**44.51%**	**271**	**46.72%**
Written (Female)	974	13.64%	45	16.61%
Written (Male)	5853	81.97%	213	78.60%
Presentation (Total)	**8900**	**55.49%**	**309**	**53.28%**
Presentation (Female)	1206	13.55%	37	11.97%
Presentation (Male)	7302	82.04%	260	84.14%
# Activities Meeting Criteria	**562**	**3.50%**	**17**	**2.93%**
Female	94	16.73%	3	17.65%
Male	436	77.58%	12	70.59%
Written (Total)	**280**	**49.82%**	**7**	**41.18%**
Written (Female)	40	14.29%	1	14.29%
Written (Male)	224	80.00%	5	71.43%
Presentation (Total)	**282**	**50.18%**	**10**	**58.82%**
Presentation (Female)	54	19.15%	2	20.00%
Presentation (Male)	212	75.18%	7	70.00%

Then, a stratified sampling approach was utilized to randomly select 580 individual data points from the original data set. A stratified sampling approach is a probability sampling technique that allows for an adequate sample by reducing error during random sampling [15]. To accomplish this randomization, each of the original data points were assigned a random number from zero to one. Data points for this study included score on the individual assignment, overall course score, gender and age.

Table 3. Minimum sample size calculation.

Minimum Sample Size Calculation		
Confidence Level	0.95	0.95
P	0.5	0.5
Error	0.04	0.04
Population size	16040	2749
alpha/2	0.025	0.025
Z-score	1.95996	1.95996
Sample Size	578.577	492.659
numerator	600.228	600.228
denominator	1.03742	1.21834

4 Results

The FMEA analysis produced some interesting findings. As was expected, the risk for failure to master learning objectives (LO) and program outcomes (PO) is elevated when students fail to submit assignments. Failure to submit all assignments, resulting in not mastering program outcomes (RPN = 125) was found to pose the highest risk to achievement of learning outcomes. Assignment weights also showed an elevated risk for student achievement of learning outcomes with an RPN of 75. As illustrated in the course breakdown, "freshmen level courses" included higher weights for presentation assignments than "senior level courses". This may communicate an inaccurate deemphasizing of these assignments by students. Again, the statistics from this study informed the rating. While this may have been a concern for the students who opted out of submitting assignments, weights did not seem to impact the majority of students in this sample. This reality impacted the probability rating for this potential failure mode (probably rating = 3).

Perhaps surprisingly though, was the RPN for delayed course completion. While this is not something that was explored in this research project, retention and attrition is a concern at the research location and could be a potential factor when considering system requirements. This should be explored further in future studies.

With an RPN of 25 each, compound learning objectives and faculty adherence to assignment submission policy in the syllabus are found to carry quite a risk to student learning outcome achievement. As discussed previously, failure of an instructor to adhere to the policy in the syllabus which states that students may fail the course if all assignments are not submitted, directly relates to the failure to meet set learning objectives and potentially associated program outcomes. Furthermore, and perhaps more indirectly, experience with a faculty member who allows students to pass the course without submitting all the assignments may contribute to mental models and inform student risk assessment which could lead to similar behavior in future courses. In addition, compound learning objectives and program outcomes make achievement difficult to measure. A compound objective or outcome includes the word "and". Including more than one criterion in a learning outcome such as "upon completion of this course, students will be able to communicate effectively using both written and oral communication skills" cannot be adequately assessed and therefore measured. For more accurate and specific assessments, associated learning outcomes must have a singular focus.

5 Conclusions

This case study set out to illuminate the facts surrounding a given situation. It was hypothesized that student decisions about submitting assignments negatively impacted their achievement of learning objectives and program outcomes given the research on student perceptions and the application of a systems engineering approach on the achievement of learning outcomes in higher education. This decision questions the mastery of stated learning objectives and program outcomes.

FMEA results found that the risk of not meeting learning objectives and program outcomes is very high when students fail to submit assignments. The analysis provided insight on various contributing factors. First, compound learning objectives and program outcomes make it difficult to adequately measure student achievement. This project justified the liability and increased risk posed by compound objectives and outcomes. It is recommended that all courses be audited to correct any compound objectives/outcome as well as to ensure the measurability of the associated assessments. Furthermore, and related to course design, the assignment weights resulted in a high RPN and so are considered potentially problematic. During the necessary assessment audit, assignment weights should be revisited to ensure proper weight is given to learning outcome assessments. Along with these tasks, college administrators should review their decision to design courses with single points of assessment to ensure this is indeed the path they want to follow. Related to single points of assessment, next, given the weak language in the syllabus and the general discretion afforded to the faculty at the research location, students can successfully pass a given course without submitting all assignments and potentially not master all learning objectives. Verbiage from the syllabus should be strengthened to avoid ambiguity. Rather than "may" it should say "will". Strengthening the language in the syllabus and providing adequate training around submission expectations for faculty and students is recommended. References to the importance of completing all assignment should be included in the Online Student Readiness Course available to all incoming students as well as reviewed in the required initial and recurrent training for faculty. This would ensure that all students complete the work that demonstrates mastery of the learning objectives and program outcomes that contributed to the course design. This policy change would support the single point of assessment decision and contribute to the successful transfer of knowledge in a higher education setting.

References

1. Angelo, T., Cross, K.: Classroom Assessment Techniques: A Handbook for College Teachers, 2nd edn. Taylor & Francis, Milton Park (1993)

2. Hart Research Agency. Falling short? College learning and career success (2015)
3. Rothwell, W.J., Benscoter, B., King, M., King, S.B.: Mastering the Instructional Design Process: A Systematic Approach, 5th edn. John Wiley & Sons, Hoboken (2015)
4. Wiggins, G., McTighe, J.: Understanding by design, 2nd edn. Pearson Education, Upper Saddle River, NJ (2006)
5. Beller, M., Gafni, U.: Can item format (multiple choice vs. open-ended) account for gender differences in mathematics achievement? Sex Roles J. Res. **42**, 1–21 (2000)
6. Gellman, E., Berkowitz, M.: Test-item type: what students prefer and why. Coll. Stud. J. **27**(1), 17–26 (1993)
7. Watering, G., Gijbels, D., Dochy, F., Rijt, J.: Students' assessment preference, perceptions of assessment and their relationships to study results. High. Educ. **56**(6), 645–658 (2008)
8. National Center for Education Statistics (NCES): Nontraditional undergraduates: definitions and data
9. Cherry, K.: Decision making strategies: how time, complexity and ambiguity influence which method we use. Verywell (2015)
10. Sternberg, R.J., Sternberg, K.: Cognitive Psychology, 6th edn. Wadsworth, Belmont, CA (2012)
11. Wilde, G.J.S.: Risk homeostasis theory: an overview. Inj. Prev. **4**(2), 89–91 (1998)
12. Kumar, A.: Doing Sociology. Lulu Press, New York, NY (2011). ISBN: 978-1-44-660793-0
13. Guerrero, H.H., Bradley, J.R.: Failure modes and effects analysis: an evaluation of group versus individual performance. Prod. Oper. Manag. **22**(6), 524–1539 (2013)
14. Leedy, P.D., Ormrod, J.E.: Practical Research: Planning and Design. Pearson, Boston (2016)
15. Salkind, N.J.: Encyclopedia of Research Design, vol. 1. SAGE Publications Inc., Thousand Oaks, CA (2010)

Participants' Perspectives on Design-Build Experience: A Qualitative Exploration

Shweta Saxena(✉), Sandeep Arora, and Rachna Khare

School of Planning and Architecture, Neelbad Road, Bhauri, Bhopal, India
{shweta,sarora,rachnakhare}@spabhopal.ac.in

Abstract. It has been established by pedagogues that for understanding and applying the knowledge, education must have both abstract and concrete components. In professional degree programs, the abstraction could be achieved through classroom teaching. However, for the longer retention of the acquired knowledge and contextually appropriate applications, the learning environment must closely resemble the work environment. A design-build studio in architecture education is one such format designed to resemble the architectural practice and to create a better learning environment for its participants. The pedagogy of design-build studios uses tools of experiential learning like role-playing, reflective practice, etc. Although, the existing literature on design-build studios present substantial elaborations on participants' learning; there is marginal documentation of the participants' perspectives on their experience. This paper presents the process of analyzing open-ended responses from sixty-six participants of a design-build project conducted at the School of Planning and Architecture, Bhopal, India. In addition, the study elaborates a major theme 'evaluating design-build learning' that emerged from the data along with five sub-themes. This qualitative exploration helped in understanding the deeper meaning of design-build studios and their effect on participants' learning. Also, it helped in evaluating the design-build pedagogy for architecture education.

Keywords: Experiential learning · Architecture education · Design-build Studio · Participants' perception · Qualitative exploration

1 Introduction

Experiential learning is a powerful tool that brings about deep levels of learning and change [1]. Based on the elements of experiential learning theory, design-build pedagogies flourished in the institutes of many countries. Architecture education is adopting this pedagogy to transform the ways of teaching and learning. Moreover, there is a need to find a teaching tool to bridge the gap between academia and the practice of architecture [2]. Design-build studios provide real experiences to the participants by immersing them in actual scenarios of designing and building. The positive learning outcomes of

S. Nazir et al. (Eds.): AHFE 2021, LNNS 269, pp. 173–179, 2021.
https://doi.org/10.1007/978-3-030-80000-0_21

design-build studios have been documented in the literature but the dialogues on the measurement of learning in the participants are limited. The documentation of the participants' perspectives on design-build experience in a studio conducted at the School of Planning and Architecture, Bhopal, India, brings to light various subthemes within the rubric 'evaluating design-build learning'.

2 Experiential Learning and Learning Theories

The relationship between experience and learning has its roots in various learning theories. *Rationalism* believed in reasons while *Empiricism* relied on senses for acquiring knowledge. Kolb placed both the learner and the experience at the center of the learning process. In such a process, students learn from their experiences and by reflecting on their experiences. Silberman clarifies the distinction of such a process from that of a lecture-dominated teaching format.

"Experiential Learning is used to signify any training that is interactive, with minimal lecture (and slides)" [1].

Dewey's educational theory supported both knowledge (theory) and action (practice) as the components of learning inquiry and was based on Realism (reality). Before Deweyan Pragmatism, theorists believed theory and practice were independent of each other. Later, many theorists developed educational theories based on Dewey's ideas. *Constructionism* believes in the tangible parts of the learning process, whereas *Constructivism* holds on to the various cognitive theories where learner makes a unique meaning of what they learn [3].

Salama's theory for knowledge integration in architecture education focuses on human behavior and people-environment research. The inquiry component of his theory uses experiential learning as the mechanism of inquiry by which knowledge is acquired [4]. Branislav mentions the connection of experiential learning with the design-build concept [5].

2.1 Architecture Education

In addition to providing knowledge and skills, education has a deeper motive of transforming the human being from within to bring positive changes in society and the world at large. Similarly, architecture education empowers the students with the knowledge and skill required to bring such changes to the users of buildings. Architecture education must provide the students with the opportunity to reflect on their identity and relate it to the responsibility towards the profession and society. To achieve deeper motives of education, participants must reflect on the experiences and gaining knowledge through the transformation of experiences [6]. Design-build studios provide the opportunity to realize such a pedagogical approach.

2.2 Design-Build in Architecture Education

Design-build in architecture education relies on the formal and informal components of teaching and learning. Where *formal is deliberate* and *informal is incidental.* In design-build studios, participants are involved in designing and building a product or structure. In such studios, sometimes participants communicate with the stakeholders, investigate materials, work individually or as a team, and make prototypes or actual structures at full scale. The scale of the project often depends on the duration and objectives of the studio, the number of participants, and the resources available.

3 The Design-Build Studio at School of Planning and Architecture, Bhopal

A three-week design-build studio was conducted at the School of Planning and Architecture, Bhopal with the intent of 'the understanding of local materials in addition to respecting the context and the site' (Fig. 1 and Fig. 2). A total of seventy-five students worked in a team of seven and there were a total of twelve teams. An open-ended questionnaire was distributed to the students on the day of the exhibition and review. Sixty-six students' responses were collected. During the initial stage of the study, open-ended questions were framed to get students' perceptions on their 'working effectiveness as a team', 'learning in group work', and to know their response on 'working in design-build group work projects'. The responses to these questions gave a fair idea of their learning experiences in design-build studios.

Fig. 1. Participants' discussing their design Photo Credit: Participants' Team.

Fig. 2. Participant chiseling the bamboo Photo Credit: Participants' team.

3.1 The Process of Qualitative Exploration

The responses received from the students were transcribed and the data was split into segments and a code was assigned to the relevant segment. To find patterns in the data the codes were repeatedly used wherever required that reinforced the perspectives among multiple participants [7]. The data set from open-ended responses were analyzed through the coding process. Later, the codes were merged to form categories and subcategories (Fig. 3 and Fig. 4). The categories that emerged from the codes helped in developing the list of design-build learning outcomes. Six categories emerged from the data; however, within the scope of this paper, only one category 'Evaluating Design-build Learning' has been presented. Five subcategories derived after merging the codes helped in understanding this category.

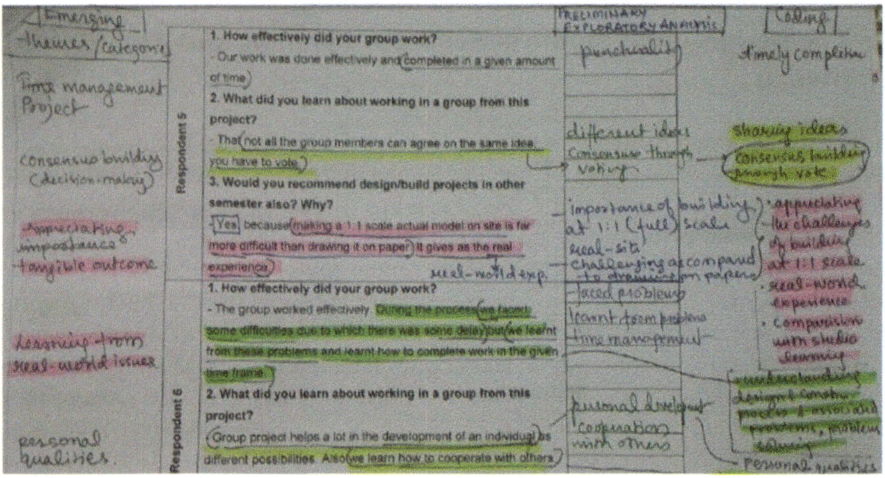

Fig. 3. Sample of manual coding

Subcategory	Code
Quality of Experience	Appreciating the challenges of building at 1:1 scale on real site
	Importance to off-studio learning
	Solving Real-problems
	Hands-on experience with the material (In Vivo)
	appreciating tangible outcome
	learning on-site decision making
	Excitement
	Maturity in work

Fig. 4. Sample of codes merged to develop subcategory

The participants' responses to which the codes were assigned and later merged to develop a subcategory are presented in the samples (Fig. 5 and Fig. 6) below:

Subcategory 1: Quality of Experience		
S.No.	Code	Participants' Response
1	Appreciating the challenges of building at a 1:1 scale on real site (Descriptive Code)	"…making a 1:1 scale actual model on site is far more difficult than drawing it on paper. It gives us real experience."
2	Importance to off-studio learning (Descriptive Code)	"Working on the field is an amazing experience and a better learning process than sheets and books."
3	Solving Real-problems	
4	Hands-on experience with the material (In Vivo)	"As it was a hands-on project, we learned from the material selection to the application of it. This could not be learned through studio work."
5	appreciating tangible outcome	
6	learning on-site decision making	"We did not plan our design on a bench/bedroom/studio. We designed it when we were in the sawmill. The design and inspiration on
7	Excitement	
8	Maturity in work	

Fig. 5. Sample of participants' response

Subcategory 2: Appreciating the Process of Working with Real Material		
S.No.	Code	Students' Response
1	Procuring Material	"…understanding the properties of materials. Techniques, joinery, etc. in a better way."
2	bringing the material to the site - tiresome	"..good exposure to field works, market survey, and material properties."
3	Ability to work with real-material	"I learned different types of joineries and even how to join bamboo."
4	Materiality	
5	Techniques	"…market survey and how to bargain."
6	Joinery	"We learn many things like how we can use the material in different ways. Mainly it was a good

Fig. 6. Sample of participant' response

3.2 Observations

The study was initiated to understand the collaborative skills that developed overtime among participants of the design-build studio. With this intent, the open-ended questions were circulated to the participants to know their perspectives and understand the impact of this pedagogy on the participants' learning. The qualitative exploration of the participants' perspectives on the design-build experience through the coding process revealed some major categories.

'Evaluating design-build learning experience' emerged as one of the major categories from the transcribed data. Five subcategories corresponded to this major category: (1)

Quality of Experience; (2) Appreciating the process of working with real-material; (3) Design-build process; (4) Suggestions; and (5) Challenges.

'Appreciating the process of working with real material' and 'design-build processes were ordinary subcategories as the researcher was expecting them in the students' responses. Whereas, quality of experience, suggestions, and challenges were the unexpected subcategories.

The findings from this study could be used to conduct a design-build studio, where the suggestions and challenges informed by the participants could help to overcome some of the constraints to conduct these studios.

3.3 Discussion

The open-ended questions initiated the thinking process of participants to reflect on their design-build experience. The experience of working with each other, investigating materials, exposure to real-world conditions, and celebrating the tangible outcome had a positive effect on participants' learning. The five subcategories define the major category 'Evaluating Design-build Learning'.

Quality of Experience: Participants' perspectives reflected the quality of design-build experience which was both rewarding and challenging. They compared it with studio-based projects and found it more realistic. The participants mentioned the moments when they felt excited about the decisions they made themselves and when their ideas took shape on the ground.

Appreciating the process of working with real material: Hands-on materials connect thinking with the making. Participants appreciated the process of material exploration. Involving in this process informed participants to relate material types and workability with space designing. Choosing the right material for construction engaged participants in understanding properties and joinery details, surveying the market for prices, and using the material in different ways.

Design-build Process: The interactive process of design-build was mentioned in the participants' perspectives. They appreciated the studio approach of working on the site. Their views on the design-build process matched with the learning theories that encourage linking theory with practice.

Suggestions: Participants' responses suggested some improvements like increasing the duration of the studio, including participants' evaluation after the project is complete, and funds to be arranged by the institute.

Challenges: Exposure to the real-world also means facing certain real challenges that are visible in students' perspectives. Depending on the scale of the project the duration of the studio is decided or vice-versa. But it is felt by many participants that due to time constraints certain hasty decisions are made. There are delays due to the unavailability of materials and budget is always a constraint as it affects the material selection and eventually affects the design. Among the group of participants, conflicts on certain decisions are there.

Despite the entire challenges and issues that arose from the participants' perspectives, it was observed that the design-build experience was appreciated for its interactive nature, be it with people, peers, materials, or tools. The fun and feeling of learning with each other and touching the real tools and materials to produce a tangible product were found exciting by the participants. The process of designing and building is not the only part of these studios; it is the involvement of the senses of the participants to give deeper meaning to their learning which was well evident through their perspectives when they evaluated their experiences.

4 Conclusion

Design-build studios provide participants with a setting that facilitates immersive engagement and interaction. It provides an opportunity for interaction between participants, clients, and stakeholders, and studio instructors to get information and find solutions to design and construction-related issues. Since 'designing', 'building', 'redesigning', and 'modifying' are parallel and on-site processes, participants can reflect on the physical expression of their ideas, and knowledge is generated through the transformation of experiences as outlined by Kolb [6]. The whole process from project inception to fabrication keeps the participants engaged. The learning in this context could be related to Higher Order Thinking as mentioned in Bloom's Taxonomy. The pedagogy of design-build studios encourages participants to 'apply their knowledge in real situations', 'analyze the changes that are needed', and 'evaluate their learning'. The studio starts with abstract components explained through lectures and discussions. As the studio advances to the development of initial ideas, with subsequent stages reinforced with the exposure to the real-world conditions, the participants move from 'abstract' to 'concrete' through the process of building. The application part of the learning process makes design-build studio different from studio-based pedagogy.

Acknowledgment. This paper presents part of the primary author's doctoral research at the School of Planning and Architecture, Bhopal. Authors acknowledge the support of studio faculty Vineet Chaddha, Apurv Shrivastava, Gaurav Singh, Abhishek Iyer, Priyanka Singh, and Ashutosh Tiwari.

References

1. Silberman, M.: The Handbook of Experiential Learning. Pfeiffer, San Francisco (2007)
2. Saxena, S., Arora, S., Shrivastava, A.: Gaps Between Practice and Education of Architect—A Study of India. In: Kantola, J.I., Barath, T., Nazir, S., Andre, T. (eds.) Advances in Human Factors, Business Management, Training and Education. AISC, vol. 498, pp. 207–217. Springer, Cham (2017). https://doi.org/10.1007/978-3-319-42070-7_20
3. Guzdial, M.: Constructivism vs. Constructivism vs. Constructionism (2018). https://computinged.wordpress.com/2018/03/19/constructivism-vs-constructivism-vs-constructionism
4. Salama, A.M.: A theory for integrating knowledge in architectural design education. Int. J. Archit. Res. Archnet-IJAR **II**(1), 100–128 (2008)
5. Folić, B., Kosanović, S., Glažar, T., Fikfak, A.: Design-build concept in architectural education. In: Creative Commons Attribution License, pp. 49–55 (2016)
6. Kolb, D.: Experiential learning. Prentice-Hall, Englewood Cliffs, NJ (1984)
7. Saldana, J.: The Coding Manual for Qualitative Researchers. Sage Publications, Upper Saddle River, NJ (2009)

Analysis of Learning Strategies and Learning Performance

Reforming Professional Education: A Case of Cognitive Human Factor/Human Element in Shipping Industry

Gholam Reza Emad(✉)

Australian Maritime College, University of Tasmania, Launceston, TAS, Australia
reza.emad@utas.edu.au

Abstract. Historically, apprenticeship considered to be one of the main methods of educating newcomers for the world of work. Apprenticeship, as the learning method, appreciate the socio-technicality of the work processes by involving the newcomers in the practice under the supervision of an expert. During the last few decades, the responsibility of training the workforce has been gradually shifted from industry to the formal education sector. This resulted in many industries such as the maritime domain to abandon most of the opportunities of learning in their workplaces. This article by focusing on the shipping industry in the maritime domain, as an exemplary case, will provide an overview of the current condition of the workforce education as related to cognitive human factor/element and provide an alternative solution to its shortcomings. The article will provide an analysis of the current educational system in place and elaborates on the opportunities and shortcomings of the formal and informal educational component of the current maritime education and training system. By utilizing the current theoretical frameworks, the paper examines the conventional and emerging workplace learning opportunities and proposes the framework of quasi-communities as an alternative approach. The quasi-communities incorporates key elements of the apprenticeship method into teaching and learning in formal settings.

Keywords: Cognitive Human Factor · Apprenticeship · Shipping industry · Maritime education and training · Seafaring · Quasi-community · Human element

1 Introduction

International Ergonomics Association (IEA) identifies three main specializations in Human Factor/Elements and Ergonomics science; namely physical, organizational, and cognitive [1]. Cognitive human factor involves with mental processes on the way it affects interactions between humans and other elements of a system. This includes topics relevant to human-machine interaction, skilled performance, mental workload, decision-making, and training and education [1]. Despite the advancement of Human Factor and Ergonomics over the recent years, there are limited studies focusing on the education and training and the development of skills and competencies of people involved with the

S. Nazir et al. (Eds.): AHFE 2021, LNNS 269, pp. 183–189, 2021.
https://doi.org/10.1007/978-3-030-80000-0_22

work processes [2]. As part of the cognitive human factor, this article provides a general description of the current workforce education and training and proposes solutions to some of its shortcomings. The paper utilizes the shipping industry as a case to be able to provide a concrete example and analysis of the current formal and informal education and training systems in place.

Shipping is considered to be a key industry as it transports more than 90% of the world trade. Human Factor and Ergonomics or as it refers to in maritime domain Human Element is recognized to be a major contributing factor to most of the incidents and casualties in shipping industry [3]. Maritime mishaps due to lack of seafarers' competencies onboard ships have been a great concern for human element professionals in the last few decades. In the maritime industry, these concerns has been addressed through the introduction of a standardised education and training system of seafarers introduced by an international convention (STCW convention). However, in spite of the extensive implementation of the international system that requires seafarers to develop the knowledge and skills and continuously upgrade them throughout their working life, maritime accidents continues to happen at a large number. The accident investigations report ultimately attribute these mishaps mainly to the human element. Years after the implementation of the STCW convention, statistics show that there is a growing number of marine accidents and incidents happens because of seafarers' lack of proper skills and competencies. The reports of leading classification societies and insurance companies after the worldwide implementation of the revised and improved version of the original convention, show that there has been an increase in number of maritime accidents [4–6]. These reports show that there are no sensible improvements in cognitive human factor/element as related to seafarers' competency development despite the effort. This raises the question that with such effort to improve the maritime education and training systems in place why we experience less success?

2 Fault Lines of Current Maritime Education System

Currently, the pedagogical practices in the maritime educational domain mainly follow the formal conventional way of schooling. The formal education system is originally intended to provide students with the underlay knowledge and know-how required by for their future jobs. Although, the central purpose of the formal education is to provide the students the theoretical knowledge and the background that they need onboard ships to perform their jobs safely, nonetheless, as discussed, the system proves to be less than perfect in achieving this goal. The maritime accident investigation reports repeatedly show that the human element is to blame due to the fact that the theoretical knowledge taught at school to the seafarers are not largely transferable to their job onboard ships.

This might be due to the fact that many schools assume that the practice and theory are separable [7]. The system presupposes that the learners would be able to perform the task only by knowing about the theory of the task. It also assumes that a task can be broken into subtasks and be taught separately. They presume that, when students are at work, they would be able to incorporate the knowledge of the subtasks they learned and competently perform the whole task. This pedagogical viewpoint fails to consider the significance of the context within which the work is done [e.g. 8, 9].

The current education system asks for a single set of curriculum to be taught in a predefined time period. This idea assumed that all students posses the average level of cognitive ability and learning style. The system disadvantages non-typical students who are either above or below the assumed standard as they either feels that the pace is slow and so are wasting their time or too fast and not being provided enough learning time. Additionally, classroom learning is mainly based on didactic method with one sided lectures and discussions about a task. This is contrary to the real world learning at work where apprentices/students onboard ships involves observing the work processes, and being coached and mentored while involved in performing the task.

As discussed, the current conventional schooling has many shortcomings and is not an effective method for educating maritime workforce. The next section discusses the concept of apprenticeship and the practical training onboard ships as part of the formal education.

3 Learning Through Apprenticeship

The recent studies on cognition, learning, and instruction provides us with the better understanding on how the workplaces afford learning environments that support the newcomers to the job to engage in interactive, meaningful, and authentic learning [e.g. 10, 7]. These research shows that the apprenticeship training promotes learning knowledge and skills in its functional and social contexts. This is different to the way formal education at schools advocates learning the abstracted knowledge, separated from its practice, in a piecemeal fashion [11]. My earlier research in the maritime field shows that for the seafarers, on-the-job training is considered to be the most effective part of their training as it helps them to learn the required skills and develop the competencies they need for their job [12].

In the traditional apprenticeship, beginners gradually develop various elements of skills while involve in observing and performing the actual work. They spend a long period of time watching how the work is performed by experts. Gradually, they will be tasked to involve with and perform the basic tasks of the job. As the apprentices learn basics and progressing, they will participate in more complex tasks of the job. Participation in these tasks is not performed as an exercise for the sake of learning but as a daily and routine part of the job. In this way they will take part in the production process and thus contribute to the final product. As they gain competencies on the specific job, they start working on different job and more complex tasks. Thus the apprentice proceeds through a "curriculum" that practices tasks and skills in their context of use. Although there is very little direct teaching, however, there are significant learning opportunities through a graded and contextually embedded practice [13]. Assessment, in this method, is an uninterrupted continuous process. The apprentice's performance being continually observed and evaluated and in case the apprentice makes a mistake it gets noticed and corrected by the mentor at the time of doing the actual task [14].

The seafaring apprentices, while on board ships, learn in a routine way, to do the job the way a competent seafarer is doing. This method also allows them to participate in the design of their own learning curriculum. They can schedule to participate in an activity, among other available activities at the time, and can redo the same task, if it is needed, the next time it is available [15]. This allows them to personalise the pace of their learning the way they feel is suitable to them. The ship provides them an environment that incorporates all the activities, problems, situations, and contexts that they may encounter as long as they remain seafarers. Thus, from the first day, they see and develop a good understanding of the kind of work life they are chosen with all its complexities. This is in contrary to the schooling system as they introduce the students to these complexities one step at a time. Through the apprenticeship there is the possibility for apprentices to quickly learn about the nature of the job and give up the career if they feel that the job they have chosen does not suit them. Those who decide to remain have a realistic idea about their future career. They live through and experience first hand the physical environment with its rewards and difficulties in addition to the culture and challenges that comes with the job.

4 The Current Apprenticeship and Onboard Training

Currently, on-the-job training is a part of the international maritime training scheme. However, the length of the time allocated by the system for the on-the-job training is limited. Additionally, learning at work is considered to be one of the elements of the formal education and training system. Thus, for the seafarers working onboard ships the training of the apprentices is not their responsibility and it is separated from their own routine everyday work at sea. Research shows that although this method of training on-the-job possess a great potential, but unfortunately, its learning outcomes at the best are unpredictable [16]. Research such as mine show that one of the main challenges with the on-the-job training is the lack of proper supervision on the training process and unwillingness of the crew members and shipping companies for full cooperation with the students' learning on board ships [16, 17]. This might be due to the fact that the training institutions have no jurisdiction over and limited access to the ships and thus inadequate control over the quality of training onboard. Furthermore, the dedicated onboard training period considers by many to be relatively short for students to being adequately trained. As a result, when compared to the traditional apprenticeship, the students have limited opportunities for exposure to the full range of activities and circumstances that an actual ship's officer needs to experience to be fully competent.

On the other hand, today's onboard ships' technology-rich work environment seems to impede many conventional authentic opportunities for apprenticeship activities [8]. This is mainly due to the invisibility of the work process of digitalized equipment. Due to the nature of technology which automate essential part of the job and limits the human involvement in the full process of work, the equipment's functions and the mental activities of the crew member who works with the electronic equipment are not observable [18]. The lack of visible trace of work processes precludes the learning opportunities through the observation of the operator at work. This profoundly changes the rich learning context which is available in the traditional apprenticeship. This resulted in the futility of the existing maritime education and training system in providing the future workforce with the entire required competencies. This calls for a new and innovative approach to the education and training system. The approach needs to provide an authentic learning environment in schools while it complements the training on-the-job. The method should provide the seafaring students with the opportunities to practice and develop their competencies in the context relevant to the onboard ships' environment.

In my continuing research in the field of the cognitive human element in the maritime industry and with the aim of improving maritime education and training system, I introduced the concept of Quasi-community as a theoretical framework for designing practitioners' learning in the formal settings [8]. The concept of quasi-community provides an instructional model based on the current understandings of how people learn for work. This concept takes benefit and utilises the principle of learning of the traditional apprenticeship to provide the opportunities for learning the cognitive skills in the school environment.

5 Quasi-community as a Novel Framework for Seafarers Training

In recent decades socio-cognitive theories such as communities of practice [10], provide insights into the elements of traditional apprenticeship and the processes of learning at work. The conceptual framework of quasi-community is inspired by these theories which provide an understanding of the learning processes in its natural settings.

The traditional apprenticeship system has historically proven to be an effective form of educating learners at the workplaces. By observing and working alongside experts, apprentices gradually learn the skills and competencies required for the job. At the same-time the students engage in the apprenticeship system while exploring the new domain would serve as resources for each other through discussing, helping and challenging one another. The process of apprenticeship has been well studied and researched by Lave and Wenger (1991) where they introduced the concept of 'communities of practice'. The communities of practice provide a valuable framework for analyzing the processes by which apprentices learn and gradually become competent and full participants in the workplace community that they enter. However, the concept developed through studying the learning processes in its natural place mainly by analysing the apprenticeship training in different work environments. Although, further studies showed the concept is applicable and appropriate for informal learning, its ability to provide an understanding of learning and knowing in the formal education at training institutions remains one of its main drawbacks [19, 20].

Additionally, there are differences between classroom learning and the workplace that hinder the development of a community of practice in a classroom. Classrooms do not provide the spatiotemporal relation which is present in a true community. They do not develop a collective memory, the same as workplaces' communities of practice, as the class members disbanded at the end of their course where the same classroom will rejoin with new members the next term. Another major difference is the fact that individual learning in the formal classroom has no effect on the collective, whereas in the true community the community memebrs' learning will change the practice during the course of time.

Inspired by the traditional apprenticeship, and utilizing the concept of communities of practice as the outset, the notion of the quasi-community has been presented as a way of replicating the critical elements of a true apprenticeship for the learner in the formal learning environment at schools [8]. The concept retains some of the dimensions of the original concept of communities of practice while discarding the inapplicable others. The quasi-community is developed to identify and ultimately improve learning in formal vocational education. The study's analysis shows many learning opportunities in the formal education and training systems that are already available or might be developed within this framework. The pedagogy based on the quasi-community allows the course participants to realize their common vocational objectives and provide an environment that makes it possible for the students to achieve those objectives.

The course curriculum, designed based on the quasi-community, is defined and co-produced by lecturers and the students based on their common objectives. To achieve that, the quasi-community praxis provides opportunities for the course participants to realize their common objectives through activities such as group work, collaboration, and knowledge sharing. This is essential for students to realize their' to realize their mutual interest that promotes the development of their quasi-community and set the pathway for the unfolding of the curriculum. This cohesive and motivated community promotes a goal-directed behavior by its participants.

One of the fundamental differences between the onboard community of practice and the classroom quasi-community is the hierarchy and distribution of expertise within the community. Quasi-community's mastery is dynamic and fluid within the community of the classroom. There the concept of expertise is relative and any member of the classroom can act as the teacher at an appropriate time depending on the proficiency needed for the matter at hand. The teacher is the one who can contribute to the problem solving for that occasion. This allows the dynamic functioning of all the learners in the classroom as they learn from or contribute to the learning of others when they function as students or teachers on different occasions throughout the course.

A quasi-community is an appropriate theorization of formal learning for the workplace, and if supported, can lead to effective learning and competency development that facilitates its transfer.

6 Conclusion

As discussed, the current practice in the training institutes failed to have a major effect on the development of skills and competencies of the maritime workforce. Additionally, today's technologically rich, complex, and ever-changing work environments deprived the traditional apprenticeships to be able to fully utilize all the aspects of authentic training that it used to offer.

This paper argues that vocational education and training would benefit from a pedagogical approach that would incorporate the conscious creation of a quasi-community of practice united by a common objective. A quasi-community can be considered as learning through participation in a shared enterprise [8]. The quasi-community would facilitate motivated, goal-directed, and behavior-oriented community towards their common objectives. The alignment of the pedagogical approach to the common objective would enable the quasi-community members to share expertise and engage productively, reflectively, and collaboratively. This paper argues that such an approach will improve the application of cognitive human factor by enabling the training to be effective and efficient in meeting the common objectives. Furthermore, it can expand the action possibilities of students thereby contributing to transferrable learning. The quasi-community has the potential to reform the cognitive human factor/element as it has the capacity to revolutionize the learning for work and professional education.

References

1. International Ergonomics Association, Definition and Domains of Ergonomics. http://www.iea.cc/whats/. Accessed 7 Jun 2018
2. Prasetyo, Y.T.: A Comparative Assessment of Human Factors and Ergonomics Courses for Industrial Engineering in the Philippines (2020)
3. International maritime Organisation (IMO) Human Element. https://www.imo.org/en/OurWork/HumanElement/Pages/Default.aspx
4. Richardsen, P.W.: Expensive Safety Hangover in the Shipping Industry. Det Norske Veritas, Oslo (2008)
5. Allianz, Safety and Shipping Review 2015. Munich: Allianz Global Corporate & Specialty (2015)
6. EMSA, European Maritime Safety Agency. Annual Overview of Marine Casualties and Incidents 2014. EMSA, Lisbon, Portugal (2014)
7. Brown, J.S., Collins, A., Duguid, P.: Situated cognition and the culture of learning. Educ. Res. **18**(1), 32–42 (1989)
8. Emad, G.R., Roth, W.-M.: Quasi-communities: rethinking learning in the adult and vocational formal education. Instr. Sci. **44**(6), 583–600 (2016)
9. Barab, S.A., Duffy, T.: From practice fields to communities of practice. In: Jonassen, D., Land, S.M. (eds.) Theoretical Foundations of Learning Environments, pp. 25–56. Lawrence Erlbaum Associates, Mahwah, NJ (2000)
10. Lave, J., Wenger, E.: Situated Learning: Legitimate Peripheral Participation. Cambridge University Press, New York (1991)
11. Collins, A., Brown, J.S., Newman, S.E.: Cognitive apprenticeship: teaching the crafts of reading, writing, and mathematics. In: Resnick, L.B. (ed.) Knowing, Learning and Instruction: Essays in Honor of Robert Glaser, pp. 453–494. Lawrence Erlbaum Associates, Hillsdale, NJ (1989)
12. Emad, G., Roth, W.-M.: Contradictions in the practices of training for and assessment of competency: a case study from the maritime domain. J. Educ. Training. **50**(3), 260–272 (2008)
13. Resnick, L.B.: Learning in school and out. Educ. Res. **16**, 13–20 (1987)
14. Lave, J.: Tailor-made experiments and evaluating the intellectual consequences of apprenticeship training. Q. Newsletter Inst. Comp. Hum. Dev. **1**, 1–3 (1977)
15. Becker, H.: School is a lousy place to learn anything. Am. Behav. Sci. **16**, 85–105 (1972)
16. Emad, G., Roth, W.-M.: Challenges of vocational education reform in maritime domain. J. Vocations Learn. **2**(1), 19–35 (2009)
17. Lewarn, B.: Seafarer training – Does the system defeat competence? In: Paper Presented at the 3rd General Assembly of the International Association of Maritime Universities, Maine (2002)
18. Asyali, E.: Impact of man-machine interface on maritime casualties. In: Proceedings of International Association of Maritime Universities (IAMU), Fourth General Assembly and Conference, Alexandria, pp. 89–90 (2003)
19. Boylan, M.: Ecologies of participation in school classrooms. Teach. Teach. Educ. **26**, 61–70 (2010)
20. Hagedorn, A., Springgay, S.: 'Making' our way through: DIY and crafting communities in Toronto. Craft Res. **4**(1), 11–30 (2013)

The Effect of Student Learning Style and Lesson Structure on Student Outcomes in an Online Learning Environment

Matthew R. Romoser(✉)

Department of Industrial Engineering and Engineering Management, Western New England University, 1215 Wilbraham Road, Springfield, MA 01119, USA
matthew.romoser@wne.edu

Abstract. Due to the COVID-19 pandemic, many instructors have been forced on short notice to prepare their lessons for online delivery. Many simply saved their lessons as slide-based HTML files and uploaded them to their course management systems. The focus of this research was to investigate how to improve the effectiveness on online learning modules in a distance learning environment. Student learning style (Passive vs. Active learners) was assessed, and students were provided on online lecture on statistical quality control. Results show active learners scored significantly higher on a post-test than passive learners. Additionally, students receiving the structured lesson rated their lessons significantly better on the attitude scale. Lesson structure and learning style significantly influence the students' ability to accurately develop a mental map that reflected how the material was related.

Keywords: Engineering education · Distance learning · Online learning · Training · Learning styles · Lesson structure

1 Introduction

The global COVID-19 pandemic has forced a paradigm shift upon those who work within the education sector. Educators have been forced to prepare courses that they have long taught in person for online delivery. As schools closed their doors to in person instruction and moved their courses online, teachers and faculty were directed to prepare to deliver their classes for online delivery. Many instructors do not possess the experience or technological skill set to do this effectively [1, 2]. As a result, many instructors merely posted their notes or PowerPoint lecture slides to their course management websites and called it a lecture. In fewer cases, instructors may have recorded a voice over to go with their slides and posted them as a video lecture. Others did their best, despite bandwidth and usability issues, to hold their lectures synchronously via web conferencing platforms such as Zoom or Skype. Students were similarly challenged as each of their instructors opted for the teaching methods that best suited their individual teaching styles. As such, students were confronted with a wide assortment of presentation modes, schedules, and

S. Nazir et al. (Eds.): AHFE 2021, LNNS 269, pp. 190–198, 2021.
https://doi.org/10.1007/978-3-030-80000-0_23

software and hardware technological requirements that they may or may not have had ownership of or skills to use at home.

When instructors – be they in K-12, higher education, or industry – design a course for delivery online, certain students seem to struggle with using the internet for learning [3, 4]. The question becomes, how can we best create online lessons to best serve students in an online environment when no instructor is readily available? Careful consideration must be given to the design of online lesson content that takes in to account the differences between students. The first issue is lesson structure. Online lectures can take the form of simple, slide-based documents with still text and pictures, or they can be highly dynamic with multimedia interactions, electronic conferencing, media, email and online submission of homework and assessments. The former is easiest to implement but does not provide much of an advantage over the traditional textbook. The latter can provide a highly active and engaging learning environment but requires considerable resources in terms of time and technological know-how to implement [5].

Depending upon the level of experience and learning style of the student, how should online lessons and modules be structured to best facilitate learning? Learning style varies from individual to individual. Students who possess a more passive learning style [6] may by nature struggle with online-based education. On the other hand, individuals who are more active learners may require less structure in an online lesson and be better equipped to seek out information on their own within the content to fill the gaps in their knowledge. One would expect it would be easier for active learners to develop their own mental structure and cognitive maps of the material.

The second issue involves learning strategy and delivery medium. Once a structure has been chosen, should the lessons contain more visual or verbal components? Research has show that students with high visualization ability often out-perform student with low visualization ability [7]. Varying the course delivery medium (combining visual, text, handouts, study-guides) has been shown to increase student performance on a task as student have more options on how to immerse themselves in the material [8].

One of the observations that has come out of the research has been that some students seem to do very well with online learning while others struggle. To have some students be more successful than others is to be expected. However, when students whose learning style favors one-on-one engagement with an instructor perform poorly when engaged online, then we need to do what we can as education designers to engineer the delivery of content so that it can be interacted with in a way that best suits the student. Online education requires flexibility, not rigidity. Though there is a lot of overlap between learning online and learning in a traditional classroom, the two arenas differ in some fundamental ways. For instance:

- Students who are typically more passive learners are forced to become active learners when learning online with contend that requires the student to dig in to the material rather than be led [4].
- Without formal structured lectures, internet-based courses force students to become self-paced learners [4].
- Basic computer skills become pre-requisite for the course, whereas before students may have managed without using a computer to do homework [4].

- Questions are more difficult to get answered. Courses are forced to be more asynchronous and the student has a virtual barrier between themselves and the instructor. Therefore, students who learn better in social situations are at a disadvantage [4].
- Online learning by its nature makes interactions between students more difficult. It becomes harder to establish study-groups and discuss questions in real time [4].

It is reasonable to presume that some students who may be successful in a traditional classroom setting may suddenly find themselves at a significant disadvantage and struggle in a class that is administered solely over the internet. Though that could also work in reverse. Some students who work better alone may do better in an online setting. However, the former case is likely the larger issue because a student who works best alone will probably elect to work alone in either setting. So, the question becomes, how can we best serve students in an online environment when no instructor is readily available? For the above stated reasons, many students struggle with online learning. The current paper attempts to address this question. In the study reported here, the effect of student learning style and lesson structure on student performance and ability to develop a mental map of the material is considered.

2 Online Learning Study

This online learning study was conducted to investigate the effect of online lesson structure and student learning style on transfer of learning and ability to construct an accurate mental map of the material's topical relationships. Subject novices do better with learning systems that are designed to reduce search and extraneous load [9]. Novices also become overwhelmed by non-linear learning systems and do better with structured systems [10].

2.1 Participants

Participants were engineering students between the ages of 18 and 20 years of age recruited from sections of a probability and statistics for engineering course. They were informed that the project would require a commitment of a single one-hour session, and they would be compensated at the end. A total of thirty participants completed the study. Of the thirty participants, there were twenty-two male and eight female students represented. Gender in engineering majors is typically heavily skewed male.

2.2 Online Lessons (Quality Control Tools)

Two different versions of an online lesson were created. These versions were the "structured" and "unstructured" versions. Both lessons contained identical content with the only difference being how the material was presented and the student's navigation options through the material. The subject of both lessons was a basic introduction to quality control tools. None of the participants would have previously had this material in class, but their current statistics course ensured they'd have the prerequisite knowledge. Lessons contained both text-based and multimedia interactive components. Interactive components in the lessons were the "show me" and "explore this" lessons that provided dynamic interaction with the material and animated demonstrations of concepts.

Participants were provided instructions stating that they had forty minutes to interact with the lesson which was designed to be completed within twenty-five to thirty minutes. Students were asked to learn the material and were aware they would be tested on their knowledge afterward.

Structured Online Lesson. The structured version of the lesson was linear in nature. It limited the navigational freedom of the participant and took them linearly through the material. The design of the structured lesson was similar to a textbook in that it was broken up in to three sequential sections with each section containing a few subsections. Experimentally, the purpose of the structured version was to limit the participant's navigational freedom and force them to follow the lesson in a sequential fashion in which the information on the current slide builds upon that contained in the previous slide. Participants were provided with linear navigation tools (back and next buttons), an index which allowed them to jump anywhere forward or backward, and section numbers. However, students were not allowed to jump ahead in the material until they'd at least visited the sections that came prior.

Unstructured Online Lesson. The unstructured version of the lesson was a node-based or hypermedia design where it was up to the participant which link to follow. It allowed considerably more navigational freedom by providing less topic to topic guidance compared to the structured version. Participants navigated by clicking on "link" icons which were labeled with the subject of the node they were linked to. The nodes contained subject headings but did not contain section numbers since the could potentially confuse the participant if viewed out of order. The icons changed color with a green checkmark included to indicate nodes that had previously been visited by the student. The unstructured version wasn't a complete free-for-all through navigation-wise. Not every node was linked to every other node. At the bottom of each page, participants were provided a list of "related nodes" to provide some guidance as to what material might be relevant to the current page. As a whole, the unstructured lesson contained exactly the same amount of material, pages and interactive content as the structured version. Experimentally, the unstructured version was intended to force participants to prioritize and decide which material should be viewed next and establish their own framework for the lesson's content.

2.3 Measures

Passive/Active Learning Scale (PALS). In order to study the effects of student learning characteristics on performance in online learning, it is necessary to find a means to measure student learning styles. One measure is the Passive/Active Learning Scale (PALS) developed by Yung-Bin Lee [6]. PALS is a 32-item assessment tool developed to discriminate between students who are "active learners" and "passive learners". Active learners tend to take initiative, explore the material to fill gaps in their knowledge, and routinely go beyond teaching objectives. Passive learners tend to follow their instructor's lead, require guidance with material while learning it and deliver just what is expected and rarely more. PALS was used for this study because it was proven to be a reliable predictor of student performance within various learning scenarios [6].

Mental Map Exercise. A mental map exercise what developed to provide a snapshot of how subject organized information related to the lesson. In order to arrive at this, participants were asked to draw, using whatever method they desired, how they internalized the information contained in the hypermedia lessons. Each participant was provided a list with the topics and sub-topics shuffled in a different random order to prevent ordering bias in the results. Participants were told that some information on the list could be considered "top level" or "broad category" while other information may be considered "lower level" or "sub-category." They were also told to create as many levels or sub-categories as they believed necessary, that they should feel free to expand upon the list, and that they did not have to use all the information provided to them. Finally, they were instructed they could represent the material in any way they wished, verbally, visually or both.

Results from this measure provided a convenient side-by-side comparison of how students of different learning styles and who receive different lesson structures internalized the information contained in the lesson. In analyzing the results, patterns and trends in how participants reported their personal "mental maps" were sought. Mental maps were categorized so that a Pearson Chi-Squared analysis could be run to determine if there were significant differences in the relative distribution between lesson structure or learning styles.

Post-Tests. A post-test was developed to assess the participant's transfer of learning with the control-charting lesson provided during the experiment. It was administered after the participant completed their mental-map sketch. It consisted of twenty-three multiple choice questions and two "higher level understanding" workout questions. Scoring was completed by an unbiased double-blinded scorer using a rubric provided by the experimenter.

2.4 Experiment One Procedure

The thirty participants were recruited from a probability and statistics course for engineering majors and were scheduled one at a time. Sessions lasted approximately one hour. Participants arrived and were provided informed consent and a brief introduction to the study. After informed consent was received, participants were administered the PALS learning style scale.

At this point participants were randomly assigned to receive either the structured or unstructured version with gender balanced between groups. They were informed that they would be interacting with an online lesson that provided a basic introduction to quality control tools for engineers and would have forty minutes to interact with the lesson and study the material after which they would be assessed on their knowledge. After the lesson was completed, participants were given the post-test assessment and then asked to sketch their mental map of the material.

3 Results

A MANOVA analysis was run using the participant's post-test score as the dependent variable. Because the distribution of scores was not normal, an inverse hyperbolic tangent transformation function was applied to the data to normalize it. The transformation equation used was as follows.

$$\text{Post} - \text{Test(Transformed)} = \text{Tanh}^{-1}(\text{Post} - \text{Test Score}/23) \quad (1)$$

A higher score on the transformed variable corresponded to a higher score on the post-test. The higher the score on the PALS, the more active the learner. Lesson structure was the fixed factor and PALS score treated as a covariate with Structure. Results show that there was a significant main effect of the participant's PALS score on post-test performance ($F(1{,}26) = 20.56$, $p << 0.001$). The more active the learner, the higher the post-test scores tended to be. Structure received did not have a significant main effect on post-test score. Figure 1 contains a scatter plot of the participants' transformed post-test score versus their PALS score, along with a linear regression equation line.

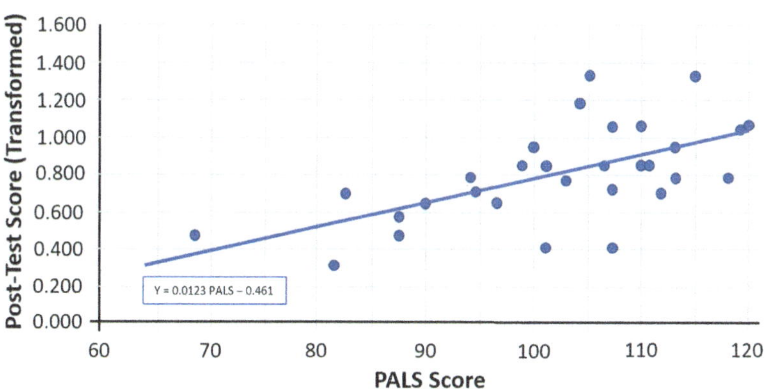

Fig. 1. PALS score vs. Post-Test score (Transformed)

With respect to the mental map task, participants reported their conceptualization of the material in a wide variety of ways. After categorizing the sketches, some interesting trends emerged. Mental maps were turned over to a double-blinded scorer for categorization in to one of five categories (Table 1).

For comparison purposes, participants were sorted in to three groups: passive (8), neutral (12) and active (10) learners, based on their PALS scores (Table 2). A Pearson's chi-squared test was run on the data in the table with the statistical question to be answered was whether the groups differ in relative distribution from each other. The results indicated that there was indeed a significant difference in relative distribution ($\chi^2_{obs} = 12.08$, df = 5, p < 0.025). The differences could primarily be seen in the "hierarchy" and "nodes" categories. There were more active and neutral learning style participants that reported those types of mental maps than passive learners.

Table 1. Mental Map categorization rubric

Map group	Criteria
Lineal/Flowchart	Presented in an order similar to how the material flows from topic to topic. Outline or flow chart form
Hierarchy	General "areas of knowledge" with categories filtering down to more specific information
Nodes	Series of interconnected informational nodes. Nonlinear
Pictures	Pictures and/or charts presented in lesson. Participant relying on pictures to describe their mental map
Loose	Information incomplete or was organized in a haphazard, difficult to interpret manner

Table 2. Mental Map frequency distribution by PALS.

Mental map category	PALS category			
	Passive	Neutral	Active	Totals
Linear/Flowchart	6	4	4	14
Hierarchy	0	0	3	3
Nodes	1	4	2	7
Pictures	1	2	1	4
Loose	0	2	0	2
Totals	8	12	10	30

Table 3. Mental Map frequency distribution by structure.

Mental map category	Lesson structure received		
	Structured	Unstructured	Totals
Linear/Flowchart	10	4	14
Hierarchy	3	0	3
Nodes	0	7	7
Pictures	0	4	4
Loose	2	0	2
Totals	15	15	30

In another analysis, the participants' mental maps were again divided in to groups, but this time according to the structure the participant received (Table 3). Again, a Pearson's chi-squared analysis revealed that the relative distribution of mental maps across lesson structure was significantly different ($\chi^2_{obs} = 18.57$, df = 5, $p < 0.005$). Those participants who received the structured lesson primarily internalized the information linearly or with some type of hierarchy. Those who received the unstructured version primarily either constructed a node-based mental map, drew pictures of images they saw or, in four of the cases, managed to structure the information in a meaningful linear fashion. The structured version also appeared to help participants recall more topic areas.

4 Discussion

From the results of this research, there does appear to be additional value for choosing a more structured version of the lesson containing mixed media interactions. Even though the structure of the lesson did not have a significant main effect on post-test scores, there is support for having a linear, textbook style structure and appropriate placement of multimedia interactions within the lesson. PALS score proved to be a significant predictor of performance, regardless of lesson structure. This suggests that active learners struggle less with online education when compared to passive learners. Limiting navigational freedom to guide passive learners through the material did not seem to be as effective as anticipated with respect to post-test score. So, other methods of making online instruction more effective for passive learners need to be considered. Regular virtual check-ins with instructors and synchronous feedback might be one method which could be investigated in future research efforts.

Structure did prove to significantly affect how participants develop their cognitive maps of the material. The structured version led to more participants developing highly linear or hierarchical mental maps of the material that more accurately reflected the progression of concept relationships in the lesson. The unstructured version resulted in less cohesive or incomplete mental maps of the material. Passive students did seem to benefit from structured lessons in developing their mental maps. Overall, it's clear that active learners are able to do well in online learning regardless of the structure of the lesson. However, providing structure did lead to more structured conceptual mental maps with more passive learners.

References

1. Basilaia, G, Kvavadze, D.: Transition to online education in schools during a SARS-CoV-2 Coronavirus (COVID-19) pandemic in Georgia. Pedagogical Res. **5**(4), em0060 (2020). https://doi.org/10.29333/pr/7937
2. Moralista, R.B., Oducado, R.M.F.: Faculty perception toward online education in a state college in the Philippines during the Coronavirus disease 19 (COVID-19) pandemic. Univ. J. Educ. Res. **8**(10), 4736–4742 (2020)
3. Eberts, R.E., Chenge, K.-C.: Learning, or not learning, through multimedia. In: Proceedings of the Human Factors and Ergonomics Society 42nd Annual Meeting, pp. 606–610 (1998)
4. Ward, M., Newlands, D.: Use of the web in undergraduate teaching. Comput. Educ. **31**, 171–184 (1998)
5. Eberts, R.E.: Computer-based instruction. In: Helander, M.G., Landauer, T.K., Prabhu, P. (eds.) Handbook of Human-Computer Interaction, 2nd edn. Elsevier, Amsterdam (1997)

6. Lee, Y.-B.B.: Effects of Learning Style and Instructional Cues on Achievement and Learning Interactivity in a Hypermedia Instructional System. Doctoral Dissertation, Purdue University, University Microfilms No. 90-08652 (1989)
7. Sein, M.K., Olfman, L., Bostrom, R.P., Davis, S.A.: Visualization ability as a predictor of user learning success. Int. J. Man Mach. Stud. **39**, 599–620 (1993)
8. Wragg, E.C.: Training skillful teachers. Teach. Teach. Educ. **1**(3), 199–208 (1985)
9. Kalyuga, K., Chandler, P., Sweller, J.: Levels of expertise and instructional design. Hum. Factors **40**(1), 1–17 (1998)
10. McDonald, S., Stevenson, R.: Effects of text structure and prior knowledge of the learner on navigation in hypertext. Hum. Factors **40**(1), 18–27 (1998)

Analysis of Generic Competences Needed Onboard a Ship

Ana Gundić(✉), Dino Županović, Dalibor Ivanišević, and Luka Grbić

Maritime Department, University of Zadar, Zadar, Croatia
{agundic,dino.zupanovic,divanisevic,lugrbic}@unizd.hr

Abstract. STCW Convention, which has been defining maritime education and training, introduced a concept of competence-based approach (into education) in 1995. The analysis of STCW Convention and related IMO Model Course 7.01 has shown the prevalence of professional competences in them whereas generic competences are only marginally present in both of them. STCW Convention and related IMO Model Courses are both missing a detailed classification of generic competences as well as an insight into generic competences needed on board ships. Real world situation review of duties aboard LNG and cruise ships has shown requirements for their inclusion within STCW Convention and related IMO Model Courses. Hence, a more explicit definition of generic competences required of the master and the first deck officer on board is the main purpose of this paper. Required generic competences have been suggested upon a thorough analysis of the master and the first deck officer's scope of duties on board LNG ships as well as cruise ships, and are classified into three categories: instrumental, individual and systemic. Additionally, each competence has been assigned with its accompanying occurrence on board a ship.

Keywords: Generic competences · STCW convention · Master · First deck officer · LNG vessels · Cruise ships

1 Introduction

IMO's convention called International Convention on Standards of Training, Certification and Watchkeeping for Seafarers (STCW Convention), adopted in 1978 and entered into force in 1984, has been regulating seafarers' education on the international level. The Convention consists of two parts – Code A, which is mandatory and consists of the minimum of regulations needed to successfully fulfil the Convention's provisions, and Code B, which consists of the recommendations and guidelines, needed to apply the provisions of the Code A [1]. The Convention itself was significantly amended several times. However, the amendments important for this paper were adopted in 1995 when a concept of competence-based education was introduced into the STCW Convention [2, 3]. Competences as defined by the STCW Convention are a combination of knowledge, proficiency and skills. Competences needed for on board duties are listed in the Code A. They were determined according to the departments included in the on board activities, namely deck department, engine department and radio department.

S. Nazir et al. (Eds.): AHFE 2021, LNNS 269, pp. 199–207, 2021.
https://doi.org/10.1007/978-3-030-80000-0_24

Seafarers' education is carried out at MET (maritime education and training) institutions that can use IMO Model Courses containing detailed teaching syllabus when planning an educational program. The Courses themselves are some sort of an educational program. STCW Convention and IMO Model Courses both aim at acquiring professional competences, whereas generic competences are only partially present in both of them. Generic competences included in the SCTW Convention and IMO Model Courses are teamwork, team management and decision-making. Recently, generic competences have been referred to as crucial when performing a certain task [4, 5]. The employers on the international market are looking not only for the employees with highly developed professional competences, but also for the employees who have generic competences as well. They are looking for the employees who can work in a team, solve problems, be flexible, take the initiative, multitask and analyze more information simultaneously [6]. Generic competences, also called core competences or key competences, are competences needed in different sectors and occupations. The analyzed sources reveal different definitions of generic competences that depend on the scientific field they have been defined in. That means that every profession needs different types of generic competences [7].

Competences analyzed in this paper refer to the master and the first deck officer's duties. To identify generic competences required of the master and first deck officer, the authors suggest the classification of generic competences used in the project called Tuning Educational Structures in Europe (Tuning project). The project differs three types of generic competences [8]:

1) Instrumental
2) Individual
3) Systemic

Instrumental competences are divided into the cognitive ones referring to the ability to understand an idea or a conception; methodological competences referring to the organizational capability; technical competences referring to the ability to use devices; and linguistic competences referring to the ability to communicate. Individual competences make social interaction and cooperation easier. Systemic competences refer to the systemic approach, that is, to abilities concerning whole systems, which should help to plan changes and improvements. They are extremely dependent upon instrumental and individual competences and cannot develop if instrumental and individual competences have not been acquired.

2 Generic Competences and Their Classification

In this chapter, it will be suggested which generic competences are required of the master and the first deck officer. The competences in question were suggested upon the analysis of on board jobs. For every competence, a situation in which its usage is necessary was determined. The classification itself was based on the Tuning project and it does not cover all the competences listed in the project report. It was modified to suit maritime profession. Situations, in which these competences are used, were determined on the basis of the interviews with masters and the first deck officers of different types of ships.

2.1 Instrumental Competences of a Master and First Deck Officer

Instrumental competences, the masters and the first deck officers should acquire for the successful completion of a task as shown in Table 1 are:

1) Effective communication
2) Time management
3) Problem identification and troubleshooting
4) Use of technology
5) Abstract thinking
6) Decision making

Table 1. Instrumental competences of a master and first deck officers.

Instrumental competencies	Situation in which its usage is necessary
Effective communication	Mooring a ship; Communication between crewmembers and with external stakeholders related to ship's voyage and cargo; Presenting documentation and processes on board to various inspections; Sending a financial report to the company; Compiling a list of ship's documentations before arriving to the port; Sending a report to the company on emergency situations; Submitting deck, engine, and cargo logbooks, minutes of meetings, health and other reports and documents to the company; Submitting documentation to the ship agent required to obtain all permits from port authorities; Monthly meetings
Time management	Planning and approving schedule of watchkeeping at sea and in port for officers and other crewmembers; Scheduling crew rotations; Planning all activities during their stay in the port; Developing a planned maintenance schedule and keeping records of all activities
Problem identification and troubleshooting	Navigating through adverse weather conditions; Equipment malfunctions on the bridge and on cargo systems; Cases of an unexpected delay of the ship; Assessing the risk of all on board operations
Use of technology	Navigation; Receiving meteorological information; Correcting and updating charts and publications; Voyage planning; Communicating with internal and external stakeholders; Planning cargo loading and unloading operations; Keeping ship documentation
Abstract thinking	Avoiding collisions; Planning cargo loading and unloading operations; Maneuvering a ship
Decision making	Making decisions regarding job allocation and use of human and material resources; Altering voyage plan; Dealing with consequences of the crew's offences; Scheduling crew rotations; Anchoring; System incompatibilities and/or system deficiencies

2.2 Individual Competences of a Master and First Deck Officer

Individual competences the masters and first deck officers should acquire for the successful completion of a task are:

1) Self-critical thinking
2) Ability to work in multidisciplinary and multicultural teams
3) Teamwork
4) Team Management

Table 2. Individual competences of a master and first deck officer.

Individual competencies	Situation in which its usage is necessary
Self-critical thinking	Conducting exercises as well as during pre- and post-exercise meetings; Navigation; Evaluating crew members; Planning and performing all on board activities that require risk assessment; Conducting incident, accidents and non-conformities investigations; Reporting illicit behavior
Work in a multidisciplinary and multicultural environment	Coexisting for several months in conditions of restricted movement with crewmembers of different nationalities and cultures; Navigating a ship between countries of different cultures in which communication with external stakeholders, i.e. agents, port employees, government officials, classification societies, charterers, customs and health authorities and other recognized official bodies, is essential
Teamwork	Anchoring operations; Navigating in a traffic separation schemes; Lowering a lifeboat; Boarding a pilot; Inspections and audits; Handling the cargo; Berthing and unberthing; Attending the meetings
Team Management	Assigning tasks from a master to a chief officer; Training other officers; Writing promotion recommendations; Evaluating a team; Reporting to a company which team members achieve good results together; Participating in various on-board meetings; Familiarizing crew members with company policies and regulations; Ensuring that every crew member has rights that are legally assigned to him/her

2.3 Systemic Competences of a Master and First Deck Officer

Systemic competences the masters and first deck officers should acquire for the successful completion of a task are:

1) Creative thinking

2) Fast adaptation
3) Training and evaluation

Table 3. Systemic competences of a master and first deck officer.

Systemic competences	Situation in which its usage is necessary
Creative thinking	Compiling a measure to improve a ship's safety management system; Navigation; Participating actively in ship budget planning; Maintenance planning; Ship maneuvering
Fast adaptation	In emergency situations on the ship and at the terminal; Avoiding collisions; On-board activities with less crewmembers than usual; Situations where non-conformities and/or findings have been identified during inspection
Training and evaluation	Conducting a training program to help the crew improve their own competences or increase their chances for promotion; Evaluating crew members and sending reports to the company

It is important to emphasize that there is a possibility of overlapping among these competences. In the process of acquiring required generic competences of the master and the first deck officer, acquiring one or more other competences could be necessary and important.

3 Tasks and Generic Competences on LNG Ships and Cruise Ship

In order to determine which generic competences are required of the master and the first deck officers on LNG and cruise ships, a documentation of the Ship Security Management System at LNG and the cruise companies was analyzed. Generic competences needed by the employers are not clearly listed in the analyzed documentation. However, a task analysis identifies generic competences required to do the job. Tables 2 and 3 outline only a part of the duties, required of the master and the first deck officer, for which generic competences are needed on LNG ships, and a part of the duties of the master, staff captain and the first deck officer on cruise ships. The appropriate generic competences refer to the tasks they are needed for (Table 4 and 5).

On the basis of the analyzed documentation, it was found out that generic competences identified as the most important ones are: working in a multidisciplinary and multicultural environment, time management, training and evaluation, effective communication, fast adjustment, decision making, identification and problem solving.

Table 4. Tasks and generic competences on LNG ships.

Master	Task	Generic competence
	Maintenance of correspondence, records and reports sent from the vessel or received by the vessel	Effective communication Use of technology[a]
	Timely delivery of deck log, engine log, cargo log, meeting reports, health reports, safety reports and other documents in the company	Effective communication Use of technology Workload and Time Management
	Providing the agents with the documents required for port authorities	Effective communication Use of technology
	Providing familiarization for the Chief Mate in understanding the Master's duties	Teaching and evaluation
	Conducting monthly meetings	Effective communication Workload and Time Management Teamwork Team Management
	Evaluating crewmembers and delivery of evaluation documents to the company	Effective communication Teaching and evaluation
	Cooperation with all government departments, classification societies, etc.	Effective communication Work in multidisciplinary and multicultural teams
First deck officer	*Task*	*Generic competence*
	Planning, implementation and monitoring of maintenance work according to maintenance schedule	Workload and Time Management Decision making
	Making a maintenance schedule and monitoring of maintenance work	Workload and Time Management Effective communication
	Supervision of the crewmembers assigned for the cargo operations	Team Management Decision making
	Ensuring professional behavior and appearance of the crewmembers assigned for the cargo watch	Team Management Teamwork Work in multidisciplinary and multicultural teams

(*continued*)

Table 4. (*continued*)

	Investigation of the cause of injuries and preparation of injury and illness reports	Team Management Teamwork Effective communication Decision making Problem Solving
	Training of other crewmembers, including CBT	Teaching and evaluation Use of technology Decision making
	Maintaining training records for new crewmembers familiarization	Effective communication
	Planning the crewmember training	Workload and Time Management

[a]In this case, the use of technology refers to the use of computers, general and specified computer programs, and web services.

Table 5. Tasks and generic competences on cruise ships.

Master	*Task*	*Generic competence*
	Ensuring proper training and proficiency of crewmembers in the operation of lifesaving, firefighting, security, navigation and environmental systems	Teaching and evaluation Teamwork
	Promoting the two-way communication between ship and the company and other stakeholders	Teamwork Team Management Effective communication Work in multidisciplinary and multicultural teams
	Carrying out ship inspections on a regular basis	Workload and Time Management Decision making Problem Solving
	Training the Staff Captain	Teaching and evaluation
	Evaluating the crewmembers	Teaching and evaluation Decision making
	Monitoring the development of junior officers	Teaching and evaluation

(*continued*)

Table 5. (*continued*)

	Participating in various shipboard meetings and maintaining personal contacts with other crewmembers	Work in multidisciplinary and multicultural teams Teamwork Effective communication
	Socializing with guests during the voyage	Work in multidisciplinary and multicultural teams Effective communication Fast adaptation
Staff captain	*Task*	*Generic competence*
	Cooperation with the officer in charge of the environmental issues, with the ship securing officer, with the firefighters and the officer in charge of the security when planning monthly activities that regard safety	Teamwork Work in multidisciplinary and multicultural teams
	Training the first officer	Teaching and evaluation
	Supervising the deck department, developing and maintaining close working relations with the departments through the first officers, non-commissioned officers and carpenter	Teamwork Team Management Work in multidisciplinary and multicultural teams
First deck officer	*Task*	*Generic competence*
	Coordinating loading and ensuring the things on board	Workload and Time Management Teamwork
	Cooperation with hotel (accommodation) department regarding cleaning and maintenance	Teamwork Work in multidisciplinary and multicultural teams

4 Conclusion

Generic competences required of the masters and the first deck officers have not been determined in detail so far. Some generic competences have only been marginally present in the STCW Convention and IMO Model Courses. These partially identified generic competences are teamwork, team management and decision-making. Other competences, discussed in this paper, are only marginally present in STCW Convention and IMO Model Courses or have not been mentioned at all.

The analysis of jobs on LNG and cruise ships has shown that generic competences should be more present in both, STCW Convention and IMO Courses. This conclusion refers primarily to acquisition of the following generic competences: usage of technology, teaching and evaluation, effective communication, work in multidisciplinary and multicultural teams, problem solving, workload and time management.

Generic competences like teaching and evaluation are mentioned in STCW Convention and IMO Model Courses only as a part of professional competences. Topics needed to acquire specific generic competences have not been determined at all. Therefore, it is suggested that generic competences required of the masters and the first deck officers should be identified and elaborated in the STCW Convention and IMO Model Courses. In that way, their acquisition would be easier and simpler to recognize.

References

1. STCW Convention, IMO, London (2010)
2. Zec, D., Maglić, L., Šimić-Hlača, M., Gundić, A.: Current Skills Needs: Realitiy and mapping, Skillsea (2020)
3. Gundić, A., Vujčić, S., Maglić, L.I., Ivanišević, D.: Methods for demonstrating a competence and criteria for evaluating a competence in STCW convention. Pomorstvo **34**(2), 245–251 (2020)
4. Michael, E.M., et al.: Global maritime professional body of knowledge. IAMU, Tokyo (2019)
5. Gundić, A., Maglić, L., Šimić Hlača, M., Maglić, L.: Process approach for determining competencies. In: Proceedings of the International Association of Maritime Universities Conference, p. 329 (2019)
6. National Centre for Vocational Education Research, Leabrook: Defining Generic Skills at a Glance. National Centre for Vocational Education Research, Australia (2003)
7. Cheetham, G., Chivers, G.: Towards a holistic model of professional competence. J. Eur. Ind. Train. **20**(5), 20–30 (1996)
8. Gonzáles, J., Wagenaar, R.: Tuning Educational Structures in Europe II: Universities' Contribution to the Bologna Process. Universidad de Deusto, Bilbao (2005)

Admission Points Score Predicting Undergraduate Performance – Comparing Quantity Surveying vs. Construction Management

Danie Hoffman(✉), Inge Pieterse, and Derick Booyens

University of Pretoria, Lynnwood Road, Pretoria, South Africa
danie.hoffman@up.ac.za

Abstract. The Department of Construction Economics, University of Pretoria, offers three-year BSc programmes in Quantity Surveying (QS) and Construction Management (CM). Both programmes have a main entry threshold of an Admission Points Score (APS) of 30, based on school performance. The need to improve first-year students' marks and the throughput rate of enrolled students brings the validity of the current threshold requirements under scrutiny. A previous study found that QS students from 2010 to 2015 with higher APS and better marks in Mathematics and Natural Science consistently outperformed students with lower marks in aspects such as average first-year marks, likelihood to qualify in three tears and to qualify with distinction. This new study compared the performance of CM students against their APS as the independent variable. The study found that for CM students, similarly to QS students, a higher APS correlated with better first-year performance. The correlation of performance was more varied, and the regression lines' gradient tends to be flatter than for QS students. The study's findings are important for the department to make informed decisions about amending future entry threshold levels. The findings also add to the existing generic body of knowledge on comparing school level performance with undergraduate performance.

Keywords: APS · Construction management · School performance · South Africa · Undergraduate performance

1 Introduction

Students starting their undergraduate studies often experience significant challenges to adapt successfully to the demands of tertiary studies. The combination of independent study discipline, the rapid pace of work progression, and the quantum and complexity of academic content often exceed students' skills and ability. Efforts from tertiary institutions to address the problem often include careful attention to the criteria used to select students, providing bridging courses, and additional support for challenging modules [1]. The South African matric pass rate keeps improving but without increasing throughput

S. Nazir et al. (Eds.): AHFE 2021, LNNS 269, pp. 208–214, 2021.
https://doi.org/10.1007/978-3-030-80000-0_25

rates at the first-year university level. Poor student performance in mathematics, physics and languages remains a big concern [2].

Construction Management (CM) students are often misconstrued to be more practical orientated with less focus on academic and theoretical aspects than their professional counterparts. CM students entering the programme face many quantitatively orientated modules as part of the curriculum to prepare them for what will be required in the industry one day. According to Harris *et al.* [3], construction management focusses on and *"addresses the effective planning, organization, application, coordination, monitoring, control and reporting of the core business processes of marketing, procurement, production, administration, accounts and finance necessary to achieve economic success and/or profitability for an enterprise or organization engaged in the provision of construction facilities."* Therefore, CM students need to be trained not only in the physical execution of projects but also in the numerous financial aspects a contractor needs to manage daily as per the definition mentioned above. Therefore, a high level of academic numeracy would be a definite requirement when applying towards entry into the program.

2 Literature

In 2007 Scott, Yeld and Hendry [4] argued that South African school curricula did not provide a solid foundation for higher education curricula and tertiary studies. Prince and Frith [5] agreed that the higher education sector does not regard prospective students' academic numeracy as proficient. In a 2020 study by Prince and Frith [6] on the academic numeracy and mathematical competence of 7467 South African university students, only 13% were classified as proficient measured against higher education expectations. A major contributor to this problem is that school mathematics often focuses on algorithms, standard forms and procedural knowledge rather than insight [7].

Schöer *et al.* [8] evaluated the South African NSC Mathematics results as a predictor of undergraduate academic performance. They found NSC Mathematics results to be 12–13% higher than the higher-grade Mathematics marks of the previous Senior Certificate system for students with similar undergraduate performance.

Similar studies to this research were done in the department on BScQS entrants from 2010 to 2015. Students with an APS of 32 and higher consistently outperformed students with lower marks in aspects such as average first-year marks, likelihood to qualify in three tears and to qualify with distinction [9]. Students with better NSC marks in Mathematics and Natural Science also achieved higher marks and better first-time pass rates in challenging quantitative undergraduate modules [10].

CM students not equipped with the underlying foundation of academic numeracy will often fail in tertiary programmes that teach advanced mathematical principles and apply them to aspects such as time-cost optimization of construction programs. Therefore, an argument is proferred for a higher APS score, especially in mathematical subjects of CM students being admitted to the programme, to ensure a higher probability that students will interpret these numeric concepts on the levels required.

3 The Methodology

The study considered the weighted average undergraduate performance of BScCM students registered from 2010 to 2015 as dependent variables against their APS as the independent variable. Aspects such as the weighted first-year marks, the throughput rate of enrolled students, and the percentage of enrolled students who qualified were considered. The data were treated and described using the following structured approach:

- Group 1 (students with an APS of <30);
- Group 2 (students with an APS of 30 or 31);
- Group 3 (students with an APS of 32 or 33);
- Group 4 (students with an APS of 34 or 35);
- Group 5 (students with an APS of 36 or 37);
- Group 6 (students with an APS of 38 or 39); and
- Group 7 (students with an APS of ≥40).

4 The Data and Findings

The cohort size of the 233 students first time entering BScCM students in the study varied between a minimum cohort of 35 in 2012 and 2015 and a maximum of 46 students in 2011.

4.1 Overall Relationships Between APS and Average First-Year Marks

The combined data set revealed a relatively weak correlation between the APS as the independent variable and the weighted average first-year marks as the dependent variable. The Pearson's correlation of 0,32019 and the resulting coefficients of determination of 0,10252 mean that only 10,25% of the dependent variables' variability can be ascribed to the variability in the independent variables. The size of the stratified groups and the average APS per group are detailed in Table 1.

Table 1. The number of first-time entering students per APS stratified grouping.

Stratified APS groups	<30	30 & 31	32 & 33	34 & 35	36 & 37	38 & 39	≥40
Number of students enrolled	44	51	46	47	32	4	9
Average APS	28.2	30.7	32.4	34.3	36.4	38.3	40.4

4.2 Weighted Average Marks in First-Year

The weighted average first-year mark of students was evaluated as the dependent variable against the average APS data as the independent variable within the different stratified groups. A simple linear regression line calculation is detailed in Table 2. A positive relationship exists between the independent and dependent variables (see Fig. 1). The slope of the simple linear regression line of 1,89 indicated a significant increase in average first-year marks for a corresponding rise in APS.

Table 2. Weighted average marks in first-year of study

Stratified APS groups	<30	30 & 31	32 & 33	34 & 35	36 & 37	38 & 39	≥40	Slope	Y-intercept	Error ϵ
Average APS	28.2	30.7	32.4	34.3	36.4	38.3	40.4			
Weighted average marks (%)	42.46	43.05	45.97	48.43	51.85	58.32	65.50	1.89	−14.04	2.665

The high Pearson's correlation coefficient of 0,9581 and the coefficient of determination of 0,9179 confirm the strong correlation. However, the fact of using stratified averaged data as an independent variable must be considered.

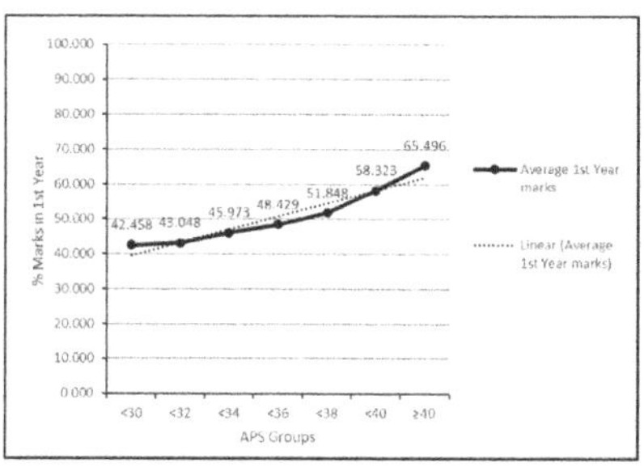

Fig. 1. Weighted average marks in the first year of study.

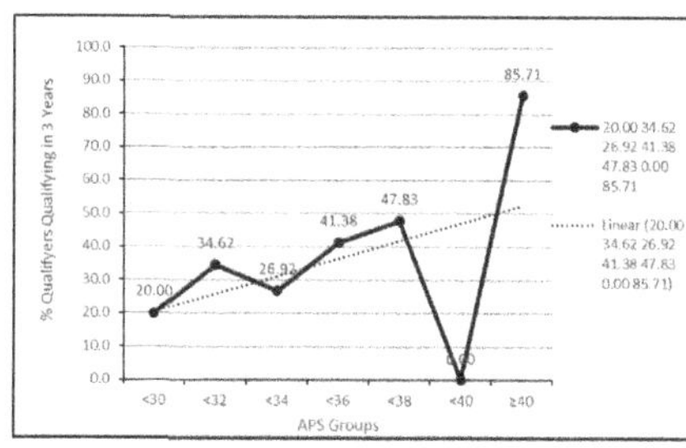

Fig. 2. % of students qualifying in 3 years.

4.3 Through-Put of Students

More students are desired to qualify in the minimum prescribed period of three years. The study, therefore, perused the % of students qualifying in three years per stratified group. The average APS per group was the independent variable, and the % of students qualifying in three years the dependent variable (see Table 3). The small group of students with APS of 38&39 did not perform well, resulting in a low Pearson's correlation coefficients of 0,4447 and the coefficient of determination of 0,1978. A positive relationship still exists between the independent and dependent variables. The slope of the simple linear regression line of 2,75 indicated a significant increase in % of students qualifying in 3 years for a corresponding rise in APS (see Fig. 2).

Table 3. % of students qualifying in three years.

Stratified APS groups	<30	30 & 31	32 & 33	34 & 35	36 & 37	38 & 39	≥40	Slope	Y intercept	Error ϵ
Average APS	28.2	30.7	32.4	34.3	36.4	38.3	40.4			
Students qualifying	25	26	26	29	23	2	7			
Students qualifying in 3 years	5	9	76	12	11	0	6			
% qualifying in 3 years	20.00	34.62	26.92	41.38	47.83	0.0	85.71	2.75	−57.99	4.230

4.4 Number of Students Who Manage to Qualify

Another measure of success is the number of students that eventually qualified. The study also considered the % of students eventually qualifying per stratified group. The average APS per group was the independent variable, and the % of students qualifying the dependent variable (see Table 4).

Table 4. % of students that qualified.

Stratified APS groups	<30	30 & 31	32 & 33	34 & 35	36 & 37	38 & 39	≥40	Slope	Y-intercept	Error ϵ
Average APS	28.2	30.7	32.4	34.3	36.4	38.3	40.4			
Registered students	44	51	46	47	32	4	9			
Students qualifying	25	26	26	29	23	2	7			
% qualifying	56.82	50.98	56.52	61.70	71.88	50.0	77.78	1.38	1377	3.903

The small group of students with APS of 38&39 again performed poorly, resulting in a low Pearson's correlation coefficients of 0,5630 and the coefficient of determination of 0,3170. A positive relationship still exists between the independent and dependent variables. The slope of the simple linear regression line of 1,38 indicated a moderate increase in % of students eventually for a corresponding rise in APS (see Fig. 3).

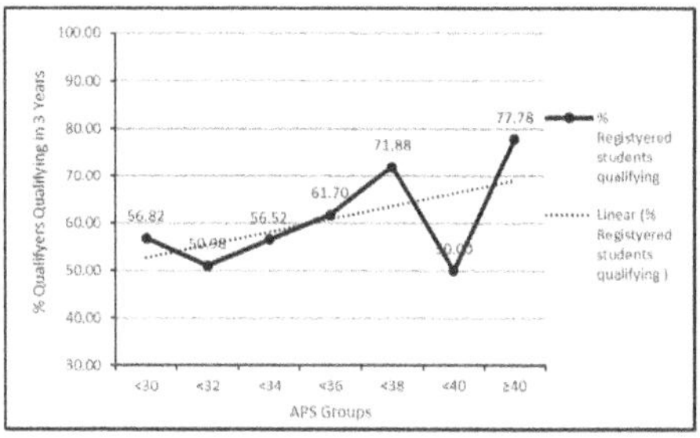

Fig. 3. % of students that qualified

5 Conclusions

The study found a constant, positive and reasonably significant correlation between the APS of CM students, structured in seven stratified groups, and their undergraduate performance. However, the study also identified that the group of students with the second-highest APS of 38 & 39 performed poorly. This group was very small, but only half of them eventually qualified, and none of them qualified in three years. The inconsistent performance of these students indicates that even though the APS may be a prominent indicator of undergraduate performance, other factors also need to be considered.

The Pearson's correlation between APS and first-year marks was high at 0,9581. However, due to the above inconsistent performance from group 6 students, the Pearson's correlation dropped to 0,4447 and 0,5630 respectively for qualifying in three years and qualifying eventually.

Compared to their QS counterparts, the QS students displayed more pronounced positive relationships between APS and undergraduate performance. When comparing the slope of the simple regression line for weighted first year marks the QS students had a 2,6% steeper slope {[(1,94 – 1,89)/1,89] × 100/1%}; for qualifying in three years; QS students had a 116,4% steeper slope {[(5,95 – 2,75)/2,75] × 100/1%} and for qualifying eventually QS students had a 132,6% steeper slope {[(3,21 – 1,38)/1,38] × 100/1%}.

The first-year marks of CM students also lent some support to the findings of Schöer *et al.* [8] that NSC marks tend to be too high. The group 7 CM students (APS $\leq$40) typically achieved average school marks of >80%, while their average first-year mark was 65,5%.

6 Recommendations

The above findings support earlier studies' recommendations that the Department of Construction Economics should pursue its strategy to adjust the admission requirements for Mathematics and Natural Sciences upwards.

It is recommended that the study be expanded to student cohorts of 2016 onwards to determine if the trends identified continue and to keep being informed of students' relative performance in the rapidly changing demography of the South African tertiary environment. The study should be extended to include other institutions nationally that offer a similar qualification. Other causes that could explain the variability in student performance should also be investigated.

References

1. Council on Higher Education: Access and Through-Put in South African Higher Education: Three Case Studies. Council on Higher Education, Pretoria (2010)
2. Fisher, G.: Improving throughput in the engineering bachelor degree. Report to the Engineering Council of South Africa (2011)
3. Harris, F., McCaffer, R., Baldwin, A., Edum-Fotwe, F.: Modern Construction Management, 8th edn. Wiley Blackwell, Hoboken (2021)
4. Scott, I., Yeld, N., Hendry, J.: Higher education monitor no. 6: a case for improving teaching and learning in South African higher education. The Council on Higher Education, Pretoria (2007)
5. Prince, R., Frith, V.: The quantitative literacy of South African school-leavers who qualify for higher education. Pythagoras **38**(1), 22–35 (2017)
6. Prince, R., Frith, V.: An investigation of the relationship between academic numeracy of university students in South Africa and their mathematical and language ability. ZDM Math. Educ. **52**(3), 433–445 (2019). https://doi.org/10.1007/s11858-019-01063-7
7. Council on Higher Education, CHE: A proposal for undergraduate curriculum reform in South Africa: the case for a flexible curriculum structure, p. 58. CHE, Pretoria (2013)
8. Schöer, M., Ntuli, M., Rankin, N., Sebastiao, C.: Hunt, K: A blurred signal? The usefulness of National Senior Certificate (NSC) marks as predictors of academic performance at university level. Perspect. Educ. **28**(2), 9–18 (2010)
9. Hoffman, D., Pieterse, I.: Academic school performance as an indicator of success in undergraduate studies in construction economics in South Africa. In: Nazir, S., Ahram, T., Karwowski, W. (eds.) AHFE 2020. AISC, vol. 1211, pp. 123–128. Springer, Cham (2020). https://doi.org/10.1007/978-3-030-50896-8_19
10. Hoffman, D., Pieterse, I.: Performance in mathematical and scientific school subjects as an indicator of success in undergraduate modules in construction economics in South Africa. In: Nazir, S., Ahram, T., Karwowski, W. (eds.) AHFE 2020. AISC, vol. 1211, pp. 56–62. Springer, Cham (2020). https://doi.org/10.1007/978-3-030-50896-8_9

Mapping Aspects for Assessing Aptitude for Architecture Education with Psychological Testing

Poonam Khan(✉) and Nikhil Ranjan Mandal

School of Planning and Architecture, Bhopal, India
{poonamkhan,nrmandal}@spabhopal.ac.in

Abstract. Architecture is considered an important domain in the making of the built environment and architecture education is as an important branch of professional education. There are more aspirants for the course than the number of seats available. Hence, the screening process through the entrance examinations is an important admission activity. The current research focuses on the expert survey opinion to map the aspects and category of aspects as derived from the study of several psychological tests developed by past researchers to test the necessary intelligence, creativity and aptitude. The process of the survey was conducted with Amabile's Consensual Assessment Technique satisfying all conditions necessary for the assessment. The results imply that there is a further scope of inclusion of aspects as derived with the already established knowledge of various psychological tests. Hence there is a further research required in the performance assessment of the above entrance examinations.

Keywords: Architecture education · Entrance examinations · Intelligence · creativity and aptitude · Psychological tests · Expert opinion survey

1 Introduction

Architecture education is an important branch of professional education with its responsibility of developing the sensitivity of students towards the growing depletion of natural resources and the degeneration of the environment [1]. With the growing number of aspirants as prospective candidates to the professional course of architecture education, it is necessary to select the students most suited for the course. There are several admission criteria prevailing worldwide for admission to the undergraduate level of architecture education viz., high school academic scores, portfolios, psychometric tests, interviews, letters of recommendation, personal goals and statements, entrance aptitude tests, etc. [2] and [3]. In India too, there are several national and state level entrance examinations conducted for the selection and shortlisting process for admission to architecture education. This paper restricts itself to the study of the contents of the two nationalized entrance examinations conducted in architecture in India. These are the Joint Entrance Examination, Mains – Paper 2 and the National Aptitude Test in Architecture [3]. As

S. Nazir et al. (Eds.): AHFE 2021, LNNS 269, pp. 215–222, 2021.
https://doi.org/10.1007/978-3-030-80000-0_26

per the comparative study conducted by the authors on the contents of both the entrance examinations, it was found that all abilities tested in the entrance examinations conducted in India can be categorized into the following abilities useful for architecture education like mathematics, imagination, three-dimensional geometry, architectural knowledge and geography with building sciences.

Simultaneous studies were carried out to find more aspects that test intelligence and creative abilities. It was found that the field of psychology is very rich with several batteries of tests developed by past researchers which are used to measure these abilities. An extensive study of literature was conducted to identify the aspects these batteries of tests are looking for.

These aspects were then grouped to form categories and sub-categories of aspects. The mode of testing was listed out and similar types were coded and classified into categories of aspects. The triarchic theory by Sternberg which says that all abilities are classified as synthetic-intelligence skills, analytical skills and practical-contextual skills were used as sub-categories for the study [3] and [4].

Amabile's Consensual Assessment Technique was employed for the assessment of sample questions from previous set of question papers. Experts from the domain of architecture education and educational psychology were selected for judging the sample questions from the selected papers.

2 The Use of Psychological Tests as Tools for Evaluating Intelligence and Creativity

There have been several methods in research to study the intellect and creative abilities of humans in the field of psychology. Researchers following the psychometric approach claim that these are standardized tools used for measuring capacities of intelligence and personality [5]. Psychological tests also known as psychometric tests have been known for their immense potential in the selection of prospective students for professional courses and for suitable candidates for jobs.

The psychological tests developed over time by several researchers is developed through several theories and models. One prominent theory is the theory of Divergent Thinking developed by the psychologist J.P. Guildford [6]. Guildford studied aspects like fluency, flexibility, originality and elaboration. The modes of testing covered in the 'contents' phase of the Structure of Intellect Model was basically, visual, auditory, symbolic, semantic and behavior. This led to further study of the divergent theory where many researchers developed several psychological tests. The Wechsler Intelligence Scale developed by David Wechsler was developed with two modes, verbal and figural tests [7]. The Lewis Terman's Concept Mastery Test was developed for testing intelligence with the verbal and semantic modes. The Getzels-Jackson's test again was developed testing semantic abilities with a word test, spatial abilities with a figural test and numerical abilities with a Math test [8]. The Getzel's – Jackson's battery consisted of semantic, figural and math tests [9]. Wallach and Kogan's Test [10] and the Torrance Test [11] was developed with verbal and figural tests again under the divergent theory to examine aspects like fluency, flexibility, originality and elaboration. The above researchers have expressed the need of further research in the field of divergent thinking.

Among other tests and theories apart from divergent thinking, researchers developed another test to evaluate intelligence, the Remote Association Test developed by Medick using verbal and semantic modes [9] which was further developed as the Functionally Remote Associates Test [12]. The Consensual Assessment Technique developed by Amabile [5] implied that both crystallized and general intelligence are important for creativity. Her technique of assessment gave three conditions where experts are asked to assess the work done. The three conditions expected were, that the experts essentially needed to be from different domains, they were expected to have a certain number of years of experience and the work to assessed needed to be other than creativity [13]. The Differential Ability (DAS) test developed by Elliot was found to be suitable for cognitive abilities to be assessed in all age groups [14]. The test was developed with the verbal and spatial modes. The Investment Theory by Sternberg and Lubart [15] relied on testing of intelligence with six factors – intelligence, motivation, knowledge, thinking styles, personality, motivation and environment. As per the Triarchic Theory, Sternberg [4] has classified all abilities in three groups – synthetic-intelligence skills, analytical skills and practical-contextual skills.

Csikzentemihayl's System Model includes the presence of three aspects in creativity i.e., the presence of the person with the creativity, the domain in which the activity holds its occurrence and the experts from the domain who accept and endorse the creative output [16]. The Das-Naglieri's Cognitive Assessment Battery was conducted in visual, auditory, verbal, figural and non-verbal modes testing intellectual capacities of individuals [17]. The Propulsion Model came ahead with a major paradigm shift in the studies involving cognitive contributions in creative initiations with a new terminology of aspects [16].

Psychological approaches in the 21st century continued. A psychological test developed by Kaufman known as the Kaufman Assessment Battery for Children (KABC) and later developed into the KABC-II which was designed to test neurocognitive intelligence [18]. These tests were again developed as verbal, figural, semantic, numerical, visual and auditory modes. The Test for Creative Thinking – Drawing Production (TCT-DP) developed by Urban included fourteen aspects like continuations, completions, new elements, humor and effectivity, etc., [19]. These tests used the figural, verbal and non-verbal modes of testing. The Creative Composite Indicator gave a scale to identify barriers in the creative abilities which can be used as occasions for enhancing cognitive abilities [20]. The Five Factor Model helped in identifying the traits of a personality [21].

2.1 Deducting Aspects with Psychological Tests

From the extensive literature study on the development of psychological tests it is seen that, many more aspects have been utilized by psychologists to test intelligence and creativity at different levels. The psychometric tests developed by different researchers in the past have used different modes of testing. On further research with the attributes of testing mechanisms, it was revealed that these modes come in various forms viz., verbal, non-verbal, figural, picture completion, word association, auditory, sensory, numerical, etc., These modes of testing were further clustered and an effort to categorize them was done. The modes of testing with examples of the type of enquiries are given in Table 1.

Each coded-category of aspects was again divided into sub-categories based on the Triarchic Theory. The sub categories are enlisted in Table 2. The aspects were then suitably divided in each category.

Table 1. Final coding of aspect identification

Final coding	Mode of testing Font size and style
Behavioral	Judgement, memory
Logical	Logical, abstract, non-verbal
Semantic	Verbal, auditory
Spatial	Figural, symbolic, perception, visual, DT tests, non-verbal, spatial, memory
Quantitative	Math, abstract, quantitative

Table 2. Sub-categories assigned to each category

Sub-category	Description
Synthetic-intelligence skills	Intellectual skills that build up with time
Analytical skills	Skills that help in analyzing
Practical-contextual skills	Contextual related and help in day-day decisions making processes

3 Aspects Tested in Architecture Entrance Exams

In India, admission in architecture institutions is basically by two entrance examinations. The National Aptitude Test for Architecture (NATA) is conducted by the Council of Architecture (CoA) in India. The test examines students for their observation and drawing skills, sense of proportion, aesthetic sensitivity, Physics, Chemistry, Math, critical and logical thinking ability [22]. The paper consists of two parts, part A is the drawing ability test and Part B is the test consisting of Multiple-Choice Questions.

Similarly, the JEE Mains paper-2 is conducted by the National Testing Agency of India under the Ministry of Education [23]. The paper examines students for three-dimensional perception and visualization of 2D drawings and 3D objects, analytical reasoning and mental ability, color theory, architectural awareness, drawing skills from memory and imagination.

Looking for similarities and differences in both the papers, the pattern of papers in NATA examination has changed over the years in comparison to the JEE Mains Paper 2.

3.1 Similarities from Both the Nationalized Entrance Examinations

With minor differences in the pattern, including the number of questions asked in both the papers, there is a similarity found in the contents and aspects both the papers look for. Table 3 below gives the similarities along with the description of the aspects as tested by both the papers.

Table 3. Similarities in aspects examined by the nationalized entrance examinations in India

Aspects	Description
Mathematical	Class 10+2 syllabus
Imagination	Drawing skills, memory drawing, free-hand drawing, three-dimensional visualization, still life drawing, logos, compositions, color theory
Logical Reasoning	Sequences, patterns, odd figures, hidden figures, mirror images, logical reasoning
Architectural - Knowledge	Knowledge about architects, monuments, materials, history, building elements
Geography and Culture	Climate and topography, type of soil, culture
Communication skills	Verbal and linguistic skills-oral and written skills

The above study of the aspects, categories of aspects and sub-categories of aspects helped in the analysis of sample questions from the set of five-year question papers.

3.2 Sampling

For the current research, question papers of five-years were taken as sample papers. The selection of the five-year question papers is done on the basis of similarity in the trend of the question paper. There were 80 number of questions in each paper; 30-mathematics based multi-choice questions and 50-aptitude based multiple-choice questions. Hence, with the collection of five-year question papers, we have 400 questions in all, taken as samples.

4 Expert Opinion Survey and Analysis

4.1 Conditions of the Consensual Assessment Technique

The expert opinion survey was conducted with the Consensual Assessment Technique as the guideline. As per this technique, in order for judgement of any activity conducted, the opinion of experts is utilized. There are three conditions set by Amabile [15]. Firstly, the selection of the experts is done on the basis of certain criterion set by the controllers. The experts should have spent a certain number of years in their respective domains. Secondly, the assessment of the work should be done independently and thirdly, the work to be assessed should be beyond creativity. For the research survey, two experts from the domain of Architecture Education and two experts from Educational Psychology were taken. Sufficing the three conditions as specified by Amabile in the CAT technique, the experts were supposed to be with more than 15 years of experience. Secondly, the experts did the assessments independently. Thirdly, the survey was beyond creativity as it included aspects from quantitative aptitude, architecture education and educational psychology.

4.2 The Survey and Results

The list of aspects, category of aspects and the sub-category of aspects along with the sample questions with the brief of the exercise was given to the experts. All 400 questions were categorized by the four experts individually, with respect to the categories of aspects. The responses were accepted where there were three similar responses. Similarly, where there was less than one similar response, the response was rejected. With this inclusion criteria, the sample questions were classified accordingly. Responses of all experts was calculated firstly, with the sample questions as pe their presence in the categories of aspects and in the sub-categories of aspects. Secondly, the frequency of questions appearing in each category and sub-category were mapped.

5 Conclusion

As a result, of the expert opinion survey, it was found that there were 24 questions which were differing in the opinions of the experts and the responses of around 376 questions matched with each other. The questions which differed from each other were discarded from the sample group and the remaining questions were used for analysis. The 376 sample questions, were further categorized into sub-categories. From the readings, as given in Fig. 1, it was found that the sample questions were found in the Logical reasoning, Spatial reasoning and Quantitative categories only. Furthermore, except for the Quantitative category, the proportions of sample questions lying in the two sub-categories of the Synthetic-Intelligence and the Analytical skills, there is a difference found in the proportions of questions asked. The trend between the two sub-categories was not identified whereas under the quantitative category, the questions range somewhat equally in the two sub-categories of skills – viz., the synthetic-intelligence skills and the analytical skills. Another observation was that there were no questions asked in the Behavioral and Semantic categories and so was the sub-category of the practical-contextual skills absent.

This implies that though the field of architecture requires a balanced proportion of all the above categories – Behavioral, Logical, Semantic, Spatial and Quantitative aspects, the categories with aspects from the Behavioral and Semantic categories can be added in the existing examinations. Similarly, the practical-contextual skills will also require further studies and may also find a suitable place in the existing entrance examinations.

The trend in the proportions of questions asked in the categories of Logical reasoning, Spatial reasoning and Quantitative reasoning needs further research. Similarly, a further research in the trend of questions asked over the years, in the two sub-categories of synthetic-intelligence skills and analytical skills needs to be identified.

The image below, shows the frequency of the responses from the Expert opinion survey conducted.

Psychometric Tests (Parameters)	Synthetic- Intelligence					Analytical Skills					Practical Contextual Skills				
	Yr. 1	Yr. 2	Yr. 3	Yr. 4	Yr. 5	Yr. 1	Yr. 2	Yr. 3	Yr. 4	Yr. 5	Yr. 1	Yr. 2	Yr. 3	Yr. 4	Yr. 5
Behavioural	0	0	0	0	0	0	0	0	0	0	0	0	0	0	0
Logical	12	11	9	15	0	1	2	5	0	5	0	0	0	0	0
Mathematical	20	20	20	15	20	9	9	9	13	8	0	0	0	0	0
Semantic	0	0	0	0	0	0	0	0	0	0	0	0	0	0	0
Spatial	0	1	0	0	14	33	31	29	33	28	0	0	0	0	0

Fig. 1. Frequency of sample questions under the listed categories and sub-categories of aspects

Acknowledgments. This research was supported by the Doctoral Research Committee, Department of Architecture, School of Planning and Architecture, Bhopal, India. The authors wish to express their appreciation to the experts who were a part of the research process and to the referees for their valuable comments on the earlier draft of the paper.

References

1. International Union of Architects: UNESCO-UIA Charter for Architectural Education, Revised edition (2017)
2. Goldschmidt, G., Sebba, R., Oren, C., Cohen, A.: Who should be a designer? Controlling admission into schools of architecture. In: Designing of Context: Proceedings of Design Thinking Research. Symposium, vol. 5, pp. 277–295 (2002)
3. Khan, P., Mandal, N.R.: An analytical study of aptitude tests for entrance to architecture education: a case of India. In: Ahram, T., Taiar, R., Gremeaux-Bader, V., Aminian, K. (eds.) IHIET 2020. AISC, vol. 1152, pp. 663–668. Springer, Cham (2020). https://doi.org/10.1007/978-3-030-44267-5_100
4. Sternberg, R.J., Universi, Y., Cutler, A.: Beyond the IQ: A Triarchic Theory of Human Intelligence (1985). ISBN: 0521262542
5. Amabile, T.M.: Social psychology of creativity: a consensual assessment technique. J. Pers. Soc. Psychol. **43**, 997–1013 (1982)
6. Batey, M.: A psychometric investigation of everyday creativity. Thesis (2007)
7. Keith, T.Z., Fine, J.G., Taub, G.E., Reynolds, M.R., Kranzler, J.H.: Higher order, multisample, confirmatory factor analysis of the wechsler intelligence scale for children - fourth edition: what does it measure? Sch. Psychol. Rev. **35**, 108–127 (2006). https://doi.org/10.1080/02796015.2006.12088005

8. Baird, L.L.: The role of academic ability in high-level accomplishment and general success, pp. 1–39 (1982)
9. Batey, M., Furnham, A.: Creativity, intelligence, and personality: a critical review of the scattered literature. Genet. Soc. Gen. Psychol. Monogr. **132**, 355–429 (2006). https://doi.org/10.3200/MONO.132.4.355-430
10. Hocevar, D.: Intelligence, divergent thinking, and creativity. Intelligence **4**, 25–40 (1980). https://doi.org/10.1016/0160-2896(80)90004-5
11. Kim, K.H.: The torrance tests of creative thinking - figural or verbal: which one should we use? Creat. Theor. Res. Appl. **4**, 302–321 (2018). https://doi.org/10.1515/ctra-2017-0015
12. Worthen, B.R., Clark, P.M.: Toward an improved measure of remote associational ability. J. Educ. Meas. **8**, 113–123 (1971). https://doi.org/10.1111/j.1745-3984.1971.tb00914.x
13. Amabile, T.: Amabile_A_Model_of_CreativityOrg.Beh_v10_pp123-167.pdf (1988)
14. Anastasi, A.: Psychological testing: basic concepts and common misconceptions. In: G. Stanley Hall Lecture Series, vol. 5, pp. 87–120 (2004). https://doi.org/10.1037/10052-003
15. Sternberg, R.J., Lubart, T.I.: An investment theory of creativity and its development. Hum. Dev. **34**, 1–31 (1991). https://doi.org/10.1159/000277029
16. Sternberg, R.J., Kaufman, J.C., Pretz, J.E.: The propulsion model of creative contributions applied to the arts and letters. J. Creat. Behav. **35**, 75–101 (2001). https://doi.org/10.1002/j.2162-6057.2001.tb01223.x
17. Naglieri, J.A., Otero, T.M.: The cognitive assessment system: from theory to practice. In: Flanagan, D.P., Harrison, P.L. (eds.) Contemporary Intellectual Assessment. Theories, Tests, and Issues, 3rd edn, pp. 376–399, xviii, 926 p. Guilford Press, New York (2012)
18. Reynolds, M.R., Keith, T.Z., Fine, J.G., Fisher, M.E., Low, J.A.: Confirmatory factor structure of the Kaufman assessment battery for children-second edition: consistency with Cattell-Horn-Carroll theory. Sch. Psychol. Q. **22**, 511–539 (2007). https://doi.org/10.1037/1045-3830.22.4.511
19. Urban, K.: Assessing creativity: the test for creative thinking - drawing production (TCT-DP) the concept, application, evaluation, and international studies. Psychol. Sci. **46**, 387 (2004)
20. Kadabadi, S., Garg, K., Parikh, R.: Creativity Composite Indicator. GTECH 2009, 4–15 April 2010
21. Allen, T.A., Carey, B.E., McBride, C., Bagby, R.M., DeYoung, C.G., Quilty, L.C.: Big Five aspects of personality interact to predict depression. J. Pers. **86**, 714–725 (2018). https://doi.org/10.1111/jopy.12352
22. COA: Council of Architecture-Practice of Architecture. Indian Institute of Architects (2014)
23. National Testing Agency, National Testing Agency (nta.ac.in)

Construction Competence Model of Music Appreciation Course Teachers in Science and Engineering University

Yingxin Guan[1], Songqi Fei[2], Yaning Zhao[1], and Yihang Du[3,4](✉)

[1] Art Education Department of Youth League Committee, Beijing Jiaotong University, Beijing 100044, China
[2] Department of Academic Research, China Conservatory of Music, Beijing 100101, China
[3] Beijing Institute of Technology, Beijing 100081, China
[4] National Academy of Chinese Theater Arts New Media Arts Department, Beijing 100073, China

Abstract. The music appreciation course is an important part of general education in comprehensive science and engineering colleges. Teacher competence is playing the big role to measure the level of education, which is a significance of factor in the cultivation of student comprehensive ability. By the factors frequency analysis, teacher competency evaluation model is constructed and subjective evaluation is carried out. Eleven professors were selected in the art education industry of science and engineering universities and the top of level musicians as participants, by semi-structured interviews to calculate weight of factors and important sequence. Four most important factors were obtained through results showed, which are teachers' moral, teaching ability, information technology ability and non-technical skills. The related conclusions can be used in teacher selection and training, which will help to optimize teachers and improve the teaching effect.

Keywords: Teacher competence · Music appreciation course · Semi-structured interview

1 Introduction

The comprehensive quality education of college students is highly valued by the Ministry of Education and various colleges and universities. On the basic curriculum of professional education technology and specialized courses, a large number of general education courses are offered for students to choose and study. General education courses are an important part of cultivating comprehensive quality of students. Music appreciation course in science and engineering universities is a general education course, which bears the responsibility of cultivating students' aesthetic sentiment and stimulating their heuristic thinking, and is conducive to promoting students' all-round development from perspective.

Music appreciation course in science and engineering universities is a kind of public elective courses for non-music majors. The teaching of music appreciation course is also

S. Nazir et al. (Eds.): AHFE 2021, LNNS 269, pp. 223–231, 2021.
https://doi.org/10.1007/978-3-030-80000-0_27

different from that of ordinary professional courses. There are still many disadvantages between teacher competence and quality of music appreciation course in science and engineering universities, and it lacks of standardized and unified standards.

Music teacher is an important factor in music appreciation course, and the cultivation of students' appreciation ability needs the correct guidance. The lack of competence of music appreciation teachers will directly affect students' learning enthusiasm and make them unable to perceive the meaning of music art. The humdrum teaching method cannot efficiently improve students' artistic accomplishment, which leads to the unsatisfactory teaching effect. At the same time, music appreciation teachers often lack the update of professional knowledge due to the development of The Times, and their teaching methods gradually fall behind, leading to the low overall teaching quality.

Therefore, based on the literature analysis and field research, effective means were adopted to identify the influencing factors of music appreciation courses in science and engineering universities. By means of quantification to extract the components of music appreciation teacher competence, it will be used to provide a powerful basis for improving teacher competence from multiple perspectives, which are helpful to the selection, training of teachers, and then the improvement of teaching effect.

2 Materials and Methods

Three sets index system were selected for semi-structured interview, including Steven Cornelius Music appreciation course project [1], AEC (European Association of Conservatoires) Music teaching plan [2], MEE (Music Education Ecosystem) Music Education Ecosystem [3], HEAD teacher evaluation system [4] and college teacher competency model [5]. The semi-structured interview was adopted to obtain the components of Music appreciation courses in comprehensive colleges and universities. The specific process as depicted in Fig. 1.

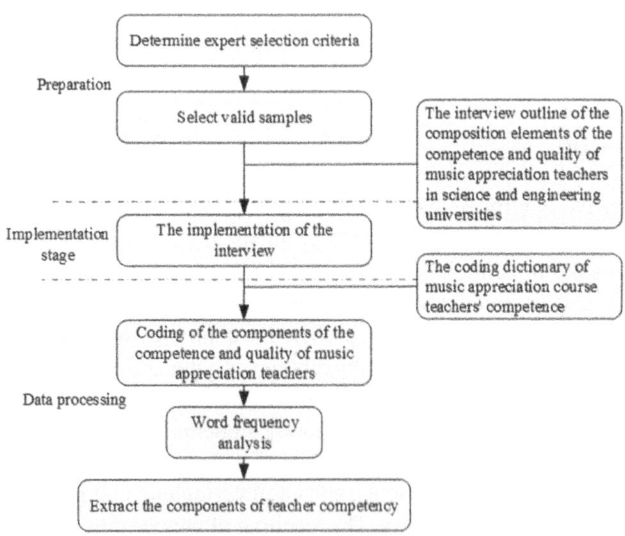

Fig. 1. Extraction of competence components for music appreciation teachers

2.1 Semi-structured Interviews

Interview is a survey method to understand the psychology of interviewees through face-to-face communication between researchers and interviewees. Its advantage is that it can accurately, deeply and effectively collect information about the attitudes and opinions of interviewees. Semi-structured interview is between structured interview and unstructured interview. On the basis of referring to the proposed interview outline, the space for interviewees to express their own ideas and opinions is reserved [6]. The semi-structured interview will be helpful to understand the actual situation of music teaching, the design of music appreciation course and the factors of teachers' competence.

In this study, six professors were selected as the interview subjects, respectively from Beijing Jiaotong University, China Conservatory of Music, Central Conservatory of Music, and National Academy of Chinese Theatre Arts. Due to the particularity of instrumental performance, musicians in national level of orchestras often undertake a lot of basic teaching work while performing. Therefore, the study selected 5 domestic National First Grade Instrumentalist as the interviewees, respectively from China National Symphony Orchestra, China Philharmonic Orchestra, and China Film Orchestra, etc. aged 37–65 years old. The specific expert information is shown in Table 1.

Table 1. Information of interviewed experts

Experts name	Major of teaching	Institution
**Bin	Art theory	Beijing Film Academy
**Tao	Instrumental teaching	Central Conservatory of Music
**Guo	Instrumental teaching	MinZu University of China
**Gang	Instrumental teaching	Central Conservatory of Music
**Qi	Instrumental teaching	China Philharmonic Orchestra
**Bing	Art Administration	Beijing JiaoTong University
**Peng	Vocal performance	China Conservatory of Music
**Xue	Instrumental performance	China Film Orchestra
**Pan	Vocal performance	China National Symphony Orchestra
**Ou	Instrumental performance	China National Symphony Orchestra
**Han	Vocal teaching	Beijing University of Technology

According to the curriculum requirements of music appreciation, the Interview Outline of the Composition elements of the Competence and Quality of Music Appreciation teachers in Colleges and universities is designed and shown in Table 2.

Table 2. Interview outline of teacher competence for music appreciation course

	Question	Purpose
1	Please introduce your teaching experience and content	Predict the direction and focus of the interview
2	Please list the difficulties encountered in music teaching in recent years, describe in detail the process, ideas, measures taken, and put forward relevant solutions and suggestions for improvement	The key stage of interview: comprehensively understand and ask for details What abilities should a music teacher have?—Understand the teaching background; What problems have you encountered in the course?—Understand the key and difficult points of music appreciation course teachers; Problem-solving thoughts (why do you do that?) —Reflect the components of teachers' competence and quality; With the development of The Times, has the teaching method been updated? How did you do it? —Existing problems of teachers' competence and quality; Strategies and suggestions to deal with problems —To find out the components of teachers' competency to be improved
3	Please list the components of music appreciation teachers' competence in science and engineering universities	The final stage of the interview: to verify and supplement the text analysis of the components of the competence of music appreciation teachers

The interview was carried out in strict accordance with the semi-structured interview process, the text of the interview results was sorted out, and the constituent elements of music appreciation course were extracted through keyword coding and analysis. The process was carried out by two researchers (coder A and coder B), and the coders' classification consistency and correlation coefficient were used as the coding reliability evaluation indexes.

Category Agreement (*CA*) refers to the proportion of the same communication skill factors identified by two coders when they independently encode the same interview data. Winter [7] gave the calculation formula of *CA*:

$$CA = \frac{2S}{(T_A + T_B)} \tag{1}$$

Where, S represents the same number of codes of the two raters; T_1 and T_2 respectively represent the number of codes of the raters.

2.2 Data Analysis

The coder analyzes the word frequency in the text, identifies and distinguishes the curriculum components in the text, and formally classifies and encodes them. Finally, the coding information of the same constituent element is classified into one category, and the name, frequency, average grade score and highest grade score of each course constituent element evaluated by each coder are counted. Taking the "Critical factors" of the events in the interview as the minimum unit of classification can best preserve the textual features of the interview [8].

3 Result

The calculated coding consistency is $CA = 89.95\%$, indicating coding consistency is good. Through the word frequency analysis of semi-structured interviews, a total of components of many music appreciation courses in colleges and universities were obtained. The distribution of keyword themes is shown in Fig. 2.

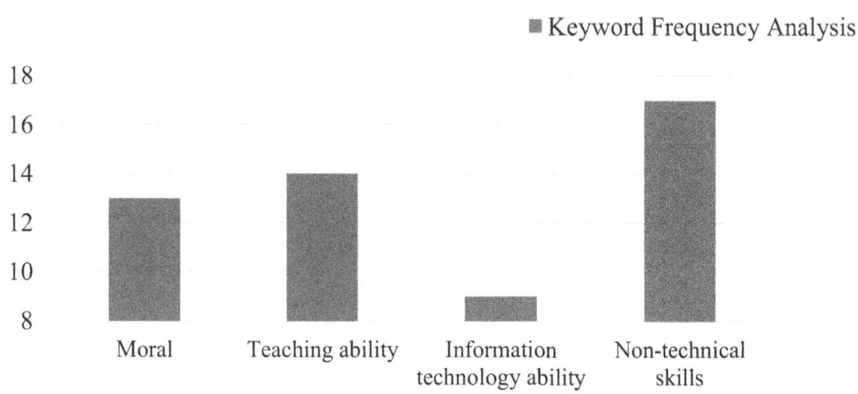

Fig. 2. Keyword frequency comparison of music appreciation teacher competency

The number of effective keywords collected in the interviews was 201, and the topic distribution of keywords was shown in Table 3. The key factors can be divided into four dimensions, that is, moral (66.5%), teaching ability (67.0%), information technology ability(64.5%) and non-technical skills (68.5%).

Table 3. Word frequency analysis of competency for music appreciation teachers

Keyword	Word frequency	Percentage	Sample
Moral	13	66.5	Teachers are not only the disseminators of music knowledge, but also the cultivators of students' morality and the promoters of all-round development, which requires teachers to constantly improve their moral quality and make perfect examples
Teaching ability	14	67.0	Teachers have effective lesson plans, content, and teaching objectives
Information technology ability	9	64.5	Teachers have diversified information technology teaching abilities and are able to use modern teaching equipment
Non-technical skills	17	68.5	Teachers can effectively convey their own emotional factors (such as leadership, decision-making ability, organizational communication, etc.) to students through language or behavior

Note: Since the variable of "Event Subject" is set as a multi-response variable, the sum of the selection frequency is greater than the total number of cases

4 Discussion

Competency quality refers to the basic conditions for a certain post, and it is the collection of professional skills and personality traits possessed by excellent job performance. The competence of college teachers generally includes moral, teaching ability, information technology ability and non-technical skills. In the 21st century modern teaching, keeping pace with The Times, the use of information equipment is also a key factor of teachers' competence, effective situational awareness, knowledge reserve, organizational communication ability, key decision-making ability, etc., also help to improve the teaching quality by improving the ability of teachers.

4.1 Moral

The word frequency of teacher morality and teacher style is 13, and the word frequency ratio is 6.5%. Teachers are excellent role models in the minds of students. Teachers' moral are mainly reflected in noble professional ethics and good personal qualities. Specific behaviors include: dignified posture, careful and serious teaching attitude, rigorous teaching style, passion and expression, etc.

Through music appreciation course, students are required to have profound humanistic quality on the basis of mastering skills, which puts forward higher requirements for professional teachers' non-intellectual factors, and requires teachers to have enthusiastic artistic feelings to drive students' learning emotions. Music appreciation course is a relatively open curriculum system, through the teacher's own exemplary role, to convey the sense of social responsibility, strong ambition and profound cultural heritage to students, to improve the moral cultivation of students play an important role.

4.2 Teaching Ability

Teaching ability is the basic factor of a teacher's teaching. The word frequency of teaching ability is 14, and the word frequency ratio is 7.0%. It is mainly reflected in the aspects of teaching design, teaching organization, teaching evaluation and teaching innovation, which directly affect the cultivation effect of students. Experts mentioned that teaching ability includes standardized syllabus, reasonable design of teaching content, clear teaching ideas, and active innovation of classroom forms, which requires teachers to have the basic teaching skills, be very familiar with the content of the knowledge being taught, and be able to flexibly apply it. The teacher will arrange interactive question-and-answer, group discussion and practical experience in class.

The training goal of music appreciation course is different from that of ordinary specialized courses, so in the process of organizing teaching, we should focus on cultivating students' interest in learning, stimulate students' initiative and creativity to the greatest extent, and realize the teaching goal in a relaxed and active atmosphere. At the same time, the teacher should pay attention to the feedback of students, adjust the teaching progress and update the teaching content in time through the dynamic and real-time teaching evaluation of students.

4.3 Information Technology Ability

Information teaching includes the use of hardware and software equipment, such as Office editing software, cameras, laptops, multimedia tools and sharing video and audio distance-learning equipment. As the teaching subject of the course, teachers control the initiative of teaching. They need to complete the construction and application of the information platform before students, that is, the object of teaching activities, including the production of teaching content, the integration of teaching resources, the production of demonstrative audio-visual images and the updating of equipment.

Through a variety of teaching tools, improve students' cognitive process, thinking mode and teaching platform experience in the class. The teacher should fully construct the communication of teaching psychology and emotion, effectively optimize the course configuration, enhance students' exploratory, analytical and experience, and achieve high quality teaching effect.

4.4 Non-technical Skills

As an adjunct to teaching ability, non-technical skills are also an important part of teachers' competence. The course of music appreciation not only requires the teacher to have an in-depth understanding of the music composition.

Experts believe that music appreciation teachers should have a rich knowledge reserve. Based on the characteristics of general education, music appreciation teachers should have the knowledge base of pedagogy, psychology and management, as well as the comprehensive knowledge of cross-culture, cross-times and inter disciplinary.

As the main of information transmission, teachers pass information to students through the teaching process, through communication and expression ability, organization and coordination ability, situational awareness, critical moment decision-making ability and so on. Through the appreciation of music, the teacher is guiding students to enjoy the joy and beauty brought by art, to achieve the purpose of edifying sentiment and perfecting personality. On the basis of achieving the teaching target, establish a good teacher-student relationship and effectively improve the comprehensive quality of students.

5 Conclusion

Semi-structured interview is used to construct the competency model of music appreciation teachers in colleges and universities of science and technology. Through the study of word frequency, it is shown that teachers' moral, teaching ability, information technology ability and non-technical skills are the four essential factors that affect the teachers of music appreciation. The further research is needed to seek the relatively important factors between teaching ability and non-technical skills, analyze these factors quantitatively and put forward feasible reform plan.

The related conclusions can be applied to the general education courses of colleges and universities of science and technology, the competency training program of music appreciation teachers. For music appreciation teachers have more clear and detailed standards in terms of morality. By the way of theory and practice, to the selection, training and optimization of teachers, and the improvement of teaching effect.

References

1. Cornelius, S., Natvig, M.: Teaching music appreciation: a cultural approach. J. Music Hist. Pedagogy **4**(1), 139–150 (2013)
2. Lehmann, A.: Using admission assessments to predict final grades in a college music program. J. Res. Music Educ. **62**(3), 245–258 (2014)
3. Said, H.: Rethinking IT education: lessons from music education. ACM Inroads **7**(1), 34–37 (2016)
4. Elena, R., Minna, H., Timo, P.: HEAD teachers managing entrepreneurship education - empirical evidence from general education. Educ. Educ. Res. **55**, 155–164 (2016)
5. Xu, A., Ye, L.: Impacts of teachers' competency on job performance in research universities with industry characteristics: taking academic atmosphere as moderator. J. Ind. Eng. Manag. **7**(5), 1283–1292 (2014)

6. Galletta, A.: Mastering the Semi-Structured Interview and Beyond: From Research Design to Analysis and Publication. NYU Press, New York (2013)
7. Dubois, D., Rothwell, W.: Competency-Based Human Resource Management: Discover a New System for Unleashing the Productive Power of Exemplary Performers. Nicholas Brealey, Hachette (2004)
8. Keaveney, S.: Customer switching behavior in service industries: an exploratory study. J. Mark. **59**(2), 71–82 (1995)

The APS and Undergraduate Performance in Construction Economics in South Africa

Danie Hoffman(✉)

University of Pretoria, Lynnwood Road, Pretoria, South Africa
danie.hoffman@up.ac.za

Abstract. The BSc in Quantity Surveying (QS) is a three-year program offered by the Department of Construction Economics, University of Pretoria, South Africa. New entrants require an Admission Points Score (APS) of 30. The Department strives for better marks in the first year of study and a better throughput rate of enrolled students. The Department needs confirmation of the validity of the APS as an entry threshold requirement. This study evaluated students' APS against their undergraduate academic performance using data from 2010 to 2015 in a stratified study. Considering the APS of students within these groups as the independent variable, the study found the following correlation factors with aspects of undergraduate performance as the dependent variable: first-year marks (0,979), the % of students who qualified (0,972), the % of students qualifying in three years (0,969) and the % of students passing with distinction (0,776). However, the study also found evidence supporting previous research that the APS is overvaluing students' ability. Compared to their APS, the top grouping of students achieved 10–15% lower marks in undergraduate studies. Some of the study data also indicate that students' non-cognitive skills need to be considered in combination with the cognitive abilities represented by the APS to explain student academic performance. The study, however, confirmed the APS's importance as an indicator of undergraduate performance. The APS should remain part of the selection process of applications from new entrants.

Keywords: Success indicator · School performance · Undergraduate performance · Quantity surveying · South Africa

1 Introduction

Legislation from 2005 and first implemented in 2009 introduced the National Senior Certificate (NSC) or final school exit level. The NSC is a 130-credit certificate made up of six subjects of 20 credits each and Life Orientation adding 10 credits. The Admission Points Score (APS) of a learner is based on the percentages for each of the six subjects in the NSC. Learners score 7 for 80–100%, 6 for 70–79%, 5 for 60–69%, 4 for 50–59%, 3 for 40–49%, 2 for 30–39% and 1 for 0–29%. All South Africa's higher education institutions use the APS as part of their admission criteria [1]. Admission to bachelor's degree studies require a threshold of 50–59% or better in four NSC subjects, but higher education institution may prescribe stricter admission requirements [2].

S. Nazir et al. (Eds.): AHFE 2021, LNNS 269, pp. 232–240, 2021.
https://doi.org/10.1007/978-3-030-80000-0_28

The BSc QS degree from the Department of Construction Economics requires an APS of 30, with English and Mathematics at an achievement level of 60–69% (APS of 5) and Physical Science or Accounting at an achievement level of 50–59% (APS of 4) [3]. The Department is considering increasing the admission threshold to 32 points to improve both the overall marks in the first year of study and enrolled students' throughput rate. This study needs to confirm that the APS closely correlates with actual undergraduate performance.

2 Literature

Higher education institutions face the challenge of selecting students with the ability to complete their chosen programmes successfully, based on marks attained in school-leaving examinations. Several international and South African studies confirmed the positive correlation between school marks in Mathematics and undergraduate performance in quantitative nature modules. Undergraduate admission requirements, therefore, include thresholds set per the above findings [4–8].

Various South African studies have assessed the NSC's ability and suitability to forecast learners' abilities to manage undergraduate studies competently. Schöer *et al.* [9] findings indicated that the NSC Mathematics results were 12–13% higher than the higher-grade Mathematics marks of the previous Senior Certificate system for students with similar undergraduate performance.

Contrary to this, a 2010 study by Wolmarans *et al.* [10] found some measure of support for the NSC in indicating the declining trend in first-year engineering students' results for Mathematics 1 and Physics 1 at the University of Cape Town from 2005 to 2009. A 2011 study at the University of KwaZulu Natal identified that lower NSC mathematics marks resulted in much higher failure rates in the engineering, health sciences, and management sciences faculties but not in all other faculties [11].

A study by Hoffman and Pieterse [12] found that students with an APS of 32 and more outscored students with an APS <32 by 23,3%, and were 29,6% more likely to pass challenging first-year modules on their first attempt. Hoffman and Pieterse [13] also confirmed that students' pass rate in challenging modules of quantitive nature improved significantly for students with higher school marks in Mathematics and Naural Science.

However, various other studies are critical of disregarding non-cognitive skills such as self-discipline, persistence, and motivation for success at university [14, 15]. Other studies acknowledge these non-cognitive skills but agreed that cognitive high-school marks were better indicators of undergraduate performance [16].

3 The Methodology

The records of entry cohorts of BScQS students from 2010–2015 were studied to explore the relationship between undergraduate performance and APS. The study evaluated undergraduate performance such as weighted average first-year marks, the throughput rate of enrolled students, the percentage of enrolled students who qualified, and the percentage of enrolled students who qualified with distinction (75%+). The study used the APS as the independent variable and the mentioned aspects of student performance as dependant variables.

A structured approach of seven groups was applied to treat and describe the data:

- Group 1 (students with an APS of <30),
- Group 2 (students with an APS of 30 or 31),
- Group 3 (students with an APS of 32 or 33),
- Group 4 (students with an APS of 34 or 35),
- Group 5 (students with an APS of 36 or 37),
- Group 6 (students with an APS of 38 or 39), and
- Group 7 (students with an APS of ≥40). extensive

4 The Data and Findings

The number of first-time entering BSc QS students from 2010 to 2015 totaled 361. The stratified group and the average APS per group are detailed below (see Table 1).

Table 1. The number of first-time entering students per APS stratified grouping.

Stratified APS groups	<30	30&31	32&33	34&35	36&37	38&39	≥40
Number of students enrolled	54	67	55	66	46	40	33
Average APS	28.1	30.6	32.4	34.4	36.4	38.5	40.9

4.1 Weighted Average First-Year Marks

Students' weighted average first-year mark was evaluated as the dependent variable against the average APS as the independent variable within the stratified groups. Details of the simple linear regression line are detailed in Table 2. A positive relationship exists between the independent and dependent variables (see Fig. 1). The slope of the simple linear regression line of 1,94 indicated a significant increase in average first-year marks for a corresponding rise in APS.

Table 2. Weighted average marks in first-year of study

Stratified APS groups	<30	30&31	32&33	34&35	36&37	38&39	≥40	Slope	Y-intercept	Error ϵ
Average APS	28.1	30.6	32.4	34.4	36.4	38.5	40.9			
Weighted average marks (%)	42.4	45.2	48.4	52.9	56.2	59.2	69.6	1.94	−13.13	2.024

The high Pearson's correlation coefficients of 0,9799 and the coefficient of determination of 0,9603 confirm the strong correlation, but the averaged stratified data used as an independent variable must be considered.

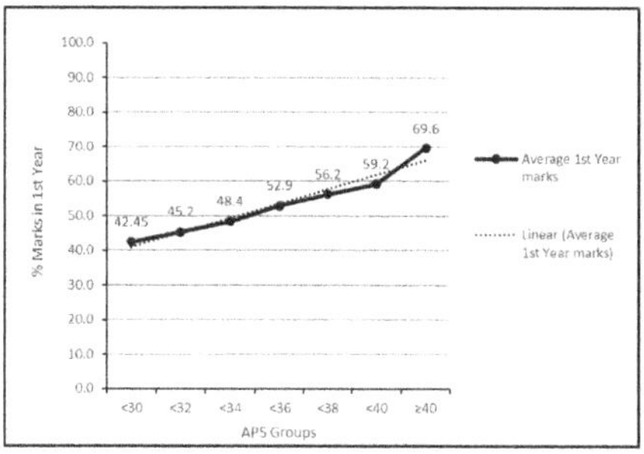

Fig. 1. Weighted average marks in first-year of study

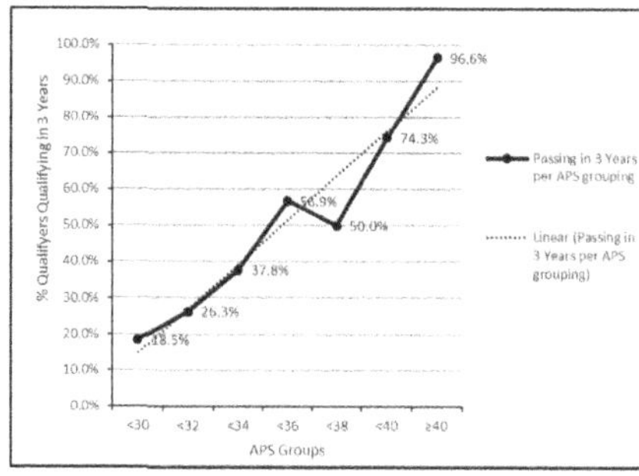

Fig. 2. % of students qualifying in 3 years

The group 7 students (APS of ≥40) achieved average school marks of 80%+, but only achieved an average first-year mark of 69,6%, lending support to the findings of Schöer *et al.* [9] that the APS may be 10–15% too high.

4.2 Through-Put of Students

The department desires better throughput rates of students. The simple linear regression line between the average APS per group as the independent variable and the % of students qualifying in the minimum period of three years as the dependent variable is detailed in Table 3. The data again indicated a positive relationship between the independent and dependent variables (see Fig. 2).

Table 3. % of students qualifying in 3 years.

Stratified APS groups	<30	30&31	32&33	34&35	36&37	38&39	≥40	Slope	Y intercept	Error ϵ
Average APS	28.1	30.6	32.4	34.4	36.4	38.5	40.9			
Students qualifying	27	38	37	51	36	35	29			
Students qualifying in 3 years	5	10	14	29	18	26	28			
% qualifying in 3 years	18.5	26.3	37.8	56.9	50.0	74.3	96.6	5.95	−153.3	7.401

The slope of the simple linear regression line of 5,95 indicated a very significant increase in % of students qualifying in three years with a corresponding rise in APS. The high Pearson's correlation coefficients of 0,9690 and the coefficient of determination of 0,9389 confirms the strong correlation.

Almost half of all students in the study (176 students or 48,8%) have an APS of less than 34, of which only 27,5% completed their studies in 3 years. In contrast, 84,4% of group 6 & 7 students (APS of ≥38) qualified in 3 years.

Group 4 students (APS of 34 & 35) outperformed the group 5 students (APS of 36 & 37), indicating that cognitive skills only explain part of students' performance. The non-cognitive skills of self-discipline, persistence, and motivation identified by research of Fraser and Killen [14] and Heckman and Rubinstein [15] may, in part, explain this finding.

4.3 Number of Students Who Manage to Qualify

The number of students who manage to qualify, even within five or six years, also constitutes some success and throughput measures. The regression line between the average APS per group as the independent variable and the % of students that qualified as the dependent variable is detailed in Table 4.

Table 4. % of students that qualified.

Stratified APS groups	<30	30&31	32&33	34&35	36&37	38&39	≥40	Slope	Y intercept	Error ϵ
Average APS	28.1	30.6	32.4	34.4	36.4	38.5	40.9			
Students enrolled	54	67	55	66	46	40	33			
Students who qualified	27	38	37	51	36	35	29			
Students who qualified (%)	50.0	56.7	67.3	77.3	78.3	87.5	87.0	3.21	−38.38	3.787

The data again indicated a positive relationship between the independent and dependent variables (see Fig. 3). The slope of the simple linear regression line of 3,21 indicated a significant increase in the % of students qualifying with a corresponding rise in APS. The high Pearson's correlation coefficients of 0,972 and the coefficient of determination of 0,9448 confirms the strong correlation (Fig. 4).

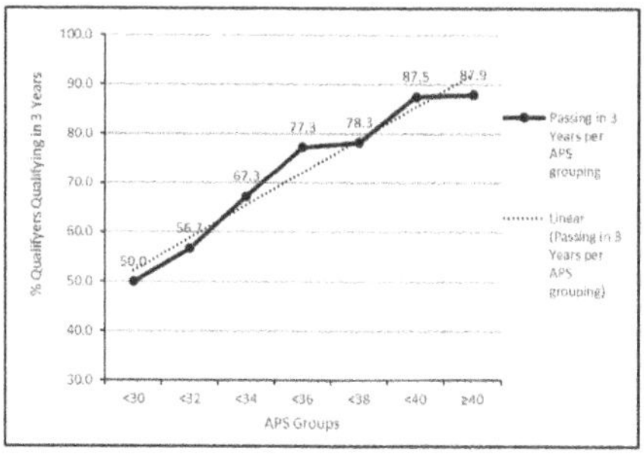

Fig. 3. % of students that qualified

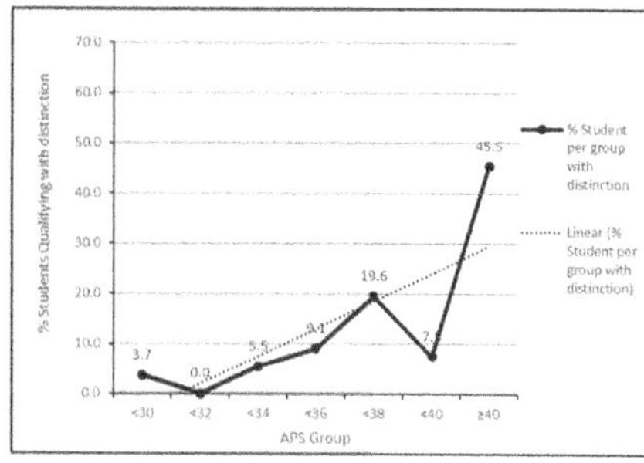

Fig. 4. % of students that qualified with distinction

4.4 Number of Students that Passed with Distinction

The number of students who qualified describes the bottom end of student performance. The number of students qualifying with distinction portrays the top end of the performance. The study also evaluated how many students managed to pass with distinction (75%+ in the final year). The simple linear regression line used the average APS per group as the independent variable and the % of students that qualified with distinction as the dependent variable (see Table 5).

Table 5. Number of students that qualified with distinction

Stratified APS groups	<30	30&31	32&33	34&35	36&37	38&39	≥40	Slope	Y intercept	Error ϵ
Average APS	28.1	30.6	32.4	34.4	36.4	38.5	40.9			
Students enrolled	54	67	55	66	46	40	33			
Students with distinction	2	0	3	6	9	3	15			
Students with distinction (%)	3.7	0	5.5	9.1	19.6	7.5	45.5	2.71	−80.33	10.76

The slope of the simple linear regression line of 2,71 indicated a significant increase in the % of students qualifying with a distinction with a corresponding rise in APS. The Pearson's correlation coefficients of 0,776 and the coefficient of determination of 0,6016 were relatively weak due to poor performance from students in group 6 (APS 38 & 39). Still, they confirmed some measure of the correlation between the independent and the dependent variables.

The study indicated that group 7 students (APS of ≥40) produced a significant share of the distinctions achieved. The poor performance of group 6 students (APS 38 & 39) also indicated some support to the studies of Fraser and Killen [14] and Heckman and Rubinstein [15], indicated that non-cognitive skills might explain part of student performance or lack thereof.

5 Conclusions

Increased throughput and better over-all performance of students are desired academic objectives of the Department. The Department required confirmation that the APS is still a reliable predictor of undergraduate academic performance. The study, therefore, perused the relationships between the APS as the independent variable and four aspects describing the academic performance of students as the dependent variables. The study found a significant correlation between the average APS of students' stratified groups and the different elements of student performance evaluated.

The weighted average marks of the students' groups varied from 42,5% (group 1) to 69,6% (group 7). The data revealed a Pearson's correlation coefficient of 0,9779, while the best-fit simple regression line had a slope of 1,94 and a standard error of the mean of 2,024. These findings confirmed a close relationship between the independent and dependent variables.

Regarding the students completing their studies in three years, only 18,5% of group 1 students complete their studies in three years compared to 96,6% for group 7. A Pearson's correlation coefficient of 0,9779 and the best-fit simple regression line with a slope of 5,95 and a standard error of the mean of 7,401 described the data. The study again confirmed a close relationship between the independent and dependent variables. Group 5 students (APS 36 & 37) produced below-par results, which indicated that non-cognitive skills also explain part of student performance.

The study also evaluated the number of qualifying students regardless of time in the programme. In group 1 50,0% complete their studies compared to 87,9% for group 7. A Pearson's correlation coefficient of 0,972 and the best-fit simple regression line with a slope of 3,21 and a standard error of the mean of 3,787 described the data. The study again confirmed a close relationship between the independent and dependent variables.

To also peruse the relationship between the APS and top students' performance, the study also included the number of students qualifying with distinction. In group 1, only 3.7% complete their studies with distinction compared to 45,5% for group 7. A Pearson's correlation coefficient of 0,776 and the best-fit simple regression line with a slope of 2,71 and a standard error of the mean of 10,76 described the data. The study confirmed a fair relationship between the independent and dependent variables. Indications that non-cognitive skills may explain part of student performance was again apparent in the data.

The study made three main findings. The primary finding was a significant correlation between the APS and all of the student performance measures under scrutiny within the stratified structure followed. The APS was found to be a reliable indicator of various aspects of undergraduate performance and should remain part of new applicants' selection process.

The study also found that students' cognitive skills reflected by the APS only explain parts of student performance. Non-cognitive skills may assist in understanding academic performance more fully.

The third finding was that the weighted average marks' analysis provided support to the conclusions from Schöer *et al.* [9] that the APS may be 10–15% too high.

6 Recommendations

The study should continue to include the 2016 and later cohorts. Future studies should include the other two academic programmes of the department. The study may also consider including other South African institutions that offer similar programmes. To repeat and expand the study will keep track of new trends in the rapidly changing demographic of the South African tertiary landscape and help identify other causes to explain the variability in student performance.

References

1. Department of Basic Education: The National Senior Certificate: A Qualification at Level 4 on the National Qualification Framework (NQF), 15 October 2009. http://education.gov.za
2. Department of Education: Minimum Admission Requirements for Higher Certificate, Diploma and Bachelor's Degree Programmes Requiring a National Senior Certificate, Pretoria, August 2005
3. University of Pretoria: Faculty of Engineering, Built Environment and Information Technology Undergraduate Admissions (2020). https://www.up.ac.za/yearbooks/2020/programmes/view/12242017. Accessed 13 Apr 2020
4. Smith, J., Naylor, R.: Determinants of degree performance in UK universities: a statistical analysis of the 1993 student cohort. Oxford Bull. Econ. Stat. **63**(1), 29–60 (2001)
5. Smith, R., Schumacher, P.A.: Predicting success for actuarial students in undergraduate mathematics courses. Coll. Stud. J. **39**(1), 165–177 (2005)
6. Parker, K.: The effect of student characteristics on achievement in introductory microeconomics in South Africa. S. Afr. J. Econ. **74**(1), 137–149 (2006)
7. Vitale, J.D., Wanger, S.P., Adams, D.: Explaining student performance in an undergraduate agricultural economics classroom. NACTA J. **54**(1), 2–9 (2010)
8. Rankin, N., Schöer, M., Sebastiao, C., Van Walbeek, C.: Predictors of academic performance: national senior certificate versus national benchmark test. S. Afr. J. High. Educ. **26**(3), 564–584 (2012)
9. Schöer, M., Ntuli, M., Rankin, N., Sebastiao, C., Hunt, K.: A blurred signal? The usefulness of national senior certificate (NSC) marks as predictors of academic performance at university level. Perspect. Educ. **28**(2), 9–18 (2010)
10. Wolmarans, N., Smit, R., Collier-Reed, B., Leather, H.: Addressing concerns with the NSC: an analysis of first-year student performance in mathematics and physics. Paper Presented at the 18th Conference of the South African Association for Research in Mathematical, Science and Technology Education, Kwa-Zulu Natal, pp. 274–284 (2010)
11. Zewotir, T., North, D., Murray, M.: Student success in entry-level modules. S. Afr. J. High. Educ. **25**(6), 1233–1244 (2011)
12. Hoffman, D., Pieterse, I.: Academic school performance as an indicator of success in undergraduate studies in construction economics in South Africa. In: Nazir, S., Ahram, T., Karwowski, W. (eds.) AHFE 2020. AISC, vol. 1211, pp. 123–128. Springer, Cham (2020). https://doi.org/10.1007/978-3-030-50896-8_19

13. Hoffman, D., Pieterse, I.: Performance in mathematical and scientific school subjects as an indicator of success in undergraduate modules in construction economics in South Africa. In: Nazir, S., Ahram, T., Karwowski, W. (eds.) AHFE 2020. AISC, vol. 1211, pp. 56–62. Springer, Cham (2020). https://doi.org/10.1007/978-3-030-50896-8_9
14. Fraser, W., Killen, R.: The perceptions of students and lecturers of some factors influencing academic performance at two South African universities. Perspect. Educ. **23**(1), 25–40 (2005)
15. Heckman, J.T., Rubinstein, Y.: The importance of non-cognitive skills: lessons from the GED testing program. Am. Econ. Rev. **81**(2), 145–149 (2001)
16. Latif, A.F., Sachs, R.A.: Can selection assessments predict students' achievements in the premedical year? A study at Arabian Gulf University. Educ. Health **14**(2), 277–286 (2001)

Teaching and Learning

MOOC Courses as a Tool for the Development of Digital Competencies of Teachers

Petr Svoboda(✉) and Lenka Mynaříková

Masaryk Institute of Advanced Studies, Czech Technical University in Prague, Kolejni 2637/2a, 160 00 Praha 6, Czech Republic
{petr.svoboda,lenka.mynarikova}@cvut.cz

Abstract. The aim of the paper is to highlight the importance of MOOC courses as a support and complement in teacher education. Online courses in the MOOC format meet the conditions to offer a possible solution to the urgent need for undergraduate and in-service teacher training in area of the digital competence that teachers have. It is necessary to develop their students' digital literacy. However, there is almost no evidence of the effectiveness and efficiency of MOOCs for developing digital competences. The paper focuses on social science teachers who lag behind others in the use of digital technologies. It is important to create MOOC courses for the development of digital competencies and for the use of digital technologies in teaching.

Keywords: MOOC courses · Digital competencies · Digital technologies · Teachers · Education · Research

1 Introduction

At present, there is little research that systematically analyses the characteristics of teachers (e.g. sociodemographic, pedagogical practice, field of study and subject taught) participating in MOOCs, specifically in terms of their professional development. So, we have no idea which teachers participate, which ones do not, what makes the courses attractive to them or what prevents them from participating. Examining the effectiveness of teaching using these courses has not received enough attention because, as [1] point out in their recent literature review, measurements of learning outcomes achieved in MOOCs are not yet very sophisticated and are often based on a single variable, such as the final grade or completion rate. More empirical and analytical research is needed to identify the potential successes and risks of MOOCs. The paper responds to these shortcomings.

The methodology of the paper is based on the analysis of the current situation in Czech secondary schools and research in this area. The study will serve as a basis for research in this area, which will be carried out at the Czech Technical University in Prague, Masaryk Institute of Advanced Studies within the research grant of the Czech Republic.

S. Nazir et al. (Eds.): AHFE 2021, LNNS 269, pp. 243–251, 2021.
https://doi.org/10.1007/978-3-030-80000-0_29

2 A New and Effective Ways of Training

Given the inevitable and urgent need to train teachers so that they can effectively develop their digital competences skills, it is necessary to look for new and effective ways of training. One possible solution is use of MOOCs for teacher education. MOOCs are online courses that allow massive participation and which can be accessed without restriction and free of charge [2]. Due to the huge proportion between students and teachers at MOOC, individual guidance and monitoring is not feasible, which is why these courses use teaching design that differs from traditional online courses to allow for extensive assessment and feedback. MOOC's educational design is a key aspect because it has a major impact on participants' motivation and outcomes [3]. Based on MOOCs, new types of online courses have emerged, such as SPOC: courses with the same characteristics as MOOCs, except that the number of participants is relatively small and access is provided only to a specific group of people. The term "MOOCs" includes all online courses with a teaching design that is characteristic of MOOCs, i.e., courses that are designed to allow massive participation, even if it actually does not have to be massive. MOOCs meet all the necessary conditions to offer low-cost solutions for the initial and continuous training of all teachers in digital competences. In fact, previous studies have shown that teachers find these courses attractive for digital competence training [4–6]. The suitability of MOOCs to address shortcomings in teacher education has not gone unnoticed by the European Union, which in 2018 led an initiative to train teachers in the safe use of the Internet through the MOOC [7].

3 Positive Social Benefits and Impacts

The findings of research in the field of social and technical sciences draw attention to the profound transformation of educational processes in society after the onset of the 4th Industrial Revolution, whose competitiveness is based on the methods, speed and quality of information processing [8]. At the same time, a profound change in the educational function of the school is expected, which will make more use of individualized learning strategies (personalized learning pathways), independent learning and peer-to-peer activities to go through the learning of pupils [9]. Basic research shows that learning management in the context of the 4th Industrial Revolution will be based on the use of digital technologies, while it will be necessary to address the problem of underdeveloped digital competencies of teachers, support their computer thinking and improve their work with digital technologies in schools [10]. The preliminary research of the project for the development of applied social and humanitarian research, experimental development and innovation [11] found that secondary schools use desktop computers, tablets, interactive textbooks and voting devices the most. Virtual reality glasses, digital assistants, and inter-active walls have been the least frequently used (or not at all) [12, 13].

Undergraduate and in-service teacher education can be provided in different ways: formal education, formal courses or workshops, mutual learning at school or in collaboration with other schools or informal events. Recently, several Massive Open Online Courses (MOOCs) on pedagogical skills have been added to this range of options, offering teachers flexibility and more training options. There is evidence of good acceptance of the MOOC format by teachers. The study showed that approximately 40% of participants in MOOC courses are former or current teachers [14]. Data from the MOOC Knowledge project confirm this high level of teacher participation in MOOCs [15].

Adequately developed digital competences are a necessity for teachers in today's information society, and given the repeatedly confirmed shortcomings in this area, but also the need and motivation to eliminate these shortcomings, the proposed solution may be an attractive choice for teachers. They can study at a time that suits them, they will receive social support from other teachers participating in the course, due to the zero financial costs for schools, it is also possible to count on the support of school management and founders. A number of significant barriers are often eliminated, which often prevent teachers from participating in further education in digital competencies [16, 17].

4 Preliminary Research

In 2019, the Ministry of Education, Youth and Sports in the Czech Republic published the Framework of Digital Competences of Teachers, which is based on the Czech translation of the European Framework of Digital Competences of teachers. The framework of digital competences of teachers in a structured and comprehensive form shows what today's teacher should know and be able to do in the use of digital technologies. It highlights activities and skills that may be underestimated or not commonly considered in connection with teaching. It focuses mainly on pedagogical competencies and provides an insight into how these competencies are influenced by the possibility or necessity of using digital technologies in teaching [18]. It should primarily motivate teachers to further develop their professional development through systematic reflection. However, it is also addressed to all teacher training schools. The framework defines 22 competencies included in six areas: professional involvement, digital resources, teaching, digital assessment, pupil support, support of pupils' digital competencies [18]. The DigComEdu framework is aimed at educators at all levels of education from pre-school to higher education and adult education, including vocational education and training, education of pupils with special educational needs and non-formal education [19].

The aim of the workshop was to acquaint participants with the DigCompEdu framework at the level that they will be able to start thinking about their digital competencies and plan their further development [19]. For example, through MOOC courses. The workshop was adapted to acquaint teachers or other interested parties with digital competencies in the form of active learning and methods of critical thinking, especially with the use of group forms of work. Thanks to this, it is possible to share diverse experiences between teachers through cooperation, discussion and mutual evaluation of different approaches and opinions. Individual work is used mainly to reflect on previous experience with the use of digital technologies in teacher education or preparation for it.

5 Implementation of Preliminary Research

We carried out the preliminary research as a workshop. The workshop was attended by a total of 16 teachers of mainly social science subjects from the Secondary Industrial

Table 1. Teachers of the Secondary Industrial School in Prague.

Competencies	Description
Digital resources	Selection of digital sources - I adapt selected and trusted sources from www pages from the Internet, search (search engines, non-text information (videos), encyclopaedias, help and documentation, English), Creation and editing of digital sources - I refer to those parts of www pages that serves in my presentation to supplement to achieve the goal of the subject topic. I professionally cooperate with the subject guarantor, Creation, storage, presentation of data - text editor (charter encoding, typography), spreadsheet (tables, formulas, graphs), graphic editor (diagrams, bitmaps, vectors), compression programs, printing (to file, on paper), videos, organization, protection, sharing of digital resources - data are organized on the auxiliary storage of the school server so that it is not possible to copy tasks to each other. When copying, I choose freely available materials with regard to their protection. Preparation for teaching - searching for new information, updating preparations, preparation of materials for teaching (presentations, worksheets, quizzes, online forms)
Teaching	Teaching - based on the assignment, students create programs. The program, i.e. software and hardware, is part of the student's workplace. In my teaching, I continuously reflect on my own further education in order to supplement or revive new applications and possibilities related to digital technologies. I take into account formative assessments to meet the needs of students and their awareness that they are making progress and improvement in my subject. Use of presentations, instructional videos, spec. SW for teaching a professional subject, use of online tools (quizzes, tests, mind maps, drawing tools, graphs, filling in online forms), conference SW for distance learning. Pupil management - in case of a pupil's questions, I visit him/her at the workplace and solve his/her problem with the help of additional questions, which encourages him/her to become aware of and achieve the fulfilment of the assigned task. Cooperation of students - cooperation consists in the fact that students respond to assigned tasks and process them as a separate project in electronic form, in which they present the studied knowledge from the manual and verified by practice in practice. Pupils' independent learning - the digital portfolio allows me to provide feedback and organize for each of the pupils and further growth of the pupil in knowledge while incorporating a lack of understanding in the following lessons
Digital evaluation	Evaluation strategy - I choose the form of evaluation from a comprehensive point of view and critically. Analysis of learning outcomes - I record in my diary the performance of students for feedback when deciding on an ambiguous decision in classification. Feedback and planning - I manually record students' progress for feedback and the possibility of consulting with a student or parent. The class teacher also has feedback on request

(*continued*)

Table 1. (*continued*)

Competencies	Description
Pupil support	Accessibility and inclusion - in my lessons I allow students with special needs to have their own laptop, tablet for recording the curriculum or their own IDE development environment for processing practical exercises in the subject of DC/practice. Differentiation and individualization - due to their individual abilities requiring different pace and level of accepted knowledge, pupils are allowed to choose the recommended learning activity (+20%, +50% according to the psychological examination). I also take the classification into account. Activation of pupils - I increase my interest in the subject so that the pupils are stimulated by the practical use in the life and operation of the company and they are thus encouraged to creatively process the assigned work, homework
Safety	Passwords, malware, phishing
Self-education	Information retrieval, online courses, webinars, continuous professional development
Communication	With pupils, with parents, with colleagues, with partner institutions, with teachers out of school, e-mail, forums, chats, conferences, video calls

School in Prague. In addition, another workshop was held with students of teaching at Czech Technical University in Prague, Masaryk Institute of Advanced Studies with a total of 16 students. The workshop devoted space to the creation of a mind map of digital competencies, which allows the introduction and evocation of knowledge that participants already have about the digital competencies of the teacher, as well as the formulation of opinions and attitudes and gradual awareness of everything in the teacher's digital competencies, what teachers use it and what they neglect.

An overview of the competencies that the participants described using mind maps can be found in table number one and in table number two.

Table 2. Students of teaching at Czech Technical University in Prague, Masaryk Institute of Advanced Studies.

Competencies	Description
Digital resources	The right choice of digital resources, creation and modification of digital repositories and resources, sharing of digital resources, their protection and organization. Searching for relevant information on the Internet - Working with the Internet, searching and evaluating resources, creating Internet resources, the ability to plan and organize (data and their collection and evaluation…). Preparation for teaching - use of educational programs, preparation of presentations, preparation of study texts, work with electronic sources of information, creation of animations. Working with data - Data backup, remote access, Sharing and Data organization/clarity

(*continued*)

Table 2. *(continued)*

Competencies	Description
Teaching	Full-time and online teaching - Ability to use hardware and software - use PC, tablet, phones, etc., be able to connect devices, solve problems with technology, knowledge of applications, know where to look for software, be able to control software, audio, projector, network connection, work with a text/spreadsheet editor, creating presentations, working with OS, working with specific programs used in the field. Working on a mobile device - Basic control of a mobile phone for searching for information and using educational applications, sharing a screen for a data projector to demonstrate working on a mobile device. PC equipment in the classroom + data projector - The need to have trained control, Presentations prepared according to recommended standards. Interactive resources used in teaching, Digital footprint when assigning or solving a task, working with interactive elements, working with prepared electronic materials, handling audio-visual equipment, working with remote and virtual laboratories, distribution of teaching materials, tablets in class, interactive pen, working with interactive textbook, google classroom. Pupil leadership, Cooperation between pupils
Digital evaluation	Assessment strategy, Analysis of learning outcomes, Feedback and planning, creating online quizzes and voting, creating online tests and assignments, working with an electronic classification system
Pupil support	Inclusion, approach to students, individualization, differentiation, activation of students. Improving pupils' digital maturity - Digital communication and collaboration, Responsibility and risks in digital. Technology, Information and media literacy, creating collaboration over creating your own content. Problem solving skills - critical thinking, algorithms, logical thinking, analytical thinking
Safety	Basics of cybersecurity, GDPR, knowledge of copyright law, classification and verification of information obtained
Self-education	Self-study and own professional development, Professional cooperation, Self-improvement
Communication	With parents, with students, online assignment, work with el. apologies, el. communication with parents and pupils, email, information system, bachelors, Bulk emails, chats, Common file storage, Student testing, Possibility of personal consultations, video conferencing, social networks, WhatsApp, Skype

6 Summary of Preliminary Research

Group 1 – Teachers of the Secondary Industrial School in Prague.
They most often use digital technologies in these activities: teaching new material, games, creating presentations, preparing digital tests, finding suitable resources for

teaching, correction of electronically submitted work of students, communication with parents, photo editing, video editing, video projection.

Software used: text editors for Arduino IDE programming, Multisim simulation program, Lego Mindstorms EV3 program, ROBOTC program for Lego, Word and Adobe Reader text program, Power Point, Google Classroom, conference software for online communication with colleagues (Skype, ZOOM, Discord) and for self-education (WebEx), applications for greater activation of pupils (Kahoot), for the development of their independent learning (Quizlet).

Change of attitude: in 10 out of 16 there was no change in attitudes towards digital competencies, in a larger number of teachers the attitude towards digital competences is rather negative, they label them nonsense or unnecessary, there is no obvious interest in expanding one's own competencies.

Group 2 – Students of teaching at Czech Technical University in Prague, Masaryk Institute of Advanced Studies.

They most often use digital technologies in these activities: preparation of study materials for self-study, creation of study texts, internet search, presentation of presentations, creation of educational presentations, communication - with parents and pupils, teachers with each other, preparation of exchange student stays, internships, organization of student competitions, work on various remote projects, searching for new information and innovations in the field of its operation, sharing a mobile screen on a data projector while showing a new application, using a tablet and applications to support electronics teaching, using animations on YouTube to represent phenomena and events in electronics use of high-speed camera for shooting fast-moving scenes, editing of video material, online test/questionnaire for teaching, editing of technical drawings.

Software used: MS Word or Adobe Acrobat, Prezi, MS PowerPoint, Google disk/Google classroom, Kahoot, Moodle, Bachelors, interactive textbooks Flex-iBook's from Fraus publishing house, Discord for online conferences, School online.

Change of attitude: 7 out of 16 did not change their attitude, the rest are more aware of the breadth of the issue and its importance. In general, in this group, the attitude towards the use and overall meaning of digital technologies in teaching is more positive, students show a greater interest in a closer understanding of the issue.

Due to the fact that two workshops were carried out with the same number of participants, it was possible to make a comparison of both groups. Part of the workshop was the identification of hitherto unstated needs of teachers in the field of digital competence development. As for where digital technologies are most often used, there is no fundamental difference between the two groups, we find mainly categories: preparation of teaching, preparation of teaching materials/study texts, preparation and evaluation of tests, integration of multimedia into teaching, communication. In terms of software, in the group of teachers we find more represented specialized software (e.g. programming, simulation, etc.), in both groups we find software for making teaching more attractive, such as Kahoot, presentation software, tools for communication with students (Google Classroom, Moodle).

7 Conclusion

Although there is a wealth of research on MOOCs [1, 20–22], examining the effectiveness of teaching using MOOCs not enough attention has been paid and the measurements of the results obtained are not very elaborate and are based on a single variable, such as the final mark or the degree of completion. Most of the literature on MOOCs has focused on topics such as course characteristics, types of MOOCs, challenges, potential impacts on education, characteristics and behaviour of participants, and certification [1, 20–22]. The existing evidence for the effectiveness of MOOCs aimed at training teachers in digital technologies is even weaker than exists for MOOCs in general. Various experiences have been described in the literature in which MOOC courses have been used to train teachers in various areas of digital technology [4, 5, 23–26]. Several of these studies did not carry out any evaluation of the effectiveness of the courses and others provided only evidence obtained through questionnaires completed by the participants themselves. Current evidence on the effectiveness of MOOCs in digital teacher competence training courses is weak and further research is needed on the possible positive impacts of MOOCs on teachers' digital competences [27].

References

1. Deng, R., Benckendorff, P., Gannaway, D.: Progress and new directions for teaching and learning in MOOCs. Comput. Educ. **129**, 48–60 (2019). https://doi.org/10.1016/j.compedu.2018.10.019
2. Siemens, G.: Massive open online courses: innovation in education? In: McGreal, R., Kinuthia, W., Marshall, S. (eds.) Open Educational Resources: Innovation, Research and Practice, pp. 5–16. Athabasca University: Commonwealth of Learning (2013). http://bit.ly/2RhB5bs
3. Castaño, C., Maiz, I., Garay, U.: Diseño, motivación y rendimiento en un curso MOOC cooperativo. [Design, Motivation and Performance in a Cooperative MOOC Course]. Comunicar **22**(44), 19–26 (2015). https://doi.org/10.3916/C44-2015-02
4. Castaño-Muñoz, J., Kalz, M., Kreijns, K., Punie, Y.: Who is taking MOOCs for teachers' professional development on the use of ICT? A cross-sectional study from Spain. Technol. Pedagogy Educ. **27**(5), 607–624 (2018). https://doi.org/10.1080/1475939X.2018.1528997
5. Gómez-Trigueros, I.M.: El MOOC como recurso para la adquisición de la competencia digital en la formación de los maestros de educación primaria. Revista de Tecnología de Información y Comunicación en Educación **11**(1), 77–88 (2017). http://bit.ly/2ZpELuC
6. Ortega-Sánchez, D., Gómez-Trigueros, I.M.: Massive open online courses in the initial training of social science teachers: experiences, methodological conceptions, and technological use for sustainable development. Sustainability **11**(3), 578–588 (2019). https://doi.org/10.3390/su11030578
7. Better Internet for Kids (ed.): #SaferInternet4EU campaign and Safer Internet Day 2018: Public report on campaign activities and successes (2018). https://bit.ly/2TmG5PL
8. Svoboda, P.: Digital technology as a significant support for the teaching process. In: Ahram, T., Taiar, R., Colson, S., Choplin, A. (eds.) IHIET 2019. AISC, vol. 1018, pp. 383–389. Springer, Cham (2020). https://doi.org/10.1007/978-3-030-25629-6_59
9. Svoboda, P.: Assumptions and reasons for the use of digital technologies in education. Studia Pedagogiczne **50**(1), 195–206 (2017). Polska Academia Nauk, Poland
10. Vuorikari, R., Punie, Y., Carretero Gomez, S., Van den Brande, G.: DigiComp 2.0: The European Digital Competence Framework for Citizens (2016)

11. Svoboda, P., et al.: Development of digital competences of teachers of social sciences at secondary vocational schools. ETA TACR project Nr. TL01000192 (2018–2021)
12. Svoboda, P., Lorenzová, J., Jirkovská, B., Mynaříková, L., Vališová, A., Andres, P.: Research of teachers' digital competences in an international context. In: Auer, M.E., Hortsch, H., Sethakul, P. (eds.) ICL 2019. AISC, vol. 1135, pp. 788–799. Springer, Cham (2020). https://doi.org/10.1007/978-3-030-40271-6_77. ISSN 2194-5357. ISBN 978-3-030-40270-9
13. Rojek, M., Leek, J., Svoboda, P.: Exploring the Virtual World of Learning Across Generations. Columbia University Press, Jagiellonian University Press (2020). ISBN 978-83-233-4830-6
14. Ho, A.D., Reich, J., Nesterko, S.O., Seaton, D.T., Mullaney, T., Waldo, J., et al.: HarvardX and MITx: the first year of open online courses, Fall 2012–Summer 2013. HarvardX and MITx Working Paper No. 1 (2014). https://doi.org/10.2139/ssrn.2586847
15. Castaño-Muñoz, J., Punie, Y., Inamorato dos Santos, A.: MOOCs in Europe: evidence from pilot surveys with universities and MOOC learners (2016). https://ec.europa.eu/jrc/sites/jrcsh/files/JRC%20brief%20MOOCs_JRC101956.pdf
16. Mynaříková, L., Novotný, L.: Knowledge society failure? Barriers in the use of ICTs and further teacher education in the Czech Republic. Sustainability **12**(17), 6933 (2020)
17. Mynaříková, L., Svoboda, P., Jirkovská, B., Lorenzová, J.: Barriers of secondary school teachers in the use of ICT for teaching. In: ICERI2019 Proceedings, pp. 2426–2431 (2019). ISSN 2340-1095
18. MŠMT: Digital Competencies Framework (2019). https://www.msmt.cz/vzdelavani/dalsi-vzdelavani/ramec-digitalnich-kompetenci-ucitele
19. Neumajer, O.: Workshop – Jak pracovat s rámcem digitálních kompetencí učitele [Workshop – How to work with the digital competencies for teachers framework] (2020). https://clanky.rvp.cz/clanek/c/Z/22351/workshop-jak-pracovat-s-ramcem-digitalnich-kompetenci-ucitele.html/
20. Chiappe-Laverde, A., Hine, N., Martínez-Silva, J.A.: Literatura y práctica: Una revisión crítica acerca de los MOOC. [Literature and practice: a critical review of MOOCs]. Comunicar **22**(44), 9–18 (2015). https://doi.org/10.3916/C44-2015-01
21. Liyanagunawardena, T.R., Adams, A.A., Williams, S.A.: MOOCs: a systematic study of the published literature 2008-2012. Int. Rev. Res. Open Distance Learn. **14**(3), 202–227 (2013). https://doi.org/10.19173/irrodl.v14i3.1455
22. Veletsianos, G., Shepherdson, P.: A systematic analysis and synthesis of the empirical MOOC literature published in 2013-2015. Int. Rev. Res. Open Distrib. Learn. **17**(2), 198–221 (2016). https://doi.org/10.19173/irrodl.v17i2.2448
23. De-La-Roca, M., Morales, M., Teixeira, A., Hernandez, R., Amado-Salvatierra, H.: The experience of designing and developing an edX's MicroMasters program to develop or reinforce the digital competence on teachers. In: Proceedings of the 2018 Learning With MOOCS conference (LWMOOCS), pp. 34–38 (2018). https://doi.org/10.1109/LWMOOCS.2018.8534697
24. Ramírez-Montoya, M.S., Mena, J., Rodríguez-Arroyo, J.A.: In-service teachers' self-perceptions of digital competence and OER use as determined by a xMOOC training course. Comput. Hum. Behav. **77**, 356–364 (2017). https://doi.org/10.1016/j.chb.2017.09.010
25. Sánchez-Elvira, A., Santamaría-Lancho, M.: Developing teachers and students' digital competences by MOOCs: The UNED proposal. In: Proceedings of the 2013 Open and Flexible Higher Education Conference, pp. 362–376 (2013). http://bit.ly/2XmEDyQ
26. Tsvetkova, M.S.: The ICT competency MOOCs for teachers in Russia. Olympiads Inform. **10**, 79–92 (2016). https://doi.org/10.15388/ioi.2016.special.09
27. Svoboda, P., et al.: Digitální kompetence učitelů středních odborných škol jako výzva současného vzdělávání. Paido, Brno (2020). ISBN 978-80-7315-272-7

Online Mentoring Platform for Developing Leaders of Character

Terence Andre[1](✉), David Huston[2], Anne-Marie Ireland[1], Andrew Stewart[1], Rod Ford[1], Christy Carr[1], and Hannah Staller[1]

[1] TiER1 Performance, Covington, KY, USA
{t.andre,a.ireland,a.stewart,r.ford,c.carr, h.staller}@tier1performance.com
[2] USAF Academy, Colorado Springs, CO, USA
david.huston@usafa.edu

Abstract. Leadership is a variable and complex task that requires many skills along with opportunities to hone those skills. The development of future leaders among current college students includes its own complexities, but these complexities are increased in depth and breadth when the college is also a military academy, shaping men and women who will be commissioned as officers and expected to lead upon graduation. In partnership with the U.S. Air Force Academy (USAFA), our team developed and tested a platform to support leader and character development through mentoring relationships. The platform, known as Advance, provided tools to support reflection, action planning, leadership portfolios, and non-academic training opportunities. Our project included an iterative development model that involved multiple rounds of testing with Air Force officers, USAFA faculty, and cadets. Through this approach we were able to refine the model, improve the mentoring process, and implement the platform.

Keywords: Leadership · Character · Development · Mentoring · Military · Reflection · Action planning · Portfolios

1 Introduction

The United States Air Force Academy (USAFA) provides an elite undergraduate educational program that integrates academic, military, airmanship, and athletic programs to develop leaders of character motivated to lead the Air Force in service to our nation. By the time cadets graduate and are commissioned as second lieutenants, they are expected to have acquired a combination of knowledge, skills, attitudes, and abilities. USAFA has described these particular characteristics in nine Institutional Outcomes.

These nine outcomes integrate into a program that is well-respected for developing quality leaders. However, USAFA leadership has identified a need to discover opportunities to improve cadet leadership development. Currently, there is a lack of transparency between the outcomes and how they are integrated into the curricula and cadet activities. This lack of transparency means that (1) cadets are unable to fully articulate their

S. Nazir et al. (Eds.): AHFE 2021, LNNS 269, pp. 252–259, 2021.
https://doi.org/10.1007/978-3-030-80000-0_30

four-year mission while students at USAFA, (2) assessment of leadership and character strengths is inconsistent, and (3) some cadet leadership activities or opportunities are difficult to identify or evaluate. Furthermore, cadets have suggested that there is a lack of consistent messages about priorities and leadership development across USAFA.

To address these challenges, our team partnered with USAFA's Center for Character and Leadership Development (CCLD) to improve leadership development among cadets through mentoring. As a professional relationship, mentoring can aid the development of cadets through multiple functions to include advising, counseling, sponsorship, visibility, protection, coaching, and role modeling [1]. While mentoring informally exists at USAFA, there is evidence to suggest that many cadets who desire mentoring struggle to begin a mentoring relationship [2]. Working with leaders at USAFA we developed the Advance mentoring platform that would (1) improve leadership development through technology supported mentoring; (2) improve integration of leadership evaluation; (3) support cadet leadership development through reflection, action planning, and mentoring; and (4) provide a leadership portfolio that reveals strengths and development opportunities and creates habits of continuous learning after commissioning.

2 Method

Our work with the USAFA CCLD was organized around two phases. In Phase I, our work focused on developing a prototype tool that supports the assessment and tracking of leadership competencies which include aspects of character and virtues. For that prototype, we built the capabilities on our Performance1 platform. Performance1 is a learning content management system (LCMS) with an architecture that is extensible for a variety of contexts. This extensibility provides opportunities to leverage the platform to adapt and mature the feature set to a variety of contexts. Performance1's architecture includes defining relationships between different types of information (e.g., leadership competencies and leadership activities and assessments) that can be used to provide and capture content in a context familiar to the user. This ability to leverage relationships between types of information also provides unique personalization opportunities that can be leveraged for the increasing responsibilities associated with leadership development.

In Phase II, and the objective of this paper, our work has focused on adding a resiliency component to the Advance platform. USAFA has seen an opportunity to develop a cadet's mental fitness as a leader of character—to respond to the growing demands that leaders in the military need to model and to help their team members grow in competencies of resiliency. Thus, our team has customized our mental fitness education experience as an additional capability within the Advance platform to support cadets at USAFA. The work in Phase II has focused on four components: (1) customize TiER1's mental fitness education program for the USAFA experience, (2) enhance USAFA's technology platform for leadership development, (3) conduct a pilot evaluation with cadets at USAFA, and (4) implement the Advance platform and training at USAFA.

2.1 Requirements Discovery

In defining the requirements for the software solution being customized for USAFA, our team used three components to aid in the analysis and documentation of the solution: user narrative, wireframe prototype, and system requirements.

User Narrative. The goal of the user narrative is to provide a story that describes the context, workflow, desired outcomes, and desired affective response to the use of the tool. For this project, our team defined two user narratives: (1) cadet-focused narrative and (2) permanent party user-focused narrative.

Cadet User Narrative. The first narrative we created focused on a cadet arriving back on the USAFA campus for their third year. Through our discovery sessions with cadets and faculty, we learned that third-year cadets start to develop a personal sense of the Air Force mission and generally approach this third year with excitement. Cadets start to recognize the investment faculty, staff, and squadron members have made in them, and look forward to learning more and beginning to invest in others. To address the user experience for a third-year cadet, we offered them the opportunity of completing a third-party 360 feedback assessment (to be completed in conjunction with Advance at least annually, with options to re-assess any time) that will help cadets begin a conversation with their mentor and enable them to create individual action plans. After completing the assessment, the cadet is provided with an overview of their strengths and recommendations on which competencies they may want to focus on over the next few months.

Permanent Party User Narrative. This narrative focused on a new faculty member who has come on board at USAFA and is beginning to learn how to mentor cadets. The dashboard for the permanent party displays notifications from cadets who have requested a mentoring relationship. Within that dashboard, the faculty member can see a cadet's recent activity and determine where a mentoring experience would be most valuable, and can initiate that experience through a meeting or by offering resources to the cadet based on the content of their reflections.

Wireframe Prototype. The wireframe prototype provides a graphic illustration of the screens identified in the narrative. This prototype helped the USAFA users visualize the tool and identify specific datasets that would be needed. We developed the prototype in Axure RP and Adobe XD, prototyping tools that allow wireframes to be linked together to give the feel of a working system. The tool also allows users to add comments throughout the iterative design process. Figure 1 shows an example wireframe prototype of the mentor dashboard, and Fig. 2 shows a mockup of what that mentor dashboard screen will eventually look like.

The wireframe prototype provides a graphic illustration of the screens identified in the narrative. This prototype helped the USAFA users visualize the tool and identify specific datasets that would be needed. We developed the prototype in Axure, a prototyping tool that allows wireframes to be linked together to give the feel of a working prototype. The tool also allows users to add comments throughout the iterative design process. Figure 1 shows an example wireframe prototype of the mentor dashboard.

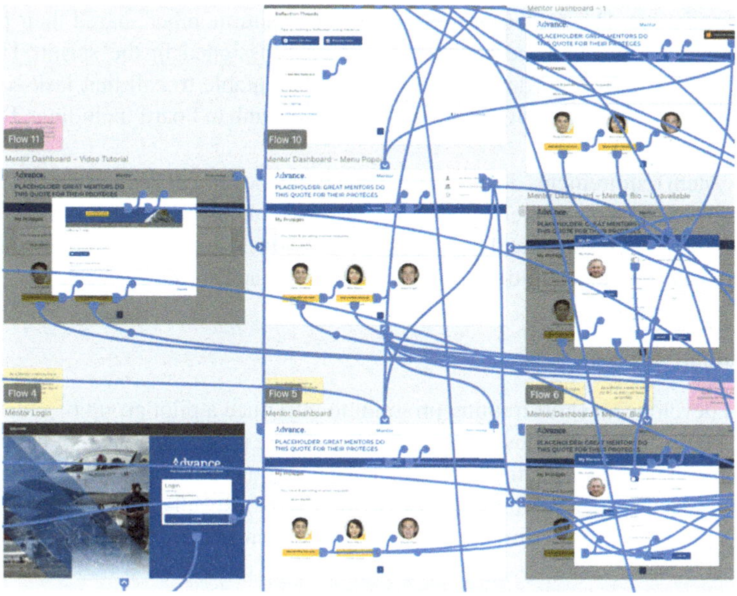

Fig. 1. Mentor dashboard wireframe.

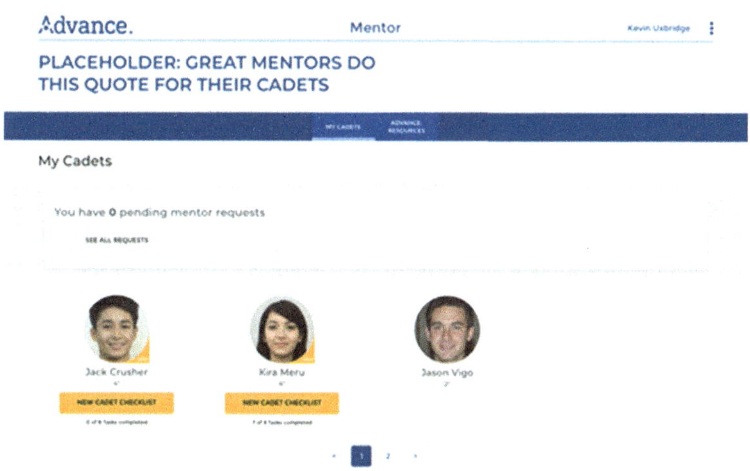

Fig. 2. Mentor dashboard mockup.

System Requirements. TiER1 used an agile framework to manage the definition of system requirements and the development process of the Advance platform. A typical development sprint included grooming the project backlog, creating a sprint backlog, estimating and assigning sprint tasks, getting development team commitment to complete assigned tasks, and then completing the work during the sprint. As the sprints ended, the team reconvened and performed the Sprint Demo, Sprint Retrospective, and Sprint

Closeout activities. During the Sprint Demo, each team member shared their progress with the team, aligned to the tasks they had been assigned in the sprint. The team discussed the progress and determined if it was acceptable to call that task complete. Each task was moved through several statuses on a Kanban board including, To Do, In Progress, and Done.

The system requirements documented throughout the project were added to our sprint management tool, Jira, where day-to-day task assignments are made, and the Kanban board is managed. Throughout the process user narratives, wireframes, and requirements are updated as new information is discovered during each sprint.

2.2 Evaluation Plan

Our team developed a short training program to introduce a pilot group of cadets to the Advance platform. We also developed a plan to evaluate the effectiveness of the Advance platform based on the Kirkpatrick [3] model of evaluation. This evaluates training at four levels: (1) Reaction, (2) Learning, (3) Behavior, and (4) Results.

At the first level, **Reaction**, we will assess what cadets thought about the training program in terms of usability, clarity, and relevance of content.

The second level, **Learning**, is intended to assess the change in knowledge achieved after cadets complete the training material. This level is highly related to the identified learning objectives for the pilot and addresses questions such as: (1) how much did they learn? (2) did they learn what they were supposed to? and (3) do they understand the habit-forming behaviors necessary to grow in their competencies?

The third level, **Behavior**, is intended to determine whether changes in on-the-job behavior have been achieved. These questions will be directed at permanent party/staff. The Behavior level addresses questions such as: (1) are they putting their learning to use? (2) are they able to teach this knowledge to others? and (3) what do you expect cadets to be able to do as a result of this training?

The fourth level, **Results**, is aimed at the organizational level and determines whether there have been organization-wide performance changes as a result of the training. This addresses questions such as: (1) what are the overall goals of the evaluation? (2) what outcomes should be achieved? and (3) was the organization positively affected by the individual behavior changes?

During the pilot study and all of Phase II, we primarily focus on levels 1 and 2. Levels 1 and 2 are intended to ensure that the training program is effective and of good quality for each individual cadet. These will be the most easily assessed levels and will serve to provide information about whether levels 3 and 4 will be effective.

To assess level 1, the reaction of the cadets, we will ask them to complete a reactionary questionnaire once they finish the training. This questionnaire was based on usability surveys and includes both Likert-type and open-ended responses. To assess level 2, the learning achieved by the cadets, we will ask them to complete an assessment developed by our instructional designers. This assessment will measure whether cadets were obtaining the appropriate knowledge and skills presented in the training. Our plan is for cadets to complete this assessment one week prior to the pilot, immediately following the pilot, and prior to the end of the year before they move on in their careers.

As mentioned earlier, the predominant focus of the pilot study is to assess cadets' reaction and learning of the training materials. Figure 3 shows a timeline of events planned for the pilot study.

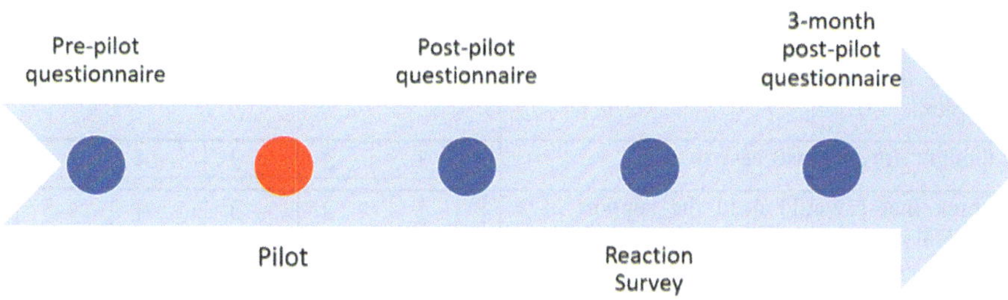

Fig. 3. Evaluation process.

Before the Pilot. Prior to participating in the Advance pilot, a sample group of cadets (17 out of 39) will be asked to complete a 360 feedback assessment. The questions will assess cadets on three identified competencies most relevant to the cadet experience. In addition, all cadets will be asked to take a survey to assess their prior experience with mentoring at USAFA (Fig. 4).

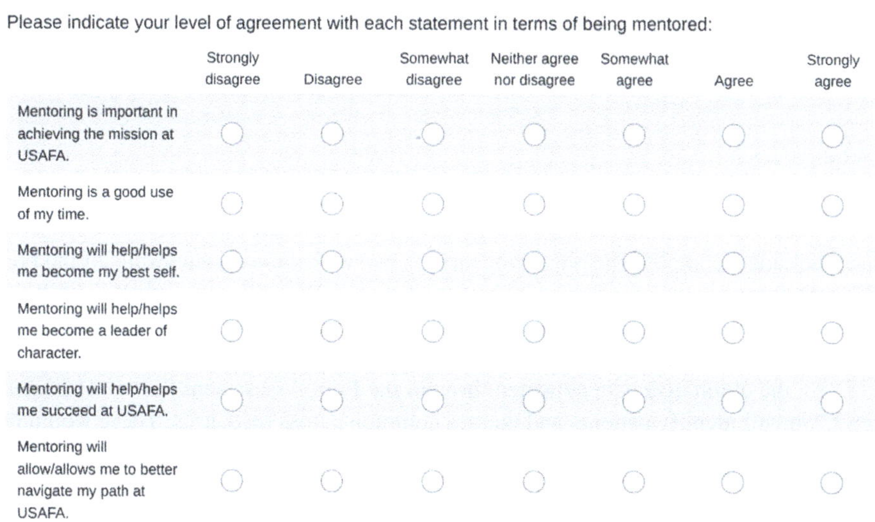

Please indicate your level of agreement with each statement in terms of being mentored:

	Strongly disagree	Disagree	Somewhat disagree	Neither agree nor disagree	Somewhat agree	Agree	Strongly agree
Mentoring is important in achieving the mission at USAFA.	○	○	○	○	○	○	○
Mentoring is a good use of my time.	○	○	○	○	○	○	○
Mentoring will help/helps me become my best self.	○	○	○	○	○	○	○
Mentoring will help/helps me become a leader of character.	○	○	○	○	○	○	○
Mentoring will help/helps me succeed at USAFA.	○	○	○	○	○	○	○
Mentoring will allow/allows me to better navigate my path at USAFA.	○	○	○	○	○	○	○

Fig. 4. Self-assessment survey.

After the Pilot. Near the conclusion of the pilot study, we will ask the same 17 participants to again complete the 360 feedback assessment. This second assessment will provide data about what cadets learned during the pilot study. Participants will also complete the pilot assessment post survey. Figure 5 shows some selected questions from that survey. This survey will assess the cadets' overall reaction to the usability of the software platform and general reactions to the content and its predicted usefulness.

I would like to use *Advance* frequently.	1	2	3	4	5
I found the *Advance* interface unnecessarily complex.	1	2	3	4	5
I thought *Advance* was easy to use.	1	2	3	4	5
I think that I would need the support of a technical person to use *Advance*.	1	2	3	4	5
I thought the various functions of *Advance* were well integrated.	1	2	3	4	5
I thought there was too much inconsistency in *Advance*.	1	2	3	4	5
I imagine that most people would learn to use *Advance* very quickly.	1	2	3	4	5
I found *Advance* very cumbersome to use.	1	2	3	4	5
I felt very confident using *Advance*.	1	2	3	4	5
I needed to learn a lot of things before I could get going with *Advance*.	1	2	3	4	5

Fig. 5. Pilot assessment post survey.

2.3 Analysis Plan

Once the cadets complete the post-pilot surveys, we will compare the results of the pre-pilot self-assessment to the post-pilot self-assessment. We will use common statistical techniques such as the two-sample, T-test or the Mann-Whitney test to determine the degree of learning the 2d Lt/cadets experienced.

From the qualitative data collected through the Pilot Assessment Survey and interviews, we will identify patterns and themes common across responses. These would be used to identify lessons learned to inform any design changes to the Advance platform.

3 Summary and Future Directions

Our team was able to successfully develop this mentoring platform and develop a plan for evaluation with users (both mentors and cadets/mentees). We aim to demonstrate the value of Advance in supporting cadets' development as leaders of character. We continue to refine the Advance platform and extend our work in three specific areas: (1) development of enhanced capabilities for the support of non-academic training to further cadets' leadership development, particularly in the area of personal and team leadership competencies to include mental fitness; (2) development of enhanced mobile device support; and (3) complete documentation and testing required in order to obtain full system authority for operating on the USAFA intraweb.

Acknowledgments. This project was supported by Air Force Contract No. FA864920C0225. Any opinions, findings, conclusions, or recommendations expressed are those of the authors and do not necessarily reflect the views of the Air Force.

References

1. Kram, K.E.: Mentoring at work: developing relationships in organizational life. Scott, Foresman and Company, Glenview (1985)
2. Huston, D.D.: An inventory of selected mentor and Protégé motivational and effectiveness characteristics at the United States Air Force Academy. Dissertations (2017). https://irl.umsl.edu/dissertation/644
3. Kirkpatrick, D.L.: Evaluating Training Programs: The Four Levels. Berrett-Koehler, San Francisco (1998)

Architectural Design for the Implementation of Learning Analytics: Case Study at Salesian Polytechnic University

Santiago Teodoro Castro Arias[1](✉), Miguel Angel Quiroz Martinez[2], Maikel Yelandi Leyva Vazquez[2], and Angela Cecilia Flores Ortiz[1]

[1] Academic Vice-Rector's Office, Universidad Politécnica Salesiana, Cuenca, Ecuador
{scastroa1,aflores}@ups.edu.ec

[2] Computer Science Department, Universidad Politécnica Salesiana, Guayaquil, Ecuador
{mquiroz,mleyvaz}@ups.edu.ec

Abstract. Learning analytics consist of measuring, capturing, and analyzing student data. Currently, the use of learning analytics in Universities constitutes a world trend that helps to make better academic decisions in the Salesian Polytechnic University with the purpose of decrease the dropout rates of the career. The institution developed tutoring processes in which help is provided to students in the teaching-learning process. The problem is that attendance at these tutorials is optional. There is a proliferation of concern about whether the students who most need it the most are the ones who attend these tutorials. Consequently, the University tries to identify the students that are most likely to drop out of their careers before it happens and offer them the help they need through these tutorials. Today, student's data are growing and are indifferent and scattered databases. This article proposes an architectural design for the implementation of learning analytics that seeks to strengthen students to achieve their personal and academic goals. The evaluation of the architecture was based on the LWPM operator and shows a high evaluation, according to the experts, highlighting the maintainability and reliability of this architecture.

Keywords: Learning analytics infrastructure · Education data · Efficient data processing · LWPM

1 Introduction

The Society for Learning Analytics recognizes the "measurement, collection, analysis and presentation of academic data and their contexts for understanding, advancing and learning strategies in the area of education" [1].

Learning analytics is a promising strategy that addresses persistent educational obstacles in Latin America, like the quality of discrepancies and school dropout rates [2]. The growth of the information that students generate when using the library, registering, enrolling, making requests, browsing the University website, interacting with collaborative learning tools, etc., have accelerated this field of study [3]. Today learning analytics plays an important role in higher education.

S. Nazir et al. (Eds.): AHFE 2021, LNNS 269, pp. 260–268, 2021.
https://doi.org/10.1007/978-3-030-80000-0_31

Most of the studies in the field of learning analytics are underdeveloped in Latin America, it is for this reason that a Data Analytics Community (LALA-Learning Analytics in Latin America) was formed and financed by the European Union. This group was created as an international group of free access formed by the Institution of Higher Education (IES) [4, 5] and today, the Salesian Polytechnic University belongs to this group.

A very important aspect in the adoption of Learning Analytics is to obtain the academic data of the students. this data must be ordered, valid, reliable, and clean to integrate them into a unified data model. Currently, the data of the Salesian Polytechnic University are obtained from the academic, financial, collaborative learning, student welfare systems, and these are stored and administered separately, as also mentioned [6] and [7]. The collection and processing of data frame a complex context which attempts the integration of this information, which is one of the most challenging aspects of the usage of learning analytics tools [8]. It is very important to protect the personal data of students, such as their health, region, sexual orientation, religion, etc. As mentioned by [4] and also [9].

The Salesian Polytechnic University, having academic data with very high failure rates, has intensely developed tutoring programs to assist students, mainly in the subjects of greater repetition, to guarantee the teaching-learning process of the students [10]. This process has helped many students to continue with their studies, however, as attendance of tutorials is optional, it raises the concern of if the students who attend these tutorials are even those who need it the most.

The Tutorials present a problem about the lack of a correct accompaniment to the students who need it the most, in this context the Learning Analytics has not been sufficiently exploited to solve this problem and identify the students who need it most. Today, the University seeks to identify students with academic problems before they request it. The objective of this work is to design something that allows adapting a learning analysis board that can predict students with the probability of dropping out, improving the effectiveness and efficiency in the institution's education, and it must also guarantee scalability to other contexts as well. They designed the LISSA project in Europe [11], and some others in Latin America. Hoping that the design of this architecture serves for the development of a Dashboard that allows the support of the conversation of a student and an advisor professor to positively motivate the student, and hold personalized conversations with each of them [12].

2 Materials and Methods

In the context of all the importance of using learning analytics tools that allow supporting an academic counseling process and analyzing the academic progress of students is undeniable. This article shows an observational, descriptive, and explanatory study to understand the existing problems in the practice of learning analytics in HEIs. This makes it possible to provide an adequate analysis of the data, to benefit academic decision-making.

2.1 Materials

Examine the Academic System (SNA), Student Welfare System (SBE), Financial System (SIGAC), and Collaborative Learning System (AVAC) to verify stored information and access to data. Base the proposed architecture based on scientific techniques and the evaluation of experts in learning analytics using the linguistic representation model based on 2-tuples. A mixed perspective was applied in the search, in which several scientific methods were used for the collection, exploration, and processing of data, among the main ones.

2.2 Methods

The learning analytics tools generated by the University of Cuenca were used as support material for the design of this architecture. That allowed a better understanding of the support used when analyzing the academic performance of students (Fig. 1).

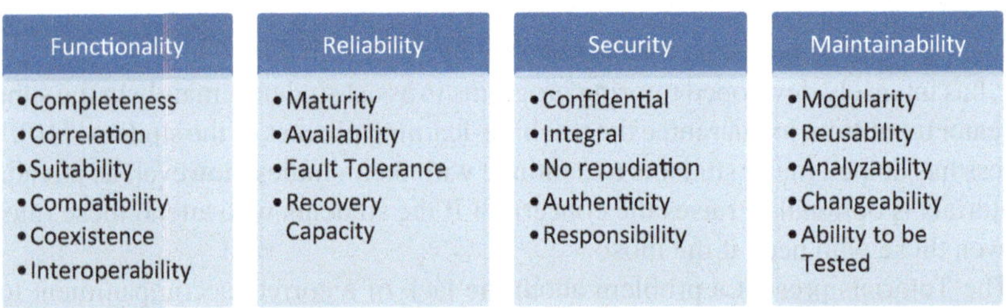

Fig. 1. Internal and external quality requirements for the development of the proposed architecture.

Architecture Evaluation. The activities included for the progress of the proposed architecture establish the feasibility, according to the experts, a framework for the evaluation is established. Workflow for the evaluation of the Architecture. In Fig. 2, the activities for the workflow are graphically displayed:

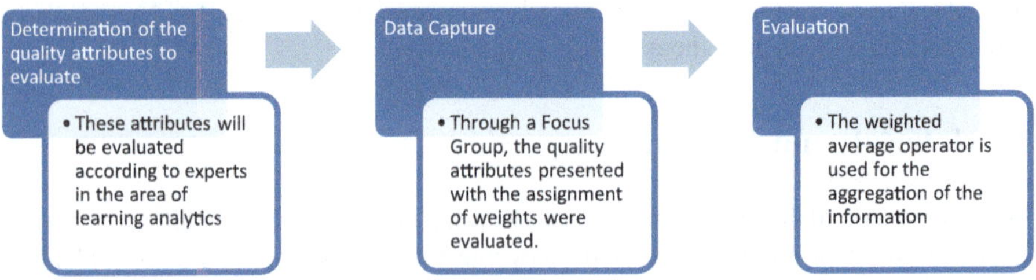

Fig. 2. Workflow activities for architecture assessment.

From the determination of the attributes, the following aggregation structure was defined (Table 1), taking into consideration the importance and the level of simultaneity of the criteria.

Table 1. Aggregation structure

Inputs	Operator		Block ID	Operator		Block ID
Completeness	0,4	C–	Functionality	C–	0,3	Architecture assessment
Correlation	0,3					
Suitability	0,3					
Coexistence	0,5	C–	Compatibility		0,2	
Interoperability	0,5					
Maturity	0,2	C–	Reliability		0,1	
Availability	0,3					
Fault tolerance	0,3					
Recovery capacity	0,2					
Confidentiality	0,2	C–	Security		0,1	
Comprehensiveness	0,2					
No repudiation	0,2					
Authenticity	0,2					
Responsibility	0,2					
Modularity	0,3	C–	Maintainability		0,3	
Reusability	0,2					
Analyzability	0,2					
Changeability	0,2					
Ability to be Tested	0,1					

3 Results

The results obtained show the design of the Architecture for the implementation of learning analytics at the Salesian Polytechnic University. Considering that the migration of educational data comes from different sources, the proposed architecture allows adding new sources to the unified data model.

3.1 Proposed Architecture

In the following architecture, predictive analysis is sought to make predictions based on the academic and financial data of the students. This process seeks to design the architecture that allows the use of the data together with analytical, statistical, and machine learning techniques to create a predictive model to predict the students with the greatest chances of graduation. The learning analytics aims to: identify the students with the highest probability of dropping out of the career. The process uses data sets in models that can generate clear results and allow acting on them to achieve the result in question.

Data Preprocessing. Clean up data by separating outliers by combining data sources, identifying missing data, or outliers to exclude. The different data sources are then combined in this case, creating unique tables for the students and their grades. Organizing the records by semesters of all students in each semester taken. Identifying all students whose approval methods are homologations or validations in a single approval method different from the normal approval method. Those students who have qualifications in more than one degree and taking only the one they are currently studying.

Training. Prediction is a confusing procedure with many variables, so branch algorithms could be used to create and train a predictive model. Perform multiple iterations with the training dataset to experiment with various approaches. After training is complete, you can test the model with new data to see how it performs.

Proposed Architecture. Once the model that accurately predicts the grade probability is found, it will be applied to the unified data model to make the analyses available to the learning analysis panel of the Salesian Polytechnic University (Fig. 3).

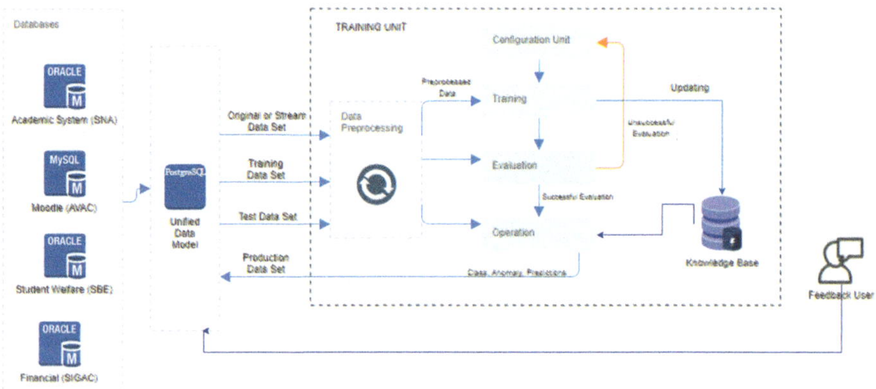

Fig. 3. Proposed architecture for LA implementation in UPS

3.2 Evaluation

The evaluation is got complete in a hierarchical aggregation process in which the logical preference scoring model (LSP) [13] is used affording to the fact that it fits the process in a more reasonable manner. The use of these aggregation operators in a hierarchical way gives suppleness to the process. The leeway of clear procurement the preferences of the specialists and their expression in the weight vectors is additional of its strengths.

Set of linguistic labels used and their respective function of belonging [14] (Table 2).

Table 2. Linguistic assessment

No	Tags	Membership functions
S0	Very Low [VL]	(0, 0, 0.25)
S1	Low [L]	(0, 0.25, 0.50)
S2	Medium [M]	(0.25, 0.5, 0.75)
S3	High [H]	(0.5, 0.75, 1)
S4	Very High [VH]	(0.75, 1)

Next, the linguistic labels and their membership functions are represented in Fig. 4 and Table 3.

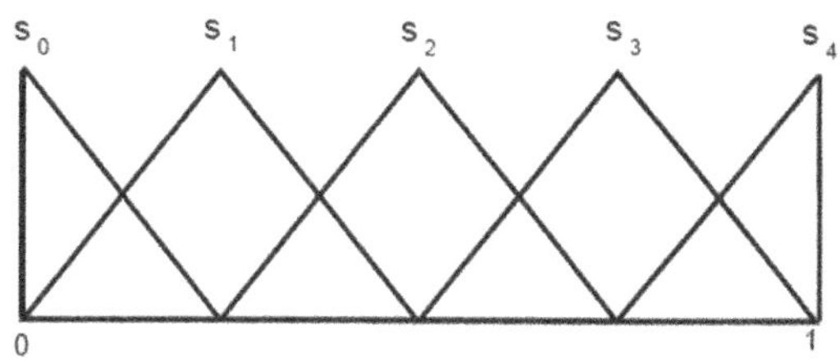

Fig. 4. Graphic representation of the labels used

Collective Evaluation of Architecture. From the evaluation provided by the experts, the following evaluation was obtained for the quality attributes selected in Table 4.

The aggregation was developed in two stages, in the first stage, the values for each of the attributes were obtained.

The result of the aggregation gives us the following linguistic information (S3, 0.35) which shows a high evaluation of the experts.

Table 3. Linguistic assessment of the criteria

Inputs	Assessment	Block ID
Completeness	S_3	Functionality
Correlation	S_3	
Suitability	S_4	
Coexistence	S_2	Compatibility
Interoperability	S_4	
Maturity	S_4	Reliability
Availability	S_4	
Fault tolerance	S_3	
Recovery capacity	S_3	
Confidentiality	S_2	Security
Comprehensiveness	S_4	
No repudiation	S_3	
Authenticity	S_2	
Responsibility	S_4	
Modularity	S_4	Maintainability
Reusability	S_4	
Analyzability	S_3	
Changeability	S_4	
Ability to be tested	S_4	

Table 4. Attribute evaluation results

Tags	Assessment
Functionality	(S3, 0.29)
Compatibility	(S3, −0.07)
Reliability	(S3, 0.48)
Security	(S3, −0.05)
Maintainability	(S4, −0.21)

4 Conclusions

The application of Learning Analytics in higher education institutions provides the ability to make better academic decisions. - The architecture proposed for the development of an application for learning analytics can be applied in a process of academic counseling, with the appropriate technology for its implementation.

The architecture proposed for the development of an application for learning analytics can be applied in a process of academic counseling, with the appropriate technology for its implementation.

The proposed architecture was evaluated using the LSP method from the quality attributes that showed a high value according to the hierarchical aggregation.

Migration of institutional databases to a unified data model will allow adding more databases if required.

As future work, the development of the architecture proposed in the article and the use of educational data mining towards the unified data model is proposed.

Acknowledgments. This work has been supported by the GIIAR research group and of the Academic Vice-Rector's Office in the Universidad Politécnica Salesiana.

References

1. Society for Learning Analytics Research (SoLAR): solaresearch (2011). https://www.solaresearch.org/about/what-is-learning-analytics/
2. Hilliger, I., et al.: Identifying needs for learning analytics adoption in Latin American universities: a mixed-methods approach. Internet High. Educ. **45**, 100726 (2020). https://doi.org/10.1016/j.iheduc.2020.100726
3. Domínguez Figaredo, D., Reich, J., Ruipérez-Valiente, J.A.: Analítica del aprendizaje y educación basada en datos: un campo en expansión. RIED. Rev. Iberoam. Educ. a Distancia **23**(2), 33 (2020). https://doi.org/10.5944/ried.23.2.27105
4. Maldonado-Mahauad, J., Hilliger, I., Pérez-Sanagustín, M., Millecamp, M., Verbert, K., Ochoa, X.: The LALA project: building capacity to use learning analytics to improve higher education in Latin America. In: Companion Proceedings of the 8th International Learning Analytics & Knowledge Conference, pp. 630–637 (2018)
5. LALA Project (2020). www.lalaproject.org
6. Lemaitre, M.J.: Quality assurance in Latin America: current situation and future challenges. Tuning J. High. Educ. **5**(1), 21 (2017). https://doi.org/10.18543/tjhe-5(1)-2017pp21-40
7. Amo Filva, D.: Privacidad y gestión de la identidad en procesos de analítica de aprendizaje con Blockchain (2018). https://gredos.usal.es/jspui/bitstream/10366/137650/1/PlandeInvestigacion-rep.pdf%0A. https://repositorio.grial.eu/handle/grial/1271
8. Sayar, F.: Towards learning analytics adoption: a mixed methods study of data-related practices and policies in Latin American universities. J. Chem. Inf. Model. **53**(9), 1689–1699 (2019). https://doi.org/10.1017/CBO9781107415324.004
9. Hilliger, I., et al.: Towards learning analytics adoption: a mixed methods study of data-related practices and policies in Latin American universities. Br. J. Educ. Technol. **51**(4), 915–937 (2020). https://doi.org/10.1111/bjet.12933
10. Solano Gallegos, S., Tobar Pesántez, L.: Repitencia estudiantil en la Universidad Politécnica Salesiana, vol. 369, no. 1 (2018)
11. Millecamp, M., Gutiérrez, F., Charleer, S., Verbert, K., De Laet, T.: A qualitative evaluation of a learning dashboard to support advisor-student dialogues. In: ACM International Conference on Learning Analytics and Knowledge, no. 2, pp. 56–60 (2018). https://doi.org/10.1145/3170358.3170417
12. Guerra, J., et al.: Adaptation and evaluation of a learning analytics dashboard to improve academic support at three Latin American universities. Br. J. Educ. Technol. **54**, 973–1001 (2020). https://doi.org/10.1111/bjet.12950

13. Dujmović, J.J., Nagashima, H.: LSP method and its use for evaluation of Java IDEs. Int. J. Approx. Reason. **41**(1), 3–22 (2006). https://doi.org/10.1016/j.ijar.2005.06.006
14. Al-Subhi, S.H.S., Mahdi, G.S.S., Alava, M.V., Pérez, M.P.Y.P., Leyva-Vázquez, M.: Operador media potencia pesada lingüística y su aplicación en la toma de decisiones [Linguistic weighted power means and its application for decision-making]. Int. J. Innov. Appl. Stud. **22**, 38–43 (2017)

Activity Design of Nature Experience Teaching

Rong-Lin Zhou(✉) and Ding-Bang Luh

School of Art and Design, Guangdong University of Technology, Guangzhou 510090, China

Abstract. Aiming at the problems of insufficient development of natural experience activities and lack of relevant theoretical research on activity teaching, the design of natural experience teaching activities is studied through literature research, observation method, and participation method. Design corresponding nature games and nature experience activities on the basis of this design method, so that educators can lead the audience to experience nature in depth through the guidance of this method, establish a connection with nature, promote the harmonious development of man and nature, and enhance the relationship between man and nature. The relationship between self and others.

Keywords: Natural experience · Activity design · Natural education

1 Introduction

With the rapid development of urbanization, the interdependence between man and nature has become a relationship between subject and object, the ethical order of man and nature living in harmony has been severely damaged, and the natural ecology is facing a serious crisis. The harm caused by detachment from nature has become a social problem that needs to be solved urgently. Nature experience education, as an important method to cure the "nature deficiency syndrome", plays an important role in improving the public's ecological environment literacy and promoting education for sustainable development. At the same time, exposure to nature also contributes to physical and mental pleasure.

Nature education in China started late, and it is relatively backward in terms of academic research and theoretical application of nature experience activities, which is manifested in the lack of experience of activity teachers, lack of professional standards, insufficient research on teaching methods and activity design methods of nature experience activities, etc., nature experience The development of the education industry urgently needs theoretical research on teaching guides and activity development.

This project aims to study the procedural methods of natural experience teaching activities design through literature research, observation method, and participation method. Through the guidance of the program, the audience leads the audience to experience nature in depth, establish a connection with nature, promote the harmonious development of man and nature, and enhance the relationship between man and nature, self and others. This research can enrich the theoretical research content of the development of natural experience education in China, provide a theoretical basis for guiding

S. Nazir et al. (Eds.): AHFE 2021, LNNS 269, pp. 269–275, 2021.
https://doi.org/10.1007/978-3-030-80000-0_32

the development of natural experience activities, enrich the content of activity teaching materials, and make up for the relative shortcomings in activity design. At the same time, in the educational sense, cultivate people's awareness of respecting and protecting nature, and cultivate responsible behavior for nature.

2 Research Status of Nature Experience

The development of nature education in China started relatively late. As of 2019, China has not issued a national special environmental education laws and regulations, but the "Environmental Protection Law" and other laws and regulations involve environmental education-related provisions. In 2010, Taiwan region of China enacted the "Environmental Education Law". In 2011, Ningxia passed the "Regulations on Environmental Education in the Ningxia Hui Autonomous Region", becoming the country's first local legislation on environmental education. Since then, Tianjin (2012), Xiamen (2013), Luoyang (2014), Guangdong (2018) and other cities have successively formulated local environmental education regulations.

2.1 The Concept of Natural Experience

Regarding the definition of the concept of natural experience, different scholars have given different views. Scholar Gu Jing proposed that nature experience education refers to activities that guide children or adults to contact, observe and explore nature through experience. By guiding children to experience the beauty and wonder of nature, it stimulates curiosity and exploration of nature and surrounding things. Desire, cultivate positive emotions of liking animals and plants, being close to nature, and caring about the environment. The course emphasizes in the real and daily nature, through sensory participation, physical experience, and game-based methods to carry out nature observation, nature exploration, nature experience, nature play, nature discovery, nature representation and other teaching activities. [1] Scholar Tao Xue defines nature experience education as: taking people as the subject and nature as the object, taking science popularization of animals, plants, natural landscapes and phenomena and participating in nature experience activities as the main form, in order to help experiencers understand and perceive nature and cultivate An educational method aimed at its natural environmental awareness and social responsibility. [2] The scholar Wu Jiahe emphasizes the education of nature around nature, by mobilizing the five senses, feeling the stimulation of the natural environment wholeheartedly, and connecting the inner world with the natural world. [3] Scholar Yue Wei believes that nature experience education is a form of education that uses nature as a place to enable students to gain direct experience through purposeful and systematic experiential activities, thereby acquiring comprehensive knowledge, improving practical ability, and establishing an emotional connection with nature.

Based on the viewpoints of the above scholars, this article interprets the natural experience as the way people touch, observe, explore, and play in the natural environment through sensory participation and physical experience, thereby establishing a connection with nature and enhancing the harmonious symbiotic relationship between man and nature. [4].

2.2 Development Status

In 2005, American writer Richard Love first proposed "Natural Deficiency" in his best-selling book "The Last Child in the Woods". In 2010, this book was released in China in Chinese. Since then, the concept of nature education has been quickly recognized in our country, and the consciousness of nature education has begun to be awakened. The mental and physical problems caused by nature deficiency have gradually attracted the attention of the society and the voice of public opinion. The nature education industry The initial start-up has also begun. The beginning of rapid development of nature education in our country can be traced back to around 2010. Before this, many institutions that carried out environmental education related work have laid out the exploration, practice, and promotion of nature education.

According to the "2018 Nature Education Industry Survey Report", newly established nature education institutions have sprung up in recent years, and their development momentum should not be underestimated. Non-formal educational institutions mainly guide people into nature and experience the beauty of nature through nature walks, nature games, and parent-child outdoor activities. China learns from the experience of other countries and has formed a natural education practice model of "teaching + nature school + nature experience". In terms of teaching, schools all over the country have begun to attach importance to the integration of the concept of natural education. In terms of nature schools and nature experience, as of 2019, the China Nature Education Commission has awarded 20 nature schools.

From the distribution point of view, most of the natural education institutions in coastal or economically developed cities such as Sichuan, Beijing, Guangdong, etc., from the site point of view, choose parks, reserves, botanical gardens, museums and other open venues to respond to the demand for venues for nature education activities. Institutions accounted for the highest proportion (35%). [3] The service types of the nature education industry include knowledge-based nature education, experience-based nature education, study tour-based nature education, research-based nature education, and interpretive nature education. From the content point of view, the proportion of experience activities/courses is the highest (35%), followed by interpretative display (17%) and travel planning (13%) [2]. It can be seen from this that nature experience education occupies an important position in nature education activities. Therefore, this article is based on the research of nature experience education, focusing on starting from the human senses and combining the relationship between people, others and nature, and putting forward suggestions on the design of nature experience activities.

2.3 Literature Research Status

China's current literature research on nature education mainly focuses on the interpretation of the thoughts of "Emile", the exploration of the concept of nature education, the development of nature education in various countries, and the reference of foreign nature education experience. The field selection, practice promotion, and organization of nature education The management and curriculum development are relatively inadequate. [5] Natural experience education is a complex systematic project, and its effectiveness depends on many factors. The lack of natural experience education in China is not superior natural education resources and hardware facilities, but the lack of advanced teaching concepts, a large number of experts focusing on natural education research, high-quality natural education teachers, and many unique and high-quality teachers. Natural education curriculum system.

In the context of the industrial revolution, developed countries have achieved rapid economic development, while environmental pollution has also followed. Compared with the environmental pollution problems that have occurred in my country due to rapid economic development in recent decades, developed countries have more To understand earlier and carry out related research. The United States, Japan, Australia and South Korea started research on environmental education earlier than China and already have a complete system of ecological and environmental protection education development programs. Foreign research has developed earlier and has relatively rich research results.

In particular, the "Flow Learning Method" teaching strategy proposed by Joseph Knell has provided ideas for the development of nature experience activities. Through the discussion of relevant literature on current nature experience courses, there are gaps in perception experience courses, including research on nature experience games and experience activities. The natural education activity plan is not perfect, and the analysis of the audience needs to be further in-depth and refined. It did not point out how the content of the course can be flexibly adjusted and appropriately applied in response to changes in the conditions of the natural place for experience education and changes in the needs of the interviewees.

Existing research shows unity in the construction of practical models. Most researchers apply the existing content and process of natural experience education activities, which makes the practical research of natural experience education the same and lacks reference value.

At the same time, the course audience focuses on infants and children, with uneven age distribution, and adults also need natural education. Many adults also lack the awareness of nature protection. The process of urbanization has reduced the opportunities for urbanites to contact nature. Contacting nature can make the busy heart live in the present, listen to the inner voice better, and make life more full. On the other hand, diversification and openness will naturally encourage social interaction and inspire social perception and knowledge, morals, and habits. In nature, it can help people sort out their thinking, look at problems from multiple angles, and relax again. From the perspective of adults, the nature experience activities discussed in this article will be more in line with the personality characteristics and actual needs of adults.

3 Research Method

This topic aims to study the procedural methods of natural experience teaching activities through literature research, network research, field research, and interview methods.

(1) Literature research method.
Through literature reading, network resources collection of domestic and foreign related natural experience activities for analysis, learn from the latest design concepts and design techniques. Lay a factual basis for the research of this article.

(2) Internet research.
Through the analysis and sorting of the questionnaire, combined with the summary of the self-made questionnaire, the research object's demand for outdoor nature experience activities is obtained, in order to explore more ideas for the design of nature experience activities.
(3) Field research.
Through a survey of existing nature experience activities in Guangzhou, this research took Maofeng Mountain, Guangzhou City, Guangdong Province, China as the location to actually participate in the teaching of nature experience activities, experience the development of immersive activities and the interactive relationship with teaching objects, To obtain relevant information, and at the same time to record the development of existing natural experience activities.
(4) Interview method.
Through interviews, we collect objective and unbiased factual materials based on the replies of the natural activity experiencers in an oral form.

Analyze the needs of the experiencer.

4 Discussion and Analysis

In the early 1980s, Joseph Knell founded the "Flow Learning" (Flow Learning), a teaching strategy designed to help people get the most out of them when they are outdoors. Also known as "Natural Teaching Method".

The basic principle of flow learning is to deepen and expand awareness. This learning provides a simple and natural structure and arranges the sequence of exercises to achieve the most profound effect.

Four-stage flow learning method: the first stage: awakening enthusiasm, the second stage: cultivate concentration, the third stage: direct experience, the fourth stage: sharing perception.

4.1 Theoretical Thinking of Activity Design

Flow learning method is a process framework suitable for most natural experience activities, but it is not enough to analyze the specific situation in depth. Analyze the flow learning method from the three perspectives of self, others and nature. The first stage of awakening enthusiasm is to open one's heart and enhance communication with others. The second and third stages are to perceive nature wholeheartedly, while accepting gifts from nature. In the process of perceiving nature, it is also a dialogue with the self. The fourth stage of shared perception is to express one's own inner perception and deepen the connection with others.

This article defines the main points of the development of natural experience activities as emphasizing the opening of the human senses: hearing, sight, smell, touch, and taste. Closely around the three relationships of self, others, and nature.

4.2 Natural Experience Activity Design Case

Starting from the main points of the activity design proposed in the previous article, using the stones in the natural world as the medium, design a nature experience game with the theme of opening the human body's tactile senses. The set natural experience way is hiking and mountaineering. The game is called "Milestone".

First choose a rock that the experiencer likes at the foot of the mountain, and take this rock to climb the mountain. This stone is today's companion who accompanies the experiencer to overcome the obstacles of the road and give spiritual strength.

This stone symbolizes the tasks and goals in the process of life. Holding this goal to the top and reaching another stage of life.

After reaching the top of the mountain, first engage in a three-minute meditation activity. Hold the stone partner in your hand, open the tactile senses, touch it with your fingers, touch its texture, and have a spiritual dialogue with it. This process requires absolute silence and concentration, and enjoy the peace of mind at the moment.

Each stone is unique, feel his life in the process of meditation, gaze, and touch.

After the meditation is over, the experiencer shares his stone with the team members, which can be its name, the content of the dialogue with it, the reason for choosing it, the experience during the hike, and so on.

In the end, the team members worked together to stack these stones and build a milestone!

Throughout the course of the activity, meditation is a dialogue with the self. Stroking a stone is a dialogue with nature. The stone stacking activity deepens the communication with the team members.

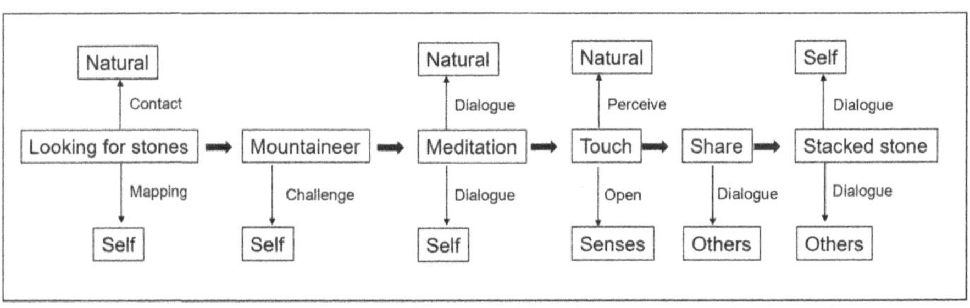

Chart 1. Milestone natural experience activity process analysis

5 Conclusion

Nature education has developed from a social problem to the formation of China's nature education industry. At this time, the research and development of teaching content, the establishment of industry standards, and the exploration of new directions will fill the industry vacancy and open up a gap in the domestic nature education field. Provide positive value to society. From the environmental protection theme of "green water and green mountains are golden mountains and silver mountains" to how to look at the integration of human beings in the natural environment, the harmonious coexistence of man and nature has become a closely related theme for everyone, and this is also an urgent concern for the entire human society. Content. The next generation is the core group of nature education. For the future, the coexistence relationship between man and nature will be more complicated, and this social value will become more prominent.

The human mind is formed through senses and perceptions, and cognition in thinking is completed through integration, judgment, and reasoning. In the process of human exploration, learning and experience in nature, the senses at all levels will become more acute and the state will become more open. Through such a series of thematic natural perception courses, participants can experience the energy charging process from relieving stress, experiencing heart flow to self-understanding, finding peace, and finally breaking through themselves and realizing life beyond the status quo. The natural experience activity development method proposed in this research focuses on providing modern people with multi-sensory experience activities and re-establishing the connection between man and nature.

Through multi-level and multi-faceted perception and experience activities, to experience and discover the connection between man and self, man and nature, and man and others. Part of the research in this article is supported by the Design Science and Art Research Center. There are still many shortcomings in this article. I hope this research can contribute to the development of nature experience education, and more and more people will benefit from nature experience activities.

References

1. Jing, G.: The positive impact of nature experience education courses on the health of preschool children. Sci. Educ. Lit. Collect. **07**, 150–151 (2019)
2. Xue, T.: Research on the Construction and Development Countermeasures of the Natural Experience Education System of Jiajin Mountain National Forest Park (Xiaojin Part) in Sichuan. Sichuan Agricultural University (2019)
3. Jiahe, W., Yoji, I., Xiangbo, C., Yuelin, L.: The practice and enlightenment of rural nature experience education——Taking Japan's GREEN WOOD Nature Experience Center as an example. Green Technol. **01**, 248–250 (2020)
4. Yue Wei, X., Fengcu, Z.: The value implication and practical logic of natural experience education. J. Guangxi Normal Univ. (Philos. Soc. Sci. Edn.) **56**(02), 115–123 (2020)
5. Yaqiong, Z., Pan, C., Yan, H., Chen, Z.: Research progress in nature education. Forestry Inventory Plan. **45**(04), 174–178+183 (2020)

Teaching of Design: Emphasis on Research of Media Characteristics

Ming Lei(✉) and Shengzhao Yu

Guangzhou Academy of Fine Arts, Guangzhou 510006, China

Abstract. The teaching aims at media characteristic, in which the common effects of various media interactions are explored, and the creative motivation of graduation design is explored. Four Diagnostic Methods Touch nonvisual shows a work whose carrier is material and texture, conveying the characteristics of pathological organs. Diseases explored by touch are selected from professional medical books, and the pathological features of these diseases are extracted. In addition, these pathological features are added through material amplification and auditory assistance. Subsequently, the audience's vision is restricted by the exhibition, which prompts the audience to interact with the exhibits with touch, to receive, understand and remember the information of the work.

Keywords: Exhibition · Design · Media characteristic · Teaching · Design process

1 Introduction

In a sense, exhibition design is a discipline that comprehensively applies media art. In the four years of undergraduate study, students have acquired the skills to explore the characteristics of various media and the common effects of multiple media. Based on this, students spend more than half a year combining their "things of interest", "things of concern" and "things they have mastered" [1].

My requirements for students who are about to start graduation design and the topic selection direction constructed with this goal are to present the works to the audience on the first floor underground of the Museum of Art in the University City with an appropriate medium, and to attract the audience to stop, to appreciate and interact with it.

2 The Topic of Four Diagnostic Methods Touch

I have been trying to transform the teaching results of the course into rational research materials for exhibition design. On this basis, we understand that the research object of exhibition design, namely the "audience", has a variety of ways to perceive and receive information, which are difficult to be thoroughly studied. "Variable research method", as a physical method, can reduce or even eliminate the interference of various factors that may affect certain object, so the research object can be simplified to be understood.

S. Nazir et al. (Eds.): AHFE 2021, LNNS 269, pp. 276–283, 2021.
https://doi.org/10.1007/978-3-030-80000-0_33

Students are always expected to continue to try to explore exhibition design without "visual sense", in other words, "the audience receives information with organs other than the eyes". In fact, this is an old proposition. Most students will make the audience perceive the information "nonvisual", which, however, is only an innovation of the "viewing form" but does not make full use of other perceptual characteristics, so the result must be disappointing.

Wei Xie and Fang Jiawa Zheng were willing to continue exploring this proposition. At first, I was hesitant about this, however, a wonderful graduation design work was displayed by the joint efforts and meticulous cooperation of them. As a supervisor, I should record it.

3 "Nonvisual"

The deliberate avoidance of vision is only an innovation of the "viewing form", so we try to find a reason to give up "vision". As we all know, in terms of the dominant characteristics of things, vision can be perceived by people faster than other perceptions, such as "size"," color", "brightness" and so on. However, some cannot be visually perceived, such as: "elasticity", "hardness", "weight". In addition, "smooth", "rough", "weight", etc. must be combined with"touch"to be accurately perceived. Interestingly, this perception experience is very common in our lives; the information people receive is gene-rated by the stimuli of images, text, sound, and digital video in daily life, and it can also activate other external stimuli [2].While people are visually stimulated by the external environment, the sense of touch is often activated. But people who are accustomed to using eyes usually attribute the credit of "touch" to "vision", which obscures our definition of perception, and is also the difficulty of this work [3].

4 "Content"

Our work really started after finding the visual "shortcomings", that is, finding a suitable "content" for the "expression" and enriching its meaning, and avoiding being reduced to "metaphysical" content later in the creative process. In my years of teaching career, it is very common to introduce the form first and then the content, and there are high-quality cases, which cannot be regarded as a violation of the design process. I believe there are two reasons: 1. The design work is difficult to apply to the standard creative process, and various factors can induce the creator's motivation; 2. The graduation design work does not participate in market competition, so there is no "content definition". More importantly, in terms of the academic inquiry of this major, starting from the form of the work, the attributes of the major of the work can be controlled more easily.

I was worried that students will take up too much time in choosing the topic, so I have discussed this question wonderfully with professor Shengzhao Yu who assisted me in the guidance, but I will not show the teaching achievements here. Finally, the content, or the "content carrier" selected by the students is: "pathology".

As the most typical form of information received visually, text has shortcomings in its receiving method, which are particularly prominent in the visceral pathology of the human body, such as: "cirrhosis", "tissue fibrosis", "hernia" "kidney calculi" and other diseases. Let's look at the two examples in medical books on the descriptions of the above in words, which explain cirrhosis:

"Cirrhosis, as a clinically common chronic progressive liver disease, is caused by one or more diseases due to long-term or repeated action to form diffuse liver damage. Most patients in our country suffer from post-hepatitis cirrhosis, and a few suffer from alcoholic cirrhosis and schistosomiasis. There is extensive hepatocyte necrosis, nodular regeneration of remaining hepatocytes, connective tissue hyperplasia and the formation of fibrous septa in histopathology, which leads to the destruction of hepatic lobular structure and the generation of pseudo lobular, and the liver gradually deforms and hardens and finally to cirrhosis. Due to the strong liver compensation function, patients may have no obvious symptoms in the early stage, and later suffer from liver function damage and portal hypertension, multi-organ involvement; patients with advanced stage often have upper gastrointestinal bleeding, hepatic encephalopathy, secondary infection, and complications such as hypersplenism, ascites, and canceration." [4] Or there are explanatory pictures like (Fig. 1):

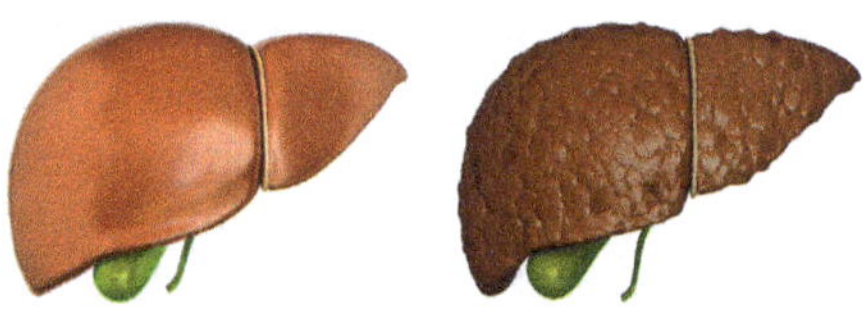

Fig. 1. Cirrhosis.

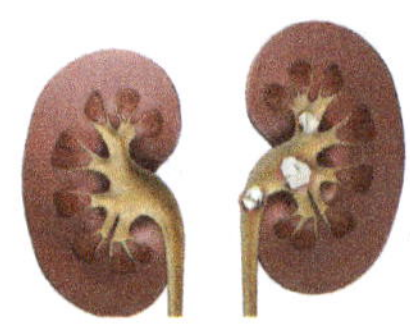

Fig. 2. Renal calculi.

Text interpretation of cirrhosis:
Renal calculi are a common and frequent disease of the urinary system. There are more male patients with renal calculi than female patients, and it is more common in young adults. There is no significant difference in the incidence of left and right kidneys. 40% to 75% of patients with renal calculi have varying degrees of low back pain. Large stones have little mobility, and the patient's waist is sore and uncomfortable, or suffers from produces pain or dull pain when the patient is vigorously exercising. The colic caused by smaller stones is a severe pain that often occurs suddenly in the waist and abdomen like a knife-like cut [5]. Or there are explanatory pictures like (Fig. 2).

It can be seen from this that regarding the pathological features of the human body that cannot be seen by the eyes, in the description of text or pictures, the combination of vision and text is less effective, and a single form of perception is not conducive to people's comprehensive knowledge of things [6] As Raymond Williams (1977: 158) said, "media is 'material social practice', rather than a specific essence determined by some elemental matters or technologies". [7] Audiences who have not suffered from these diseases will find it difficult to understand and resonate in front of large sections of text and pictures; if these materials are translated into a more direct and delicate language, not only will the audience who appreciate this work have more desires to watch it, but also understand the pathology of internal organs of the human body in a more direct way. In addition, people's health awareness will be improved, which is of academic value and practical significance.

5 "Designing Process"

5.1 Classification of Source Information

After everything is prepared, our work begins. After studying this pathology book, more than one hundred common diseases are listed, and 32 of them are selected by us as suitable source information objects. These are intestinal ulcer, intestinal obstruction, intestinal polyps, liver cancer stage I, liver cancer stage II, liver cancer stage III, end stage liver cancer, pulmonary fibrosis, alveolar enlargement, increased lung texture, pulmonary cyst, hernia, renal calculi, polycystic kidney disease, bladders that hold water, gastric ulcer, gastric perforation, bone spurs, tachycardia, obsessive-compulsive disorder, tropophobia, feelings of fatigue, hemangioma, arteriosclerosis, osteoporosis, vascular hyperplasia, wisdom teeth, high pressure, fever and ichthyosis, which can all be detected with touch alone.

We collect and collate 32 kinds of pathological data and extract meaningful information. Data on volume, weight, and texture are collected. Take cirrhosis as an example:

Volume: Organs in different locations have different volumes. Taking the liver as an example, the length, width and height of a normal-shaped liver are 25 cm * 15 cm * 7 cm, which are equivalent to a stack of B5 papers with a height of 7 cm. The morbid liver, such as fatty liver, is larger and its texture is more like "powder".

Weight: Taking the bladder as an example, the average value of a normal male bladder is 470 ML [256–810 ML (after urination and when holding water)], which is equivalent to the weight of a medium volume of Yibao mineral water (550 ml).

Texture: The texture of a healthy heart is like a steak, while the heart of an obese person is larger with more "water", that is, soft and weak.

The texture, weight, and volume information of all kinds of morbid organs collected are converted into graphical information drawings, and the difference between healthy organs and morbid organs is compared (Fig. 3). The audience judges in the comparison with their senses.

Fig. 3. Sketch of s ource Information classification.

5.2 Materials Selection and Production

The choice of materials is the key to "touch". The whole work not only explores the texture of different organ materials, but also shows the contrast between different parts of the health and disease, as well as the texture and even the temperature (unfinished), which brings the audience touch experience. After collecting, sorting, and classifying various pathological systems, the two students found multiple materials to judge whether their texture is consistent with the texture of various pathologies described in medical books.

Take "arteriosclerosis" as an example. The description of arteriosclerosis in pathology: "Arteriosclerosis, as a noninflammatory lesion of arteries, is a general term for degenerative and hyperplastic lesions of arterial wall thickening, hardening, loss of elasticity, and narrow official cavity. Common types are: atherosclerosis, calcification of arteries, and arteriosclerosis. At the same time, the members of the group go to the hospital to register and interview the doctor. The doctor's description of arteriosclerosis is: simply speaking, the human artery is blocked without the elasticity."

The hidden adjectives that are screened out were: "thick", "hard", "elastic", "narrow", and "blocked". Then these words need to be translated to be converted into the sense of touch, so the text information is translated into tactile sensations.

We find that the core differences between healthy blood vessels and hardened blood vessels are: "hard" and "soft", "unblocked" and "blocked". This process requires a "tubular" material as the "artery", which requires us to find multiple materials to find the most appropriate one (Fig. 4). In the end, the No. 2 material with elasticity and plasticity is selected as the most suitable one, and the center is injected with material. After we inject gypsum into the No. 2 material and harden it with water, a blocked hardened section is formed inside the soft hose (Fig. 4).

Fig. 4. Selection of materials for arteriosclerosis.

Part of the No. 2 material is not infused with gypsum (Fig. 4). When the work is exhibited, the audience can strengthen the understanding of "arteriosclerosis" through a comparison of tactile sensations with a profound sensory experience. These 32 diseases including intestinal ulcer, intestinal obstruction, intestinal polyps, liver cancer stage I, liver cancer stage II, liver cancer stage III, end stage liver cancer, pulmonary fibrosis, alveolar enlargement (Fig. 5), increased lung texture, pulmonary cyst (Fig. 5), hernia, renal calculi, polycystic kidney disease, bladders that hold water, gastric ulcer (Fig. 5), gastric perforation, bone spurs, tachycardia, obsessive-compulsive disorder, trypophobia, feelings of fatigue, hemangioma, arteriosclerosis, osteoporosis, vascular hyperplasia (Fig. 5), wisdom teeth, high pressure, fever and ichthyosis are translated from "text description" to "touch sense". We search for hundreds of materials to carry out comparative experiments and select the most vivid and alternative materials related to haptics.

However, the tactile characteristics of some diseases are difficult to be reflected with ready-made materials, so we must use some craftsmanship to make them. As shown in the figure below, taking "polycystic kidney" and "intestinal polyps" as examples, we constantly try to sprinkle foam balls on the silicone material after heating (Fig. 6), and then after the temperature drops, we pick out the balls, thus forming a texture on the silicone (Fig. 6), so the touch we want is obtained.

Finally, the tactile texture of these 32 diseases is classified and placed in the corresponding small box (20 cm *20 cm *20 cm), which is also to improve the audience's viewing experience (Fig. 6).

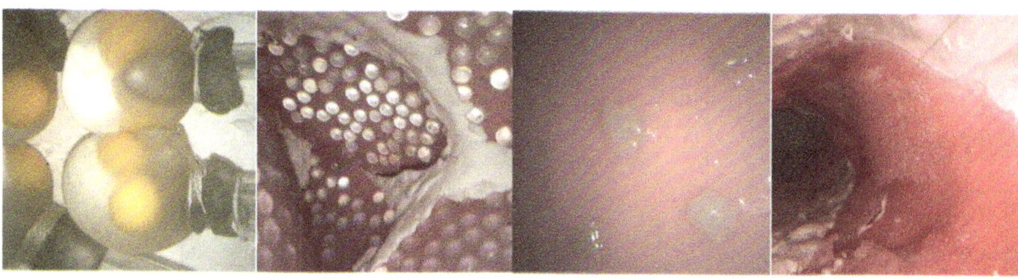

Fig. 5. Alveolar enlargement/pulmonary cyst/gastric ulcer/vascular hyperplasia.

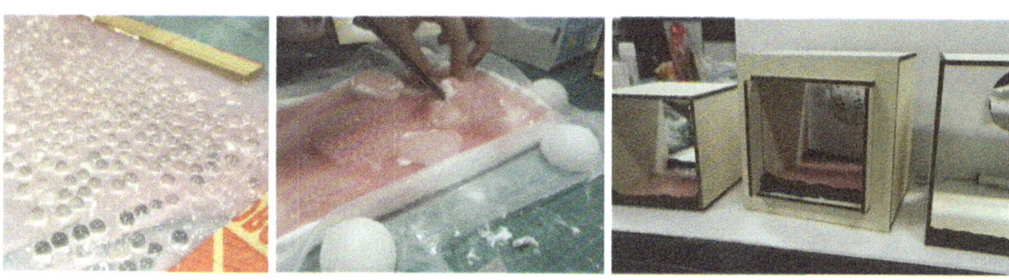

Fig. 6. The production process of polycystic kidney/intestinal polyps.

5.3 The Conception of the Exhibition Form

Our bold vision is to make the audience "look" at the work without eyes as much as possible. Most lesions of internal organs cannot be seen, and people can only understand the lesions through text messages and form blurred images in the brain.

Furthermore, this work aims to explore the texture of materials, create a touch like the texture of pathological organs, reshape the form and sense of touch, and even exaggerate its characteristics. In addition, this work exchanges information with art and sensibility, so the audience does not need to use their eyes. In addition, this work prompts the audience to perceive with their hands, and naturally leads to "audience interaction", which is dreamed by the designers of the exhibition.

We set a lot of circular holes with a diameter of ten centimeters throughout the installation to inform the audience that there is something can be "read" but cannot be seen, which forces the audience to "use their hands" (Fig. 7). This is the action we are expecting, "pull", which is an interaction that can be subdivided into a coherent movement of touching, feeling, gripping, grabbing, and pulling, which attracts the audience with curiosity and active participation in the interaction. Compared with vision, this process brings a subtle, delicate, magical feeling and comfort in the private communication of the mind. This uncommon opportunity brings more real and profound feelings to the audience [8]. In addition, infrared sensor voice prompts are used in the works to inform the audience whether the organ being touched is normal or sick, thus, the eyes can now rest.

Fig. 7. Exhibition

6 Conclusion

The standard for evaluating the quality of works from the perspective of professional exhibition is the participation of the audience, including number of audiences, reading degree, memory, resonance. This is what we require students to do during graduation guidance teaching. The work or a certain part of the work immediately attracts the audience, which is the first and critical step of the work and should be understood by students. Next, what does the audience want to see? What do you want to show them? Only when the audience's exhibition needs are precisely met, they will receive the information and may produce the later exhibition memory and the rare "emotional resonance".

The audience even spend a long time to experience the touch of 32 diseases with serious expressions. For us, this expression symbolizes the success of reading information. In other words, the detailed textual description of these 32 diseases will be a thick medical textbook, which is disliked by most ordinary audiences, so the audience will not "read" it, and "understanding", "memory" and "resonance" will also not exist, which means that the information transmission has failed.

References

1. Rare Book Division: The New York Public Library. "Fundamentals of exhibition design" The New York Public Library Digital Collections. 1939–12-1940–01
2. Pehlivanides, G.: Exploring cultural information interaction design: a case study of a multimedia exhibition based on customizable user interfaces. In: Bozanis, P., Houstis, E.N. (eds.) PCI 2005. LNCS, vol. 3746, pp. 177–187. Springer, Heidelberg (2005). https://doi.org/10.1007/11573036_17
3. Mitchell, W.J.T.: There are no visual media. J. Vis. Cult. **4**(2), 257–266 (2005)
4. Qin, L.: Nursing of ascites due to liver cirrhosis with integrated Chinese and western medicine. Inner Mongolian Med. **33**(25), 155–156 (2014)
5. Ye, Y., Jiang, H.: The importance of kidney . Health Rev. **11**, 4–9 (2016)
6. Zhong, M., He, R., Zhao, D.: Narrative perception in the exhibition space-studying of multimedia technical device design. In: Ahram, T., Karwowski, W., Pickl, S., Taiar, R. (eds.) IHSED 2019. AISC, vol. 1026, pp. 936–941. Springer, Cham (2020). https://doi.org/10.1007/978-3-030-27928-8_140
7. Witmore, C.L.: Vision, media, noise and the percolation of time: symmetrical approaches to the mediation of the material world. J. Mater. Cult. **11**(3), 267–292 (2006)
8. Geng, L.: Research on tactile design in graphic design . Hundred Sch. Arts **30**(04), 250–251 (2014)

Distance Learning during COVID-19 Era

Online Classes: Lessons Learned During the Pandemic

Angela Caridà, Maria Colurcio(✉), Ambra Altimari, and Monia Melia

Department of Law, Economics and Sociology, University Magna Graecia of Catanzaro, Campus Universitario 'Salvatore Venuta', Viale Europa, 88100 Catanzaro, Italy
{angela.carida,mariacolurcio,ambra.altimari, monia.melia}@unicz.it

Abstract. This work frames the factors that influence students' perceptions of online teaching and learning during the global pandemic. We submitted the COLLES Survey to management students and we conducted an exploratory factor analysis to identify the nature of the latent factors underlying students' perception of online classroom environment. Three main factors explain most of the variability; according to previous studies, we labelled them as (i) Corse design and development, (ii) Instructor's characteristics and role, and (iii) Learners' characteristics. Moreover, we found that some differences occur between latent factors patterns between undergraduate and postgraduate classes. Although the existing literature acknowledges that learner characteristics, course design, and instructor role may influence students' perception of the online teaching/learning experience, there are no studies linking them to one or more of these specific factors. Furthermore, there is limited research on the similarities and differences in perceptions of the online teaching/learning experience between undergraduate and graduate students.

Keywords: Online learning · Online teaching · COLLES · Students' perspective · Factor analysis

1 Introduction

COVID-19 has forced Italian schools and universities to learn from novel circumstances and adapt quickly to changes by moving from face-to-face (F2F) to online teaching. It has accelerated the digitalization and transformed teaching practices in higher education in Italy.

Online teaching is not a new issue in higher education [16], but before COVID-19 the use of digital technologies for educational purposes was occasionally limited to traditional universities, offering only a few courses and some specific activities (e.g., seminars), to specific online education providers (e.g. online universities), and/or to specific geographical areas – emerging countries among others [5].

The physical closing of universities and schools worldwide [30] and the consequent digitalization of teaching framed a new system of higher education in which rules,

S. Nazir et al. (Eds.): AHFE 2021, LNNS 269, pp. 287–298, 2021.
https://doi.org/10.1007/978-3-030-80000-0_34

practices and institutional arrangements suddenly seem inadequate to ensure the creation and dissemination of value.

By moving online, teachers are forced to adapt their teaching strategies to the new educational scenario to deliver educational content through the use of social connection platforms (such as Zoom, Meet, Teams, etc.) along with the university-specific e-learning platforms (such as Moodle, Ilias, Olat). In the current global pandemic, teachers have been forced to redesign their courses to ensure the motivation and rhythm of teaching and to prevent the lack of concentration and interaction with and among students in the classroom [12].

Social connection platforms and the Internet ecosystem enable the convergence of many elements of learning (e.g., text, audio, and video) into the same communication channel [32], therefore, they can facilitate the online learning process by making teaching more dynamic, interactive, and effective [17].

Of course, the success of online teaching and distance learning is influenced by several factors that shape the new educational context, such as the role of the professor as a facilitator of learning (e.g., moderator, participant, and observer), the degree of student autonomy [6], which affects the level of interaction with the professor and peers, and the new classes format, which must address the new and specific needs of students in this difficult and challenging time.

Due to the recent and emerging nature of the phenomenon, few studies have been conducted on online teaching as a new educational strategy imposed by contingent situations, thus transcending the intrinsic motivation, skills, and attitudes of professors and students in using digital platforms for educational purposes.

This work aims to understand the factors that influence students' perceptions of online teaching and learning during the global pandemic. Specifically, this work addresses two main questions: i) are there significant factors that influence students' perceptions of the virtual classroom environment during the pandemic, and ii) are there significant differences in students' perceptions across grade levels?

To this end, we used the Constructivist Online Learning Environment Survey (COLLES survey students) [29] to analyze the perceptions of online teaching/learning of a sample of Italian undergraduate and graduate students who had their first experience with the virtual classroom from March 2020 to June 2020. This work contributes to the existing literature on online education, which needs more research on evaluation and quality in online courses [10, 16, 29], especially in the current period when both instructors and students, many of whom have never experienced the virtual classroom, have not chosen but are forced to practice online learning/teaching.

The rest of the paper is organized as follows: The second section focuses on the literature background. Then, we present the research methodology and discuss the main findings.

2 Literature Review

There is ample evidence of the explosive growth of online teaching/learning during the pandemic COVID-19 and of its key role in higher education in the immediate future as well.

Since the 1990s, research in distance education has focused primarily on the characteristics of learners and their interaction in courses, and on course design and the characteristics of instructors. These two topics define elements that influence strategies to increase interaction with others, especially peers and instructors [8], as well as active learning [3, 16].

According to [16], learner characteristics refer to demographic, academic, cognitive, affective, self-regulatory, and motivational characteristics of online students. For example, studies on online learning found a strong and positive relationship between learner self-regulation and autonomy with communication and collaboration [2] and active interaction with others [7].

Self-regulation (or self-regulated learning) refers to students who are able to self-generate thoughts and learning by creating productive social relationships and work environments [25] rather than being passive recipients of information [26].

Artino [1] found that graduate students were more likely to use critical thinking skills than undergraduate students and showed that grade level was related to students' self-regulation for interacting with others [8].

Therefore, understanding learners' characteristics is critical to the success and quality of the course as it is closely related to course design issues and different ways of engaging learners to better meet their needs.

Course design and development depends on the availability and use of digital learning technologies (e.g., social connection platforms, chat - chat room, videos, discussion forums, etc.). The characteristics and roles of the instructor in the online classroom, are critical to promote learners' interaction with others [18, 24], and the pedagogical techniques (group and individual work, small and large group discussions, role-playing, case study analysis and discussion, etc.) [12] to engage online learners and build a sense of community [13], are the main factors that characterize the course design and development.

In the virtual classroom, the teacher's role shifts from instructive to supportive and facilitative [19]; the teacher acts as a connector to build relationships between learners and help them collaborate despite the physical distance. Different roles the teachers play, and the different course designs allow for different types of engagement which are crucial to communicate and interact with learners (e.g. learner-teacher interaction) [15], to stimulate their participation/involvement and collaboration (e.g. learner-learner interactions) [21, 28], and to incentive student's presence in online courses [22].

Interaction, collaboration, and students' continuous presence in online courses are some predictors of successful online course experiences. Researchers have developed specific evaluation models to measure the factors that influence online learning environments [9, 20, 29, 31]. Among these, the COLLES survey is the most popular survey [19] to assess students' perceptions of the online learning environment [29]. The COLLES survey assesses students' perception using six constructs (e.g., relevance of course, reflective thinking, interactivity, cognitive demands, affective support, and interpretation of meaning) and fits with the notion that students' perceptions of online courses are a multidimensional concept influenced by many factors [23].

Although the existing literature acknowledges that learner characteristics, course design, and instructor role may influence students' perception of the online teaching/learning experience (e.g., the six constructs listed above), there are no studies linking them to one or more of these specific factors. Furthermore, there is limited research on the similarities and differences in perceptions of the online teaching/learning experience between undergraduate and graduate students. Therefore, understanding the factors that explain how undergraduate and graduate students perceive online courses, especially during the pandemic, is critical to improving support for online students who are forced to take distance learning experiences and to providing guidelines for effective online teaching.

3 Methods and Tools

3.1 Participants

The participants for this study included a sample of 267 students from a public university in the southern Italy (the University Magna Graecia of Catanzaro). Of these students, 88% were undergraduates and 12% were graduate students.

Participants attended management courses offered by the departments of Law, Economics and Sociology (Diges); all courses were offered entirely online through Google Meet and the Moodle Platforms from March 2020 to June 2020 and from September 2020 to December 2020. All students had experience with face-to-face lectures, as the first semester courses of AY2019/2020 were taught in the face-to-face modality.

The management courses involved in the survey are: Business Management (Beginners), Marketing, Innovation Management (intermediate), Business management and strategy (master).

Courses have different focus, different level of analysis and different instructors. Courses present some differences in terms of design, as some of them include laboratories and applied projects to real cases, other courses are mainly theoretical courses.

3.2 Data Collection

The Constructivist On-Line Learning Environment Survey (COLLES) [29] was submitted online through the Moodle Platform, it was active between June and August 2020. The COLLES is an analytical tool of analysis focused on online courses, and in particular on students' perception of the online classroom environment [29].

The COLLES aims at investigating aspects that support students in developing themselves as collaborative and reflective learners in an online classroom. To this end COLLES includes 48 questions divided in six thematic areas:

1. **Professional Relevance**. Whether the virtual classroom environment and course content are relevant to professional practice.
2. **Reflective Thinking**. These questions assess whether the online class stimulate the development of reflective thinking.
3. **Interactivity**. Whether the virtual environment offers opportunities for interaction.

4. **Cognitive Demand.** Whether the instructor communicates effectively and offers stimuli to the students.
5. **Affective Support.** Whether peers' support occurs in the virtual classroom environment.
6. **Interpretation of Meaning.** Whether students and tutor communicate and co-construct a good reflection and analysis of the topics.

Within each thematic area there are four questions. Responses are given on a Likert scale with five gradations from 1 ("Almost Never") to 5 ("Almost Always").

3.3 Exploratory Factor Analysis

Exploratory factor analysis (or EFA) is a statistical technique that examines the possible existence of underlying factors influencing a set of measures, and the relationship between each latent factor and observed variables [11].

We conducted an EFA using the XLStat software to identify the nature of the constructs underlying responses in the questionnaire. Factors were extracted using the principal factor extraction method and selecting the varimax rotation to the first three factors [4, 14]. This method allowed us to characterize students' perceptions of the virtual classroom environment with the three principal traits: course design and development, instructor's characteristics, and learners' characteristics.

4 Results

4.1 Students' Perceptions

Our dataset contains 24 variables and 267 observations. The general perception of the online classroom environment was positive, as the aggregate mean values by area are always higher than 3, but in the Interactivity area. All classes have in common the general pattern in which Interactivity and Affective Support show mean scores lower than the other areas.

4.2 Exploratory Factor Analysis

We first conducted an EFA to verify how many latent factors can be identified. The EFA identified 14 factors, that explain a the 67.504% of cumulative variability. The first three factors explain the 52.062% of cumulative variability. Thus, we ran a second EFA selecting a varimax rotation applied to the first three factors. The Table 1 shows variables and factors.

The first factor ($\alpha = 0.916$) groups variables related to the Interactivity, to the Affective support, and two of the variables belonging to the Interpretation of meaning (specifically the two variables related to other students). The second factor ($\alpha = 0.868$) groups variables from the Relevance, from the Cognitive demand area, and the two variables of the Interpretation of meaning related to the instructor. The third factor ($\alpha = 0.831$) corresponds to the Reflective thinking.

Table 1. Factor pattern after varimax rotation

			D1	D2	D3
Professional relevance	V1	My learning focuses on issues that interest me	0.099	**0.308**	0.201
	V2	What I learn is important for my professional practice	0.082	**0.597**	0.129
	V3	I learn how to improve my professional practice	0.183	**0.580**	0.147
	V4	What I learn connects well with my professional practice	0.220	**0.571**	0.153
Reflective thinking	V5	I think critically about how I learn	0.021	0.184	**0.726**
	V6	I think critically about my own ideas	0.049	0.252	**0.696**
	V7	I think critically about other students' ideas	0.106	0.001	**0.670**
	V8	I think critically about ideas in the readings	0.032	0.183	**0.741**
Interactivity	V9	I explain my ideas to other students	**0.695**	0.008	0.289
	V10	I ask other students to explain their ideas	**0.705**	– 0.017	0.282
	V11	Other students ask me to explain my ideas	**0.741**	– 0.059	0.274
	V12	Other students respond to my ideas	**0.756**	0.052	0.295
Cognitive demand	V13	The tutor stimulates my thinking	0.068	**0.707**	0.175
	V14	The tutor encourages me to participate	0.057	**0.767**	0.132
	V15	The tutor models good discourse	0.063	**0.781**	0.086
	V16	The tutor models good critical self-reflection	0.140	**0.635**	0.114
Affective support	V17	Other students encourage my participation	**0.769**	0.100	– 0.083

(*continued*)

Table 1. (*continued*)

			D1	D2	D3
	V18	Other students praise my contribution	**0.853**	0.140	– 0.095
	V19	Other students value my contribution	**0.828**	0.169	– 0.093
	V20	Other students emphatise with my struggle to learn	**0.802**	0.177	– 0.094
Interpretation of meaning	V21	I make good sense of other students' messages	**0.455**	0.370	0.052
	V22	Other students make good sense of my messages	**0.597**	0.322	– 0.008
	V23	I make good sense of the tutor's messages	0.199	**0.613**	0.084
	V24	The tutor makes good senso of my messages	0.285	**0.557**	0.056

Values in bold correspond for each variable to the factor for which the squared cosine is the largest.
Source: our elaboration.

To test the existence of a different path in the factors, between undergraduate and postgraduate students, we selected two subsets of our sample: i) students attending the master class in Business management and strategy; ii) students of the undergraduate class in Business management (beginners). The two classes share topic and tutor and differentiate in the level and the course design. By comparing the results from the two classes, we can focus on the effect of class design and student characteristics (in terms of experience, background knowledge, team building, and communication) and hold constant the effects of instructor facilitation and topic.

We ran an EFA and selected the varimax option for the rotation that was applied to the first three factors.

The Table 2 shows that the factor pattern in the postgraduate class is in line with the factor pattern in the full sample. Nevertheless, the factor patterns in the postgraduate sub-sample and in the undergraduate sub-sample are different. In the undergraduate sub-sample, the first three factors explain the 57.478% of variability. In the postgraduate sub-sample, the first three factors explain the 58.284% of variability.

Two are the main differences in the factor patterns between undergraduate and postgraduate classes. More specifically, as already pointed out, factor pattern in the postgraduate class is in line with factor pattern in the full sample, instead, the EFA on the undergraduate class shows a different pattern in the Interactivity, Relevance and Reflective thinking areas.

Table 2. Factor pattern after varimax rotation

	Undergraduate			Postgraduate		
	D1	D2	D3	D1	D2	D3
V1	– 0.088	0.060	**– 0.236**	0.043	**0.705**	0.004
V2	– 0.050	**0.667**	0.219	0.009	**0.783**	0.048
V3	0.128	**0.503**	0.288	0.340	**0.801**	0.171
V4	0.164	**0.549**	0.319	0.095	**0.669**	0.357
V5	0.341	**0.433**	– 0.227	0.134	0.195	**0.745**
V6	0.112	**0.365**	– 0.201	– 0.078	0.380	**0.440**
V7	**0.556**	0.449	– 0.190	– 0.050	– 0.050	**0.593**
V8	0.405	**0.549**	– 0.158	0.204	0.089	**0.705**
V9	0.540	0.120	**0.628**	**0.468**	0.142	0.436
V10	0.346	0.043	**0.876**	**0.535**	0.126	0.520
V11	0.566	0.054	**0.607**	**0.808**	– 0.112	0.331
V12	**0.756**	– 0.086	0.423	**0.769**	– 0.025	0.500
V13	0.168	**0.830**	– 0.294	0.439	**0.475**	0.211
V14	0.157	**0.821**	0.073	0.272	**0.791**	– 0.043
V15	0.185	**0.755**	– 0.047	0.521	**0.609**	0.059
V16	0.305	**0.682**	0.079	0.447	**0.532**	0.135
V17	**0.705**	0.078	0.283	**0.739**	0.324	0.176
V18	**0.802**	0.183	0.231	**0.852**	0.274	– 0.058
V19	**0.905**	0.083	0.095	**0.832**	0.217	– 0.092
V20	**0.886**	0.063	0.045	**0.688**	0.451	– 0.002
V21	0.136	**0.209**	0.100	0.420	0.371	**0.470**
V22	**0.770**	0.405	0.237	**0.448**	0.418	0.316
V23	– 0.020	**0.789**	0.200	0.059	**0.618**	0.265
V24	0.017	**0.802**	0.156	0.247	**0.523**	0.039

Values in bold correspond for each variable to the factor for which the squared cosine is the largest.
Source: our elaboration.

5 Discussion

The current study examined the perceptions of the online classroom environment of undergraduate and graduate students enrolled in online management courses during the Covid 19 pandemic.

Although students were forced to quickly adapt to the online classroom and despite their lack of experience with online technologies and learning, they demonstrated generally positive perceptions of online teaching/learning. The Cognitive Demand theme, which relates to the role of the instructor and specifically whether he/she communicates effectively and provides incentives to students, showed the highest mean score. This result is not surprising and is consistent with the fact that the instructors involved in the online courses we considered already had experience with online courses. They adopted a variety of roles, tools, and instructional practices (e.g., creative lab, video and audio content, online survey, etc.) to adapt traditional instructional formats and interactive group discussions to the new online environment. As previous studies have shown [1, 12], teachers played both an instructional and a supportive/promotional role in enabling students to easily use technology for learning purposes and in engaging their classmates in meaningful interactions, i.e., encouraging, acknowledging, and reinforcing their contributions. In contrast to students' attitudes toward using online technologies to interact and build relationships with peers, we were surprised to find that interactivity and affective support, which reflect opportunities to interact and build relationships with peers in the virtual environment, had the two lowest scores.

Not surprisingly, our results show some interesting differences between grade levels; for example, Marketing and Innovation Management show optimal scores in all six domains, both of which relate to intermediate level courses that use teaching methods and tools from Business Management (1st year undergraduate programme) and Business Management and Strategy (2nd year postgraduate programme). Learners taking these courses already have experience with business topics and know peers who are part of their learning community.

In Business Management differences in Reflective Thinking mean scores are evident, with actual scores higher than preferred scores. This is consistent with studies that refer to undergraduate as students being less self-regulated and autonomous, and therefore unable to master the area of critical thinking to generate and learn thoughts on their own [1]. Finally, Musiness Management and Strategy show different values in the Professional Relevance; preferred perception is higher than actual perception. This is probably due to learners' career expectations and experiences, as most master's students have been doing accounting internships for at least two years and therefore expect their learning to be directly related to accounting careers rather than management careers.

To explore and better understand the latent factors that influence the six thematic domains that deepen students' perceptions of the virtual classroom environment [29], we conducted an EFA. The results of the EFA reveal the existence of three main characteristics, which we code as course design, instructor characteristics, and learner characteristics, following the existing literature [15, 16, 19, 21, 28].

The first factor (D1) groups variables related to Interactivity (V9, V10, V11, V12), to Affective Support (V17, V18, V19, V20), and two of the variables belonging to Interpretation of Meaning (specifically the two variables related to other students, V21 and V22). All these variables can refer to the process of co-construction of new knowledge [29], which implies reflective collaboration between learners. We code the latent factor influencing them as course design and development. This is consistent with Cho [8] who relate interaction with other students to course activities, including group projects, online discussions, and individual projects with peer feedback. This is confirmed by students attending Marketing, Innovation Management and Business Management, and strategy courses. Because these courses combine theoretical lectures with creative labs and teamwork, case study analysis, and group discussions, students tend to interact more actively with other students than with the instructor.

The second factor (D2) groups variables related to the Professional Relevance (V1, V2, V3, V4), the Cognitive Demand area (V13, V14, V15, V16), and the two variables of the Interpretation of Meaning (V23, V24) related to the instructor. All these variables can relate to the instructor's role, which ranges from imparting expert knowledge to being a facilitator and observer. Instructors act to enable learners to better regulate their own learning in online contexts as well as connect it to their scope and actual professional practice. We code as instructor characteristics the latent factor that influences them.

Finally, the third factor (D3) corresponds to Reflective Thinking (V5, V6, V7, V8), we code it as learner characteristics. This is consistent with other studies [1, 8] that refer to critical thinking strategies as the inherent and individual ability of students to apply prior knowledge to new situations or to make critical evaluations of ideas. Very interestingly, the results from our general sample indicate that Reflective Thinking has no correlations with other factors.

This finding is consistent with previous studies of distance and online education, which have focused primarily on learner characteristics and course interaction, as well as course design and instructor characteristics. However, when we turn to the comparison between undergraduate and postgraduate courses, only the factor pattern in the postgraduate subsample confirms the general results we found in the whole sample. The EFA we conducted in the undergraduate subsample shows a different factor pattern: D1 groups Affective Support (V17, V18, V19, V20), one item of Interactivity (V12), and one item from Interpretation of Meaning (V22).

Group D2 is the most miscellaneous as it explains three items on the Professional Relevance (V2, V3, V4) and on Reflective Thinking (V5, V6, V7, V8), on Cognitive Demand (V13, V14, V15), and three items on Interpretation of Meaning (V21, V23, V24). Finally, D 3 groups one item of Professional Relevance (V1) and three items covering Interactivity (V9, V10, V11). Since D2 links the same variables we found in the general results with (as additional variables) the Reflective Thinking variables, we can assume that in undergraduate studies the role of instructor is crucial for the development of some personal characteristics of the learner (e.g. critical thinking about ideas and learning). Students in the first year of undergraduate program usually lack self-regulation and autonomy, which are the basis of Reflective Thinking and of interaction: The more autonomous students are in their learning process, the more they interact with the professor and peers. Therefore, they need more explicit support and structure from the instructor.

In contrast to the results in the total sample, the variables belonging to interactivity load to a different factor in relation to the affective support variables. This difference could be due to the lack of prior academic experience of the learners and thus the lack of social presence and sense of community in the virtual classroom.

In general, this result is quite controversial, it does not allow us to clearly codify the factors that influence students' perception of the online environment. Further research is needed to better understand the reasons behind these findings. Identifying the factors that explain how undergraduate and graduate students perceive online courses, especially during the pandemic, is significant for improving support for online students who are forced to take distance learning experiences and for providing guidelines for effective online teaching.

References

1. Artino, A.R., Jr., Stephens, J.M.: Academic motivation and self-regulation: a comparative analysis of undergraduate and graduate students learning online. Internet Higher Educ. **12**(3–4), 146–151 (2009)
2. Barnard, L., Lan, W.Y., To, Y.M., Paton, V.O., Lai, S.L.: Measuring self-regulation in online and blended learning environments. Internet Higher Educ. **12**(1), 1–6 (2009)
3. Berge, Z.L., Mrozowski, S.: Review of research in distance education, 1990 to 1999. Am. J. Dist. Educ. **15**(3), 5–19 (2001)
4. Bollen, K.A.: Structural Equations with Latent Variables. Wiley, New York (1989)
5. Caridà, A., Colurcio, M., Russo Spena, T., Kandampully, J.: Service innovation in emerging economies: an inclusive perspective. Sinergie Italian J. Manag. **37**(3), 11–38 (2019)
6. Çebi, A., Güyer, T.: Students' interaction patterns in different online learning activities and their relationship with motivation, self-regulated learning strategy and learning performance. Educ. Inf. Technol. **25**(5), 3975–3993 (2020). https://doi.org/10.1007/s10639-020-10151-1
7. Cho, M.H., Jonassen, D.: Development of the human interaction dimension of the self-regulated learning questionnaire in asynchronous online learning environments. Educ. Psychol. **29**(1), 117–138 (2009)
8. Cho, M.H., Kim, B.J.: Students' self-regulation for interaction with others in online learning environments. Internet Higher Educ. **17**, 69–75 (2013)
9. Clayton, J.: Initial findings from the implementation of an online learning environment survey. Int. J. Cyber Soc. Educ. **4**(2), 127–138 (2011)
10. Daumiller, M., Rinas, R., Olden, D., Dresel, M.: Academics' motivations in professional training courses: effects on learning engagement and learning gains. Int. J. Acad. Dev. **26**(1), 7–23 (2021)
11. Decoster, J.: Overview of factor analysis. http://www.stat-help.com/notes.html
12. Gonçalves, S.P., Sousa, M.J., Pereira, F.S.: Dstance learning perceptions from higher education students—the case of Portugal. Educ. Sci. **10**(12), 374 (2020)
13. Jiang, W.: Interdependence of roles, role rotation, and sense of community in an online course. Dist. Educ. **38**(1), 84–105 (2017)
14. Kline, R.B.: Principles and Practice of Structural Equation Modeling, 3rd edn. The Guilford Press, New York, USA (2011)
15. Kuo, Y.C., Walker, A.E., Schroder, K.E., Belland, B.R.: Interaction, Internet self-efficacy, and self-regulated learning as predictors of student satisfaction in online education courses. Internet Higher Educ. **20**, 35–50 (2014)
16. Martin, F., Sun, T., Westine, C.D.: A systematic review of research on online teaching and learning from 2009 to 2018. Comput. Educ. **159**, 104009 (2020)
17. Mukhtar, K., Javed, K., Arooj, M., Sethi, A.: Advantages, limitations, and recommendations for online learning during the covid-19 pandemic era. Pak. J. Med. Sci. **36**, S27–S31 (2020)
18. Mullen, G.E., Tallent-Runnels, M.K.: Student outcomes and perceptions of instructors' demands and support in online and traditional classrooms. Internet Higher Educ. **9**(4), 257–266 (2006)
19. Patak, A.A., Naim, H.A., Ma'ruf, A., Ghafar, M.N.A.: Design and validation of online learning environment questionnaire. Int. J. Adv. Sci. Eng. Inf. Technol. **6**(3), 334–338 (2016)
20. Pearson, J., Trinidad, S.: OLES: an instrument for refining the design of e-learning environments. J. Comput. Assist. Learn. **21**(6), 396–404 (2005)

21. Phirangee, K., Epp, C.D., Hewitt, J.: Exploring the relationships between facilitation methods, students' sense of community, and their online behaviors. Online Learn. **20**(2), 134–154 (2016)
22. Phirangee, K., Malec, A.: Othering in online learning: an examination of social presence, identity, and sense of community. Dist. Educ. **38**(2), 160–172 (2017)
23. Roberts, T.S.: Computer-supported collaborative learning in higher education, pp. 1–18. IGI Global, Hershey (2005)
24. Ryan, A.M., Patrick, H.: The classroom social environment and changes in adolescents' motivation and engagement during middle school. Am. Educ. Res. J. **38**(2), 437–460 (2001)
25. Schunk, D.H.: Self-efficacy, motivation, and performance. J. Appl. Sport Psychol. **7**(2), 112–137 (1995)
26. Schunk, D.H., Zimmerman, B.J.: Motivation and Self-regulated Learning: Theory, Research, and Applications, Routledge, New York (2012)
27. Tallent-Runnels, M.K., et al.: Teaching courses online: a review of the research. Rev. Educ. Res. **76**(1), 93–135 (2006)
28. Tawfik, A.A., Giabbanelli, P.J., Hogan, M., Msilu, F., Gill, A., York, C.S.: Effects of success v failure cases on learner-learner interaction. Comput. Educ. **118**, 120–132 (2018)
29. Taylor, P., Maor, D.: Assessing the efficacy of online teaching with the constructivist online learning environment survey. In: 9th Annual Teaching Learning Forum, 2–4 February Perth, WA (2000)
30. UNESCO School closures caused by Coronavirus (COVID-19). https://en.unesco.org/COVID19/educationresponse(2020)
31. Walker, S.L., Fraser, B.J.: Development and validation of an instrument for assessing distance education learning environments in higher education: the Distance Education Learning Environments Survey (DELES). Learn. Environ. Res. **8**(3), 289–308 (2005)
32. Zhu, X., Chen, B., Avadhanam, R.M., Shui, H., Zhang, R.Z.: Reading and connecting: using social annotation in online classes. Inf. Learn. Sci. **121**(5/6), 261–271 (2020)

Ergonomics of Virtual Learning During COVID-19

Lu Yuan(✉) and Alison Garaudy

Southeastern Louisiana University, Hammond, USA
Lu.Yuan@southeastern.edu

Abstract. Virtual learning has been widely adopted in many higher education institutes. This is especially necessary and important during the ongoing COVID-19 pandemic. In order to identify the ergonomic issues that college students faced while learning remotely during the pandemic, a five-question survey was conducted in October 2020. While 60% of the respondents reported more or most comfortability while working/studying at home, 56% of the students claimed more stress and 64% of the students reported average or less-than-average productivity. In terms of using ergonomic postures while learning remotely, 56% of the survey respondents answered that they used them occasionally and another 20% reported that they never used them. Overall, the survey results are in line with the findings that were reported by other studies. The study results may contribute to the knowledge about the ergonomic challenges that virtual learning presents, especially during this unusual time.

Keywords: Virtual learning · Survey · COVID-19

1 Introduction

Virtual learning has been widely adopted in many higher education institutes. This is especially necessary and important during the ongoing COVID-19 pandemic, which was declared by the World Health Organization (WHO) on March 11, 2020. COVID-19 spread across the globe rapidly. "Stay at home" orders and safety and health concerns led to many restrictions with tons of businesses closing down temporarily or permanently and a lot of students and workers being sent home. Virtual learning presents unique ergonomic challenges for students. The home environment may be more comfortable in general; yet, the setup of ergonomic work/study stations and the use of ergonomic postures may not always be applicable. In addition, the increased screen time, the lack of in-person communications, and the technical difficulties and challenges, can all contribute to the increased stress which may lead to long-term health issues.

The right environment helps students stay focused, comfortable, and productive. This can be challenging with the distractions in the home. Although home offices are not exactly rare, those who have never worked or studied from home may find themselves lost, uncomfortable, and unprepared. Those who do not have home offices may find themselves embracing inappropriate postures and uncomfortable positions. For example,

S. Nazir et al. (Eds.): AHFE 2021, LNNS 269, pp. 299–306, 2021.
https://doi.org/10.1007/978-3-030-80000-0_35

it is tempting to do work while sitting on the couch or lying in bed. Plus, other seats in the home are generally selected with cost and appearance as factors rather than ergonomics.

Since physical distance is being enforced due to the pandemic, technology is being used now more than ever. The average person already spends a decent amount of time online each day. In addition to more computer time, being at home also gives people increased access to televisions and phones at all times, even when working. Plus, with restrictions, social distancing, and businesses continuously to adjust capacity or temporarily close, screens are increasingly being used for entertainment as well.

Not everyone has computers at home, so laptop use may be necessary. Because they were originally meant for short, convenient periods of use, they are not the best for a full work day. They are even more likely than computers to cause back pain, as well as eye, wrist, and neck strain because of their smaller screens and keyboards. A study [1] conducted by the Division of Occupational Therapy at the Medical University of South Carolina, which focused on the benefits of laptop ergonomics education to graduate students, stated that "many common laptop habits can have severe physiological effects on the user ranging from eye strain, poor posture, upper extremity pain, and overuse injuries."

Being even smaller than laptops, mobile handheld devices and smartphones can cause even more neck pain and headaches. One study found that musculoskeletal complaints related to the neck have the highest prevalence rates ranging from 17.3% to 67.8% among mobile device users worldwide [2].

Not only can working from home be physically difficult on the body, but it can also be stressful for everyone involved for a number of reasons. Everyone handles stress differently. The Internet is not always reliable, which can lead to missing meetings or deadlines. There have been several instances where people have forgotten that they are on camera during a meeting or have been interrupted by others in the house. If people are not tech savvy, it may be even harder to adjust, especially if they do not have their own computers or the same programs that they used at work. They may feel pressured to work overtime to make up for their loss in productivity.

Furthermore, adapting to communicating solely online can be difficult. Some people have more resources available to them at home, and some simply adapt to change better. The book, *The 21st Century at Work: Forces Shaping the Future Workforce and Workplace in the United States* by Karoly and Panis [3], states that "there are indications that home-based work or telecommuting may be disruptive to family relationships and be socially isolating," and that "telecommuting and face-to-face interactions are complements rather than substitutes."

Not being able to see coworkers, friends, or extended family can have a big impact on one's life, especially if a loved one is in the hospital or at high risk of dying from the coronavirus in general. Additionally, some have kids and other family members around the house. Other people can be noisy and distracting. Plus, having kids at home can be double the work, especially if they need to be homeschooled. On the other hand, the isolation may be just as hard or harder on those who live alone and are accustomed to visiting others and chatting at work.

In the article "Overcoming COVID-19: What can human factors and ergonomics offer?" Gurses et al. [4] stated that "Science-based approaches that consider human cognition and behavior while working in complex work systems, such as [Human Factors and Ergonomics], are needed to improve pandemic management." Implementing the correct solutions leads to an increase in productivity and comfortability, which is highly beneficial in the present but also in the future. Habits formed now, whether positive or negative, can and will affect the body forever. Therefore, ergonomic efforts are critical for health and wellbeing.

In order to identify the ergonomic issues that the students who are enrolled in the Occupational Safety, Health, and Environment (OSH&E) program at Southeastern Louisiana University faced while learning remotely during the pandemic and help them improve the ergonomics of virtual learning, a five-question survey was conducted in October 2020. The purpose of this study was to explore the ergonomic needs during the COVID-19 pandemic as more people had been staying at home and using technology to work and communicate. The specific aims of the study were to: 1) research how being at home during the pandemic has affected people ergonomically; and, 2) identify how to prevent the ergonomic issues from causing discomfort, injuries, or illnesses.

2 Methods

The five-question survey consisted of two multiple-choice rating questions and three short-answer ones. The specific questions include:

1. Using a scale of 1-5, with 1 being the least and 5 being the most, please rate the fol lowing in regards to your experience working/studying at home.

   - Your comfortability

     1- the least
     2-
     3- average
     4-
     5- the most

   - Your stress

     1- the least
     2-
     3- average
     4-
     5- the most

- Your productivity

  1- the least
  2-
  3- average
  4-
  5- the most

2. How often do you use an ergonomic posture (sitting comfortably, parallel to floor, back supported, feet flat) while working/studying from home?

- never
- occasionally
- frequently

3. What has been the biggest ergonomic challenge (comfortability, stress, etc.) you have had to face in regards to working/studying from home? [short answer]
4. How are you balancing personal and school/work demands? [short answer]
5. Is there anything, positive or negative, that you would like to share about your experience working/studying from home? [short answer]

The study was approved by the institutional review board at Southeastern Louisiana University. The survey was conducted via Survey Monkey in October 2020. It was sent to the 86 junior and seniors who are enrolled in the OSH&E program at Southeastern Louisiana University.

3 Results

A total of 25 students responded to the survey. The response rate was 29.1%. Figure 1 shows the results of the survey respondents' answers to the first question. While 60% of the respondents reported more or most comfortability while working/studying at home, 56% of the students claimed more stress and 64% of the students reported average or less-than-average productivity.

The results of answers to the second question are shown in Fig. 2. 56% of the survey respondents answered that they used the ergonomic postures occasionally and another 20% reported that they never used them.

When asked what has been the biggest ergonomic challenge during virtual learning, there were an array of answers. However, there were some common ones. Seven people (28%) stated that their posture was the biggest challenge; 2 of those 7 said it was because they struggled with getting comfortable in their chair, and 1 said that they simply favored the comfortability of their bed. Next, 5 people (20%) said their challenge was focusing. One explained that this was due to having to continuously watch their laptop for almost an hour. Another 4 respondents (16%) answered "stress" with one due to noise and one with keeping up with assignments and teacher expectations. Similarly, 3 other responses (12%) said teachers because of trying to understand what they expect,

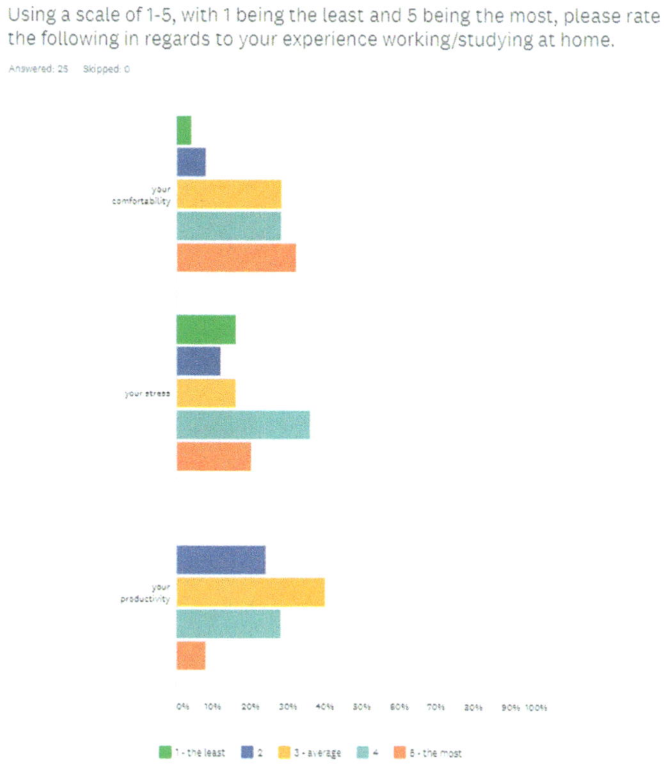

Fig. 1. Results of survey respondents' answers to question 1.

figuring out when/where class is, or just running through lecture slides. Three (12%) said no challenges, although one mentioned not having a printer was an inconvenience. Two (8%) said time management was an issue, and other challenges included staying motivated outside of the classroom, productivity, and organization.

In response to question 4, "How are you balancing personal and school/work demands?" there were a few simple answers and then some very insightful ones. Some students entered one-word or simple-phrase responses such as "poorly," "good," "ok," "It's hard," "not well," and "great" twice. Two respondents said that they were taking it "one day at a time." Three students (12%) said that the workload is much heavier this semester, and one even stated that "The teachers have loaded us up with homework and that's literally all I have time for." In contrast, another student who said s/he is graduating this semester answered "I have one class this semester, so it's very easy." Most of the respondents gave answers that utilize similar tactics, saying that they handle it by setting a schedule, practicing time management, prioritizing school work, getting into a routine, using a reminder app, and keeping a mental checklist or to-do list.

The last question asked if there was anything positive and/or negative the respondent would like to share about their experience. While 4 people declined to say anything further, the other 21 respondents provided some helpful and interesting information. Several pointed out the pros of being home, from a simple "The comfort of your home!"

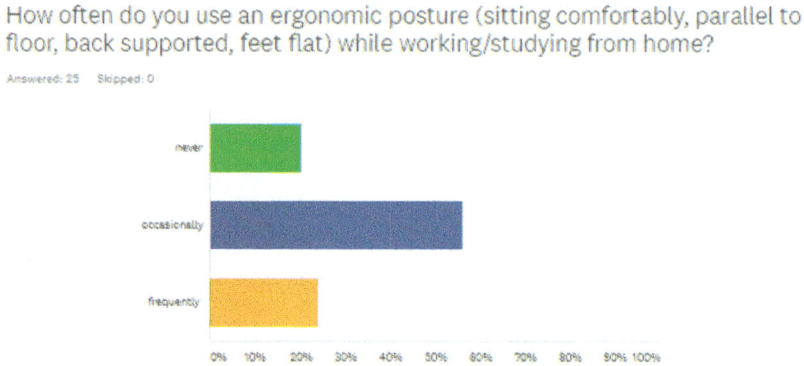

ANSWER CHOICES	RESPONSES	
never	20.00%	5
occasionally	56.00%	14
frequently	24.00%	6
TOTAL		25

Fig. 2. Results of survey respondents' answers to question 2.

and "reduced my stress levels" to more comprehensive considerations: "Has allowed me to work more and not have to travel all the way to Hammond continuously, causing less spending and more saving, overall has helped me a lot" and "My grades have improved regardless of my surroundings. I think it has to do with having more free time at home to study." Others pointed out their cons that are fairly consistent with answers to the previous questions, including more procrastination, more information being given, and more noise. One respondent also pointed out a different kind of problem they ran into – the difficulty of finding a job because of COVID-19. Some gave a mix of positive and negative, such as enjoying more sleep but favoring in-class learning, finding school work easier to do at home but dealing with some teachers who do not know how to use the technology, liking the flexibility and not worrying about commuting but missing campus and seeing everyone, and being glad they do not have to be on campus with the virus but only getting taught by PowerPoints.

4 Discussion

Overall, the survey results are in line with the findings that were reported by other studies. A study conducted by the city of Aspen, Colorado, found an average of 6.9 out of 10, with 0 being very difficult and 10 being very easy, when asking employees who were working remotely how easy or difficult it is to work effectively during the pandemic [5]. The Institute for Employment Studies in the UK conducted a Homeworker Wellbeing Survey in March 2020, and they found that 50% of respondents were not happy with their current work-life balance [6].

A number of recommendations to improve the ergonomics of virtual learning were presented in the study. One of the most important ergonomic principles to remember is the ergonomic posture. When sitting, this consists of keeping the feet fat on the floor or a footrest, wrists straight, hands at or below the elbow level, thighs parallel to the floor, and the knees about level with the hips. Adjustable office chairs with armrests that support the elbows close to the sides and a seat back that supports the natural curve of the spine when sitting up straight or leaning back slightly are the best way to achieve this. A standing ergonomic position uses some of the same principles but with the head, neck, torso, and legs in a line. At home, alternating between a standing ergonomic position and a less-than-ideal sitting position is much better than sitting the whole time.

If the equipment an employee or student is using at home is not adjustable, there are products and methods that can help. For example, if a chair does not have arm rests, the keyboard can be pushed back slightly to allow the work surface to support the forearms. A seat pillow that is molded for better all-day sitting can increase comfortability and height if needed. If a chair does not have lumbar support, a rolled-up towel or pillow can be used. Additionally, if the feet cannot rest flat on the floor and the chair does not have a footrest, a foot rest can be placed on the ground. These are available for sale, or a stack of books, a box, or any other firm surface that allows the feet to be flat and knees not extended can be used. Anyone who might be feeling antsy may prefer a foot rest with a curved bottom that allows their feet to be propped up on a surface that allows them to rock back and forth. Better yet, an under-the-desk mini-bike can help maintain healthy joints and improve circulation while providing some exercise similar to walking.

An adjustable mobile laptop desk is a great tool for those who would like to bring their laptop to different areas in the home. Research from New York University and University of Miami found that having new, diverse experiences every day is linked to positive emotions and enhanced happiness, so a change in scenery may be valuable [7]. This tool also provides an easy transition from sitting to standing to vary one's posture throughout the day, effectively avoiding back, neck, and shoulder pain.

Prolonged digital screen use may cause eye strain and excessive fatigue. During this time, Californian optometrist Jeffrey Anshel's 20-20-20 rule can be implemented [8]. Every twenty minutes one should take at least twenty seconds to look at an object at least twenty feet away or at least twenty seconds to close their eyes. The Japan Human Factors and Ergonomics Society recommends setting an alarm for every 20 min to remember to take a break, blinking frequently to prevent dry eye, and switching positioning from sitting to standing to prevent general health problems.

Some ways to reduce stress include getting ready for online meetings in advance and dedicating time to the family outside of work time. Checking connections before meetings, making sure other people in the house know when one needs a quiet space to focus, and remembering to ensure that the camera and microphone are turned off at appropriate times can help ease the mind. Webcam covers cannot only help provide extra security measures, but they can also prevent accidental broadcasting. Headphones with a microphone can minimize external noise and improve sound quality on both ends of a call. Committing time to family members and personal health outside of normal work hours can help those working at home to maintain a healthy work-life balance and to improve their focus on the tasks at hand.

Teachers and students should be able to maintain a system of open communication and clear expectations, and schools should ensure that students have access to mental health care and anything required to complete their coursework. It would also be a good idea for schools and teachers to provide students with information on ergonomics since research suggests that receiving ergonomic education can positively influence awareness of body mechanics [1].

The present study has several limitations. The response rate (29.1%) is in line with the average response rate to online surveys and was enough to get some variation and insight, but further research and a larger number of participants would improve the study. It is also important to consider the time of the year. The survey was conducted about seventh months into the COVID-19 pandemic, which posed some unique challenges. It is very likely that the responses would have been drastically different if the survey had been conducted in the period of uncertainty and adjustment that dominated the spring semester. On March 12, 2020, Southeastern Louisiana University began transitioning all classes to remote learning, and a week later, students were informed that classes would remain online for the rest of the spring 2020 semester. Juniors and seniors in the OSH&E program would have experienced this online learning, and the fall semester was also already halfway completed. By the time of the survey, students had gotten some time to grow accustomed to the circumstances of the pandemic and the semester.

5 Conclusion

The present study conducted a survey to identify the ergonomic issues of virtual learning, especially during the ongoing COVID-19 pandemic. The results may help shape future actions regarding ergonomics while working and studying at home. Setting up an effective space that can be used for safe and productive work is a challenge most have had to face. Implementing the correct solutions leads to an increase in productivity and comfortability, which is highly beneficial in the present but also in the future. During this pandemic and whenever we work or study at home, it is important that we adapt workplace ergonomics and implement the practices wherever applicable.

References

1. Bowman, P.J., et al.: Benefits of laptop computer ergonomics education to graduate students. Open J. Ther. Rehab. **2**(1), 25–32 (2020)
2. Xie, Y., Szeto, G., Dai, J.: Prevalence and risk factors associated with musculoskeletal complaints among users of mobile handheld devices: a systematic review. Appl. Ergon. **59**(A), 132–142 (2017)
3. Karoly, L.A., Panis, C.W.A.: The 21st Century at Work: Forces Shaping the Future Workforce and Workplace in the United States. RAND Corporation, Santa Monica (2004)
4. Gurses, A.P., et al.: Overcoming COVID-19: what can human factors and ergonomics offer? J. Patient Saf. Risk Manag. **25**(2), 49–54 (2020)
5. Ott, S.: COVID-19 Officevibe Survey Results - Aspen, CO. https://bit.ly/3s2zJDj
6. Bevan, S., Mason, B., Bajorek, Z.: Homeworker Wellbeing Survey. Institute for Employment Studies. https://bit.ly/2LedWbc
7. Heller, A.S., et al.: Association between real-world experiential diversity and positive affect relates to hippocampal–striatal functional connectivity. Nat. Neurosci. **23**, 800–804 (2020)
8. Japan Human Factors and Ergonomics Society: Seven Practical Human Factors and Ergonomics (HF/E) Tips for Teleworking/Home-Learning Using Tablet/Smartphone Devices, 1st edn., The IEA Press, Paris (2020)

Information Competency as a Success Factor in Distance Learning

Tatiana Shaposhnikova, Alexander Gerashchenko(✉), and Dmitry Romanov

Kuban State Technological University, Moskovskaya Street 2, 350072 Krasnodar, Russia

Abstract. The information competency of teachers and students is a fundamentally important success factor in distance learning, the relevance of which has largely increased due to the COVID-19 pandemic. Distance learning provides an opportunity to exclude or limit direct personal contact between teachers and students. Nowadays it relies heavily on information technology (information systems), being connected with informatization of education in general. This research concentrates on the analysis of the correlation between the successful application of information systems in distance learning and the information competency of the subjects of such learning. The results of the research show that the success of the operational development and application of information systems in distance learning (both by teachers and students) is determined both by the system of knowledge and skills of the individual (in the field of information technology), and by the previous personal experience of using technical information systems in general life activities.

Keywords: Information competency · Distance learning · Success factor

1 Introduction

Nowadays, nobody doubts the perspectives of such information and educational technology as distance learning: [1–13]. Distance learning is attractive due to many reasons, including the variety of its forms (portals with electronic educational resources, online courses, electronic universities, etc.). In addition, in a complicated epidemiological situation (currently caused by COVID-19), distance learning is an important factor in ensuring the continuity of the educational process in educational institutions of any type (both general and vocational education institutions).

At the same time, the introduction of distance learning technologies requires all kinds of prerequisites: in online distance learning, as well as in offline in-class learning, the main thing is not the use of technical means, but the successful solution of social and educational problems (first of all, ensuring the quality and efficiency of the educational process). For example, the implementation of telecommunication technologies for distance learning is impossible without a reliable high-speed telecommunications network, appropriate software and hardware, while network technologies of distance learning are unavailable without electronic educational resources of the proper quality,

S. Nazir et al. (Eds.): AHFE 2021, LNNS 269, pp. 307–314, 2021.
https://doi.org/10.1007/978-3-030-80000-0_36

which are located on appropriate portals, such as the YaKlass (www.yaklass.ru) digital educational resource for schools.

The most important aspect of any technology is not instrumental but social: [4, 10, 14, 15]. With regard to information and educational technologies, these are teachers and students with the proper level of information competency, as well as other cognitive factors that determine the success of work in the information and educational environment.

Information competency is the readiness of an individual to use information technologies (methods and means of working with information) not only in everyday life but also in educational, professional and other activities: [1, 5, 14, 16–18]. According to modern views, information competency is a personal and professional quality, which includes the following interrelated components: operational (knowledge and skills related to information technology); motivational-value (motives for the use of information technology, value attitude to information and information technology); behavioral (personal experience of using information technology in life); diagnostic (the ability to self-manage the development of personal information competency). Analyzing the manifestations of information competency in life (particularly in educational and professional activities), the authors of this article have suggested the criteria for the formation of its behavioral component.

Modern specialists have concluded that one of the most important socio-cultural factors in the formation of students' information competency (at all stages of the lifelong education system) is the informatization of education with the systemic (and not episodic) use of computer information systems in the educational process: [1–13]. At the same time, the events associated with the complex epidemiological situation caused by COVID-19 forced the accelerated (forced) introduction of distance learning technologies into the teaching and learning process at educational institutions of all types. The question that inevitably arises is "What factors determine the success of using distance learning?"

A necessary condition for the successful application of such educational technologies is proper equipment. Hardware, software and information support are provided depending on the types of distance learning technologies used and by the didactic tasks solved. Online training related to video conferencing needs reliable, high-performance telecommunication networks, as well as software for organizing multipoint video conferencing (for example, Zoom and Skype), while network or case technologies presuppose the presence of such powerful information and methodological support as electronic educational resources.

At the same time, distance learning information systems cannot be used successfully without appropriate cognitive prerequisites – first of all, the information competency of teachers and students (the main players in social and educational interaction). Since distance learning technologies are information technologies, the readiness for their use means the readiness to use information technologies in educational activities (which are professional for teachers), i.e. a certain level of information competency.

Despite the undoubted importance of distance learning (especially in a complex epidemiological situation) as the most advanced information and educational technology, the relationship between the information competency of subjects of the educational process and the success of their use of information systems in distance learning is still studied insufficiently.

This research concentrates on the analysis of the correlation between the successful application of information systems in distance learning and the information competency of the subjects of such learning. The object of the research is the use of distance learning information systems by subjects of social interaction in education: teachers and students. The subject of the research is the relationship between the level of teachers' and students' information competency and the degree of their success in applying information systems for distance learning.

The objective of the research is to develop information-probabilistic models of the success of distance learning. The significance of research is connected, on the one hand, with its role as the basis for further scientific understanding of the transformations that arise due to informatization of education and, on the other hand, with its contribution to the guidelines for the development of the students' information competency at all levels of continuing education.

2 Methods

Methodological foundations of the research are the systematic approach (that considers distance learning in the context of a systemic process – the informatization of education), the sociological approach (that considers the information and educational environment as the most important socially determined factor in the development of a student's personality), the competence-based approach (that considers information competency as a systemic set of knowledge and skills in the field of informatics, the motivational-value attitude to information technologies and personal experience of their application, as well as a factor in the successful use of distance learning information systems), the process approach (that considers the use of distance learning systems as information and educational interaction on a relevant portal), the qualimetric approach (that proclaims the need for a multi-criteria assessment of both the success of training and its factors), and the probabilistic-statistical approach (that considers educational activities, including the use of distance systems learning as a stochastic process).

The methods of research are the analysis of scientific literature and best practices in the organization of distance learning, modeling, methods of probability theory and mathematical statistics, methods of qualimetry (theory of latent variables), pedagogical experiment, and survey methods.

3 Results and Discussion

The results of the research show that the success of the operational development and application of information systems in distance learning (both by teachers and students) is determined both by the system of knowledge and skills of the individual (in the field of information technology), and by the previous personal experience of using technical information systems in general life activities.

Let N different information systems be used sequentially in distance learning. As observed in this research, distance learning for students (particularly Russian ones) can involve the following: e-mail (more precisely, e-mail exchange systems), the YaKlass educational portal, the Zoom video conference system, the VK social networking service, the Microsoft Team platform, etc. For distance learning of engineering students, e-mail, e-department and modular object-oriented dynamic learning environment (Moodle) can be applied. The probability of successful mastering by the subject of social and pedagogical interaction of all technical systems, according to the theorem on the probability of independent events (where p_i is the probability of successful mastering of the i-th technical system):

$$\rho = \prod_{i=1}^{N} p_i. \tag{1}$$

This formula is correct if the development of technical systems has no logical connections. However, logical connections are often present. For example, if an individual has successfully mastered the social network VKontakte, he or she will also master the Microsoft Team platform as the logic of the work is largely the same. Someone who has mastered Skype (for video conferencing) can also successfully master the Zoom system (for multipoint video conferencing), as well as the Electronic Department system (its module that provides online classes). In this case, we apply the conditional probability formula (where $p_i|p_{i-1}$ is the probability of successful mastering of the i-th system, provided that the i-1-th system is successfully mastered):

$$\rho = p_1 \prod_{i=2}^{N} \left[(p_i|p_{i-1})p_{i-1}\right]. \tag{2}$$

What determines the likelihood of successful development (use) of a particular information system by an individual in distance learning? Obviously, if the individual had previous experience with this particular system, then

$$p = \eta \cdot \kappa. \tag{3}$$

In this expression, η is the reliability of the individual's interaction with the information system and κ is the forgetting coefficient, $\kappa \leq 1$. If not a long time has passed since the last use of the information system, then $\kappa \approx 1$.

How to determine the reliability of an individual's work with a specific information system? If the number of cases N of this work coincides with the number of cases of successful work, then

$$\eta = \frac{N}{N+1}. \tag{4}$$

If the number of cases N of this.

Does not coincide with the number of cases of successful work, then

$$\eta = \frac{M}{N}. \tag{5}$$

Here M is the number of cases of successful work.

Let us address to the situation when there is an extrapolation (transfer) of experience from one information system to another, as well as the use of well-established skills and knowledge when working with an "unfamiliar" information system. To model the probability (more precisely, to simplify), let us make an assumption: the number of cases N – either of successful work with a familiar information system or of the development of a particular skill – is so great that $\eta \approx 1$, i.e. experience or skill is considered firmly established. For example, before using the distance learning system, students can work out sending emails (including those with attachments) a huge number of times, and it will not be difficult for them to exchange messages with their tutors when working at forums in an information learning environment (such as Moodle), to submit their homework etc. Another example is as follows: a very standard action when working with software, especially websites, is authentication (login with a username and password). Here is an example of extrapolation: at first, students worked with social networks such as VK (for example, sending homework to teachers and exchanging messages during online lessons), and then they easily switched to the Microsoft Team platform.

Let the degree of difference (on the scale of relations) between the information system equal Z, then the probability of independent mastering of the new information system by the individual is

$$p = \frac{1}{z + 1}. \tag{6}$$

As it can be seen, if the difference is zero, then the probability is 1.

If we take into account the level of extrapolation E of the individual (on the scale of relations: the assessment method is presented in [3]), then

$$p = \frac{E + 1}{E + Z + 1}. \tag{7}$$

Let us consider the most difficult (more precisely, the most general) cases. Let S be a set of knowledge and skills objectively necessary for working with a technical system, s – a set of knowledge and skills actually available to the individual, G – the difficulty of working with an information system (for example, the complexity of a sequence of steps, or actions, when managing an information system), H – the level of the individual's learning capacity on a logarithmic scale (the ability to successfully apply existing knowledge in various types of activity: the assessment method is presented in [3]), then

$$p = \partial(S - s).\frac{e^{H-G}}{1 + e^{H-G}}. \tag{8}$$

$\partial(S - s) = 1$ if $S - s = \emptyset$ and $\partial(S - s) = 0$ if $S - s \neq \emptyset$, where the symbol $\emptyset$ represents the empty set.

At the same time, the individual can independently master the missing elements of knowledge and skills. In this case, the probability that the individual will not master the technical information system is

$$p = \left(1 - 1.\frac{e^{H-G}}{1 + e^{H-G}}\right).\alpha(S - s) = \left(\frac{1}{1 + e^{H-G}}\right).\alpha(S - s). \tag{9}$$

In this formula, α is the probability that the individual will not be able to master the missing elements of knowledge and skills from the set $S - s$.

According to the theorem on the probability of independent events (although this assumption is very conditional, since there can be a connection between the elements of knowledge and skills),

$$\alpha(S - s) = \prod_{i=1}^{p(S-s)} \beta_i. \tag{10}$$

Here P is the cardinality of sets and β_i is the probability of non-mastering the i-th missing element of knowledge and skills.

Now let us analyze a real-life case of distance learning, using the example of Kuban State Technological University (KubSTU), a higher education institution located in Russia, in the city of Krasnodar. Due to the complicated epidemiological situation resulting from the COVID-19 pandemic, KubSTU, as well as other educational institutions of Russia, shifted entirely to distance learning for the period from the middle of March to the end of June 2020. For this purpose, the university used its information systems of distance learning, namely the Electronic Department system on elkaf.kubstu.ru and the Moodle-based information learning environment on moodle.kubstu.ru).

A stating experiment conducted at the KubSTU Department of Physics showed that, in general, the use of distance learning information systems did not cause difficulties for the majority of students, and the transition to distance learning did not lead to either an increase or a decrease in the quality of educational activities of the first- and second-year undergraduate students. The department conducted a survey on teaching physics in a distance mode that involved 2083 students of KubSTU, of whom 986 enrolled in 2018 and 1097 enrolled in 2019.

The following questions were asked: 1) What difficulties did you face in using distance learning systems? 2) What information systems, in addition to those offered by the university (the Electronic Department and Moodle of KubSTU), would you like to use in distance learning? 3) Is it advisable to use distance learning?

The answers to the first question were distributed as follows: none (771, or 37%), unstable or unproductive functioning of the Internet (1183, or 56.8%), non-optimal interface of the proposed systems (227, or 10.8%), insufficient knowledge of information technologies (368, or 17.6%). The sum of the numbers exceeding 100% is explained by the fact that a number of respondents identified several difficulties. Among the students who noted the reason "insufficient knowledge of information technologies", only 95 (25.8%) are second-year students, and 273 (74.2%) are first-year students. Indeed, in their second year, the overwhelming majority of students have formed information competency, at least at the literacy level, which makes it possible to use distance learning systems.

The answers to the second question were distributed as follows: none (733, or 35.2%), e-mail via well-known sites (1341, or 64.3%), well-known universal videoconferencing systems (429, or 20.5%), social networks, universal platforms (982, or 47.1%), interactive software and methodological complexes, computer systems for educational purposes (264, or 12.6%). The sum of the numbers exceeding 100% is also explained by the fact that a number of respondents chose several options. In other words, the majority of respondents are in favor of the variety of technical means used in distance learning.

The answers to the third question were distributed as follows: there should be no place for distance learning at all (297, or 14.2%), a full transition to distance learning is necessary (259, or 12.4%), it is necessary to combine full-time in-class learning with distance learning (1394, or 66.9%), it is hard to answer (133, or 6.3%). As you can see, the overwhelming majority of the respondents are in favor of combining in-class (face-to-face) education conducted offline with distance learning conducted online, while most of those who support this combination claim that distance learning is promising in the independent work of students (1179 out of 1394, or 84.5%).

Besides, the respondents supporting the optimal combination of face-to-face and distance learning were asked an extra question: has the university done enough to optimally combine face-to-face and distance learning? The answers were distributed as follows: it is hard to answer (241 out of 1394, or 17.2%), enough (188, or 13.4%), enough rather than not enough (307, or 22%), not enough rather than enough (395, or 28.3%), clearly not enough (263, or 18.8%). The respondents, according to whom there is not enough work done at the university to combine full-time in-class and distance learning (there are 658 of them), argue that there is a lack of the necessary diversity in the use of information systems, as well as electronic educational resources.

It is clear from the survey results that though there is still a lot of work to be done in order to develop and promote distance learning technology, the university is ready for it.

4 Conclusion

Distance learning conducted online can be considered the most promising information educational technology, especially in continuous education (adult education), part-time learning, independent work of students, and emergency situations (for example, in a complicated epidemiological situation).

Nevertheless, for the use of such technology, students must have cognitive prerequisites provided by full-time in-class learning conducted offline.

Therefore, students will benefit most from a rational combination of offline in-class learning and online distance learning.

Acknowledgments. The research was carried out with the financial support of the Kuban Science Foundation in the framework of the scientific project No. IFR-20.1/36.

References

1. Abdurazakov, M.M.: Teacher-student interaction in information-related educational environment: the culture of knowledge – cognition – informational communication. Pedagogika **9**, 39–46 (2018)
2. Dubakov, A.V.: Development of Instructional competence of a foreign language teacher in a distance refresher course. Soc. Sociol. Psychol. Pedagog. **4**, 94–98 (2018)

3. Chernykh, A.I., Shaposhnikova, T.L., Khoroshun, K.V., Romanov, D.A.: Monitoring the Quality and Efficiency Of Continuous Professional Education. Kuban State Technological University, Krasnodar (2016)
4. Polupan, K.L.: Interactive intellectual environment as a digital technology of continuous education. Vysshee Obrazovanie v Rossii **11**(27), 90–95 (2018)
5. Sedunova, A.S.: Cognitive prerequisites for interaction in the information environment among university students. Soc. Sociol. Psychol. Pedagog. **9**, 63–66 (2018)
6. Sorochinsky, M.A.: The analysis of teachers' readiness for activities in a university e-learning environment. Soc. Sociol. Psychol. Pedagog. **8**, 103–106 (2018)
7. Shafranova, O.E.: Additional education of teachers as the condition of risk minimization in informatization of higher education. Sci. Supp. Syst. Adv. Train. **2**(35), 16–24 (2018)
8. Yakusheva, I.V., Demchenkova, O.A.: Innovative Learning environment in university: meeting international requirements. Pegagogika **11**, 55–61 (2018)
9. Altaweel, N.A.: Integration of mobile technology into evidence-based practices for students with emotional and conduct disorders in classroom. Problems Educ. 21st Cent. **1**(77), 195--208 (2019)
10. Cox, A.M.: Space and embodiment in informal learning. High. Educ. **6**(75), 1077–1090 (2018)
11. Iskrenovic-Momcilovic, O.: Using computers in teaching in higher education. Mediter. J. Soc. Sci. **4**(9), 71–78 (2018)
12. Kurti, E.: Methods of teaching and their impact on learn. Mediter. J. Soc. Sci. **1**(7), 489–500 (2016)
13. Palomino, M.C.P.: Information and communication technologies and inclusive teaching: perceptions and attitudes of future early childhood and primary education teachers. Probl. Educ. 21st Cent. **3**(76), 380--392 (2018)
14. Grigorash, O.V.: The modern approaches to assessing the level of competence among engineering students. Soc. Sociol. Psychol. Pedagog. **10**, 106–110 (2018)
15. Reyes, G.E., Govers, M., Ruwaard, D.: A mathematical and conceptual model regarding social inclusion and social leverage. Mediter. J. Soc. Sci. **3**(9), 9–16 (2018)
16. Kazakova, E.I., Tarkhanova, I.: The measurement of the development level of university students' universal competencies. Pedagogika **9**, 79–83 (2018)
17. Martynov, V.G., Golunov, N.N., Makarova, E.D.: Open educational online courses of Gubkin University as an innovation in education. Pedagogika **5**, 37–44 (2018)
18. Shaposhnikova, T.L., Vyazankova, V.V., Romanova, M.L.: Qualimetry evalyation of students informational competence. Open Dist. Educ. **2**(62), 35–39 (2016)

Evaluation of Distance Learning Effects on Concentration and Relaxed States by EEG and HRV

Yuri Nakagawa(✉), Peeraya Sripian, and Midori Sugaya

Shibaura Institution of Technology, 3-7-5 Toyosu, Koto-ku 135-8548, Tokyo, Japan
{al17080,peeraya,doly}@shibaura-it.ac.jp

Abstract. The demand for distance learning is increasing due to the recent expansion of the COVID-19 pandemic. In distance learning, less communication level has been pointed out that it would reduce student's concentration level in the study. However, these were subjective evaluations based on questionnaires, therefore, there was a lack of criteria to evaluate this problem. In this work, we propose a method to objectively evaluate distance learning using the biometric information to understand the difference of the student's concentration level during the study. We measure the student's electroencephalography (EEG) and heart rate variability (HRV) to evaluate the change of student's emotion based on different communication condition during distance learning. In the experiment, we measured EEG and HRV of the students, while turning on and off face-to-face video camera. The analysis of results show that different communication condition effect student's concentration level. We found that it could be due to their preferences of the subject.

Keywords: Distance learning · COVID-19 · Electroencephalography · Heart rate

1 Introduction

Nowadays, distance learning is being promoted from the perspective of realizing diverse learning and highly specialized classes [1]. Distance learning can help maintain and improve the quality and equality of educational opportunities for students who, for various reasons, have difficulty commuting to school to receive education [2]. In addition, under the circumstances of expansion of the COVID-19 pandemic, distance learning has made it possible for students to receive education in an isolated environment even if it is difficult for all students and faculty to commute to school.

Meanwhile Tagami et al. [3] pointed out that distance learning suffers from the lack of a sense of realism and tension, which could affect the quality of the class. Croft et al. Kudo et al. [5], also pointed out that the lack of communication between learners would decrease their concentration on the class.

According to the survey regarding online classes during 2020 summer semester [6], it was found that about 78% of the respondents answered that they would like to adopt online classes as one of the forms of instruction in the future. This suggests that online

S. Nazir et al. (Eds.): AHFE 2021, LNNS 269, pp. 315–323, 2021.
https://doi.org/10.1007/978-3-030-80000-0_37

classes will continue in the future. On the other hand, a relatively high percentage (about 66%) of respondents felt that the disadvantage of taking online classes was the lack of communication with other students.

Although online classes have many advantages, but the lack of communication and the feeling of lifeless could be it's major issues. However, conventional research has not sufficiently evaluated the psychological impact of the lack of communication on the learning state, and there is a lack of objectivity in this evaluation.

In recent years, measurement methods using biometric data such as EEG and pulse have been proposed as an objective method of measuring psychological states. Ohkura, et al. investigated the evaluation of the excitement feeling by measuring changed in biometric signals measured in autonomic nervous system [7]. Also, study by Sakamoto et al. showed that the amplitude of alpha waves increased during relaxation, while the amplitude of alpha waves decreased, and beta waves appeared during tension [8]. These objective evaluation methods have not been used for the evaluation of online classes yet. In this study, we objectively compare and evaluate the learners' state by using biometric data. The purpose of this study is to clarify the objective psychology of the learners.

2 Issues and Proposals

2.1 Issues

Studies in [4–6] pointed out the problem of poor communication in distance learning. However, it is not clear to what extent the effects on learner's concentration level exist between different communication levels in distance learning. Furthermore, there are few evaluation methods provided to objectively detect the effects between different conditions of distance learning. These evaluations of the actual results in educational settings are usually mainly based on questionnaires [3–6]. Questionnaires could provide subjective reflections, but they do not provide an objective assessment of learning situations such as concentration in class. It is important to clarify the effects of these factors for the better operation of online classes.

2.2 Hypothesis

In the traditional classroom, learners and teachers share a physical space. Online class, in the contrary, learners do not share the physical space with other learners and teachers, and the class is held in a space isolated from other learners and teachers. This is the major difference between online classes and traditional classes. There is no point of contact between the two parties in the physical space, and it is difficult to exchange verbal and non-verbal information, which is necessary for communication. Therefore, the difference of the amount of information exchanged could be considered as the difference of the communication level in the online class.

If the classroom is assumed to be the space shared by learners and teachers in a traditional class, the "classroom" in an online class can be replaced by a virtual space with online meeting services that are already being used in various situations (Fig. 1).

When we step into a space where others exist, we actually share the space with them, if it is a physical space. However, when we simply participate in a virtual space, it is

Fig. 1. Traditional class and online class

only a collection of personal spaces, not the sharing of physical space. If the difference between a personal space and a shared space is the opportunity for communication, then in order to share a space in a virtual space, it is necessary to explicitly indicate each personal space on the virtual space, and then use the function that provides the opportunity for communication between personal spaces.

In our study, we considered that the most effective communication features of virtual web meeting services are voice communication and video communication. The voice call function enables verbal communication, and the video function enables non-verbal communication. In this study, we first used the video function, which is considered to enable non-verbal communication, to provide an opportunity for communication by showing an image of one's own face to the other person.

2.3 Proposals

As for evaluation methods, recent studies [7–9] proposed the use of biometric information to evaluate one's own state, which is difficult for others to evaluate objectively. In particular, electroencephalogram (EEG) evaluates the state of concentration, and heart rate variability (HRV) reflects the emotional response of the body. In this study, we propose an objective assessment using these methods to clarify the effects of different states of concentration between different levels of communication in online classes.

Based on the hypothesis, we compared and evaluate the biometric data of the participants during distance learning in different communication condition. Different communication condition is performed by turning on and off the video function of an online communication tool, indicating with and without nonverbal information.

3 Evaluation Indicators

3.1 EEG Indicators

In general, alpha waves (α) indicated the relaxing component and beta waves (β) indicated the concentrating component [8] (Table 1).

Because the frequency range of alpha waves and beta waves is wide (7.5–11.75 Hz and 13–29.75 Hz, respectively), and it is difficult to separate these values among individuals. In this study, we also define concentration state as the High α and Low β are dominant and non-concentration state as the Low α is dominant based on the analysis of EEG signals in Shiraiwa et al.'s work [11], which studied the effects High αas the

Table 1. Value ranges and psychological states of each frequency component

EEG	Measurable data (Hz)	Psychological state
Low α	7– 9.25	Relaxed, peaceful, conscious but not lazy
High α	10–11.75	relaxed but focused
Low β	13–6.75	Thinking, self and environmental awareness

indication for learning motivation. The higher the spectrum of Low α, the more relaxed, i.e. non-concentrated state, and the higher the spectrum of High α and Low β, the more concentrated state.

3.2 Heart Rate Variability Indicators

The heart rate variability index pNN50, which is a measure of heart rate variability, is used to evaluate relaxation and tension. pNN50 is considered to be parasympathetically dominant and relaxing when pNN50 is large, and sympathetically dominant and tense when pNN50 is small. In this experiment, the tension state is also evaluated by the change of the value of pNN50.

3.3 Subjective Evaluation Indicators

In order to use the subjective evaluation as a reference for the analysis, the participants were given a questionnaire before and after the experiment (for each communication condition). In the pre-experiment questionnaire, the participants were asked to answer "yes" or "no" to whether they liked studying English, the subject used in the experiment. For each communication condition, after the experiment, the participants were asked to answer a questionnaire about their concentration during the simulated class. They were asked to answer the questionnaire on an eight-point scale, with 1 representing "not at all" and 8 representing "very much". The questionnaire consisted of the following items based on the discussion from the professional point of view of education (Table 2).

Table 2. Questionnaire about the participant's concentration during the simulated class.

Question number	Detail
Q1	Do you feel lonely while solving the problem
Q2	Were you concentrated while solving the problem
Q3	Were you motivated while solving the problem

4 Experiment

The purpose of the experiment was to evaluate the effect on the state of concentration when the communication opportunity of the face-to-face video showing in the online class was set up. Two male in their 20s (1 pair) participated in the same meeting room on Zoom from their private rooms at home. The participants answered a questionnaire before performing the experiment.

4.1 Experimental Procedure

The following procedures were performed for the video-off and video-on condition.

- Step 1: Baseline measurement for 60 s by wearing electroencephalograph/heart rate monitor.
  *Only for the video-on condition, the participant would turn on the video after the baseline measurement.
- Step 2: The participants listened to 3 English listening questions given by the professor.
- Step 3: The professor explained the answers to the questions answered in Step 2.

A pre-test questionnaire was administered before the start of the experiment for both conditions, and a subjective questionnaire was administered after the end of the measurement for each condition to assess whether the participants had been concentrating.

4.2 Results

The subjective questionnaires for each condition showed that loneliness decreased in both participants. In addition, the subjective evaluation results showed that A's concentration and motivation decreased, while B's concentration and motivation increased (Fig. 2). The opposite results were found for both participants. In addition, in the pre-questionnaire, A answered that he disliked English, while B answered that he liked English. The results suggest that the students' preferences may have been influenced by the subjects.

As for the EEG indices, there was a significant difference in each frequency component of A when the video was turned on. When the video was turned on, the non-concentration component, Low alpha wave, increased, while the concentration components, High alpha and Low beta, decreased (Fig. 3).

There was no significant difference in the frequency components of the EEG of B (Fig. 3). This may be due to a slowing down of the response caused by sleep deprivation, or to the fact that they liked English so much that turning on the video did not affect their concentration.

Next, the results and discussion of the heart rate variability index pNN50 are described. For A, there was a significant difference in the change in the heart rate variability index pNN50 when the video was turned on (Fig. 4). The pNN50 decreased when the video was turned on, suggesting that the nervousness of the participants increased when the video was turned on.

The time series graph is shown in Fig. 4. The values were smaller when the video was turned off and larger when it was turned on, indicating that the biometric response was different in each condition. These results suggest that the use of the video on and off may be able to promote relaxation and tension during learning.

Next, there was a significant difference in the change of the heart rate variability index pNN50 by turning on the video for B (Fig. 5). The fact that the pNN50 increased with the video on suggests that B may have relaxed.

The time series graph is shown in Fig. 5. On the whole, the differences in displacement are small, as can be seen from the comparison of the means, but there is a tendency towards relaxation and relaxation of tension.

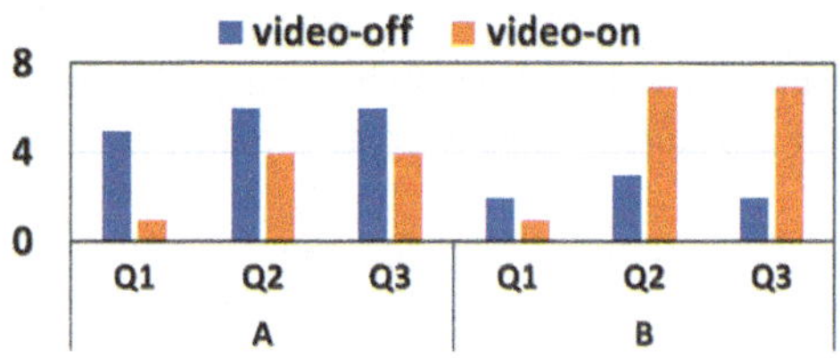

Fig. 2. The comparison of post-experiment questionnaire results by 2 participants.

The questions Q1-Q3 are described in Table 2.

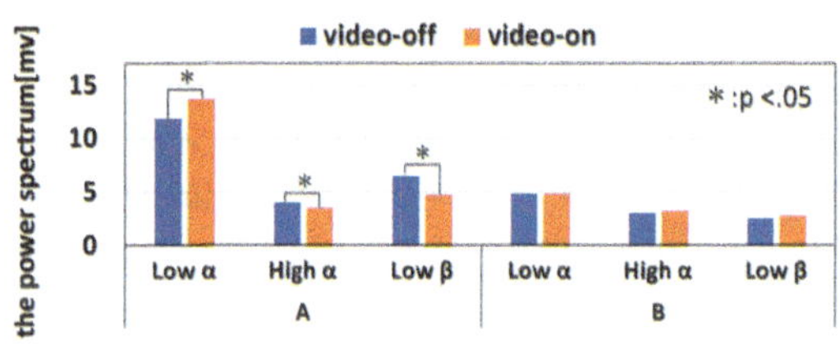

Fig. 3. Comparison of EEG (the power spectrum) of each of the two participants

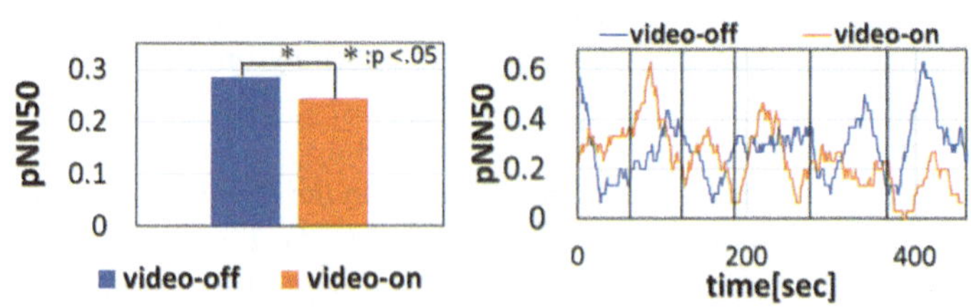

Fig. 4. Comparison of participant A's HRV (pNN50) in video-off and video-on condition. Left: average value of pNN50. Right: the time series analysis of pNN50.

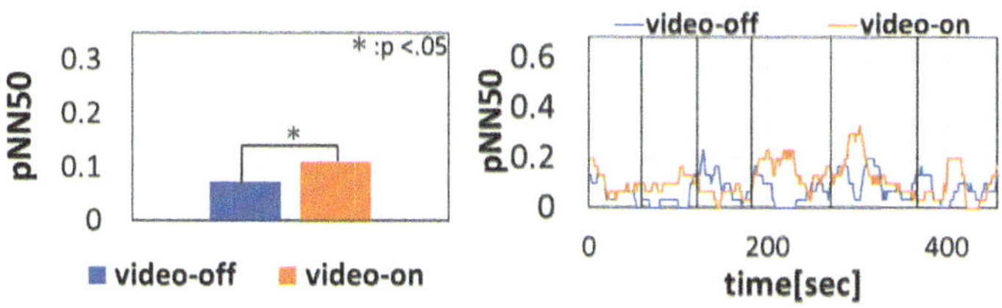

Fig. 5. Comparison of participant A's HRV (pNN50) in video-off and video-on condition. Left: average value of pNN50. Right: the time series analysis of pNN50

4.3 Discussion

From the results of the questionnaire, it was found that the feeling of loneliness was reduced when the participant is given the communication opportunity of turning on the video. Meanwhile, the results of concentration and motivation were different among the participants. These results indicate that the loneliness can be reduced and improved by providing communication opportunities. However, the result suggested that concentration and motivation in class may not simply be due to the difference in communication opportunities.

The results of the EEG and the subjective evaluation of the questionnaire were the same, indicating that the EEG could be used to evaluate the state of concentration. In addition, the results of the heart rate indicated that the communication opportunity may or may not affect the tension state during the course, suggesting that there may be individual differences among learners.

These results indicate that the use of biometric data can reveal individual psychological states. This suggests that, for learners who are not good at the subject matter, such as A, we should try not to interfere with their concentration and let them concentrate on deepening their understanding of the learning content by avoid presenting the existence of others. For learners who like the subject, such as B, we should actively give the presentations that make them feel the presences of others in order to further improve their learning motivation. In this way, it is possible to make online classes more effective for learning. In addition, since the tension of the students changes depending on whether the video is on or off, we can consider a system that encourages students to relax when the video is on.

Online courses have only been implemented in earnest since the impact of COVID-19. Therefore, there are many problems, such as the fact that effective implementation methods have not yet been established.

However, online classes have various merits such as relatively less time and effort to commute to school, equal distance between teachers and students unlike classroom seating, and the ability to concentrate on learning without excessive tension in the classroom space. We believe that online classes have various advantages. In particular, the objective evaluation of the learning state of the learner using biometric information, as we have done in this study, can be applied in various ways.

In this study, we compared the difference in learning status between learners who turned on the video and who communicated with each other using biometric data. However, we have not been able to evaluate communication in which linguistic information is shared in a virtual space. In order to clarify the effects of online communication, it is necessary to evaluate the effects of speech and dialogue opportunities. If it becomes possible to distinguish between the effects of different types of communication opportunities, it will be possible to apply them to various situations in online classes. In order to make them more practical, it is necessary to pursue the real-time evaluation. However, we have not yet established the technology for real-time application. Therefore, it is necessary to consider the evaluation of communication that shares linguistic information and the system of feedback in real time, which will lead to the development of online classes.

5 Conclusion

In this study, we assessed the concentration state of students in an online class using EEG indicators and found that the effect of turning on the video in an online class may depend on their preferences for class reinforcement.

In addition, it was suggested that the concentration state did not necessarily decrease or increase depending on the communication opportunity. However, for more abstract psychological states such as motivation to learn, the current EEG and heart rate variability assessments do not provide sufficient correlations. Therefore, it is important to clarify the relationship between EEG and heart rate variability and to objectively evaluate the psychological aspects of learning. In addition, since the number of experiments was limited, it is necessary to increase the number of experiments. These are the future tasks.

References

1. The Third Basic Plan for the Promotion of Education of the Ministry of Education, Culture, Sports, Science and Technology. Decided by the Cabinet in June 2018 [Translated from Japanese]. https://www.mext.go.jp/content/1406127_002.pdf
2. Ministry of Education, Culture, Sports, Science and Technology. Policy for the Promotion of Distance Education [Translated from Japanese] https://www.mext.go.jp/a_menu/shotou/zyouhou/detail/__icsFiles/afieldfile/2018/09/14/1409323_1_1.pdf
3. Tagami, H.: Necessity of eyes correspondence in remote teaching and a solution of the problem. J. JSiSE **25**(4), 394–402 (2008)
4. Croft, N., Dalton, A., Grant, M.: Overcoming isolation in distance learning: building a learning community through time and space. J. Educ. Built Environ. **5**(1), 27–64 (2010)
5. Kudo, N., et al.: Building interaction support system for participants in the Internet based distance class. In: Internet Conference, pp. 97–104 (2003)
6. Taura, K.: Introduction of questionnaire results regarding online classes (2020) [Translated from Japanese]. https://www.nii.ac.jp/event/upload/20200904-06_Taura.pdf
7. Ohkura, M., et al.: Measurement of Wakuwaku feeling of interactive systems using biological signals. In: Fukuda, S. (ed.) Emotional Engineering, pp. 327–342. Springer, London (2011)
8. Sakamoto, Y., et al.: Comparative analysis of state thinking by learning a simple electroencephalograph. In: Multimedia, Distributed, Collaborative and Mobile (DICOMO2012) Symposium, pp. 724–729 (2012)
9. Urabe, N., Sugaya. M.: An emotion classification method for individuals using EEG and heart rate data and deep learning. In: Proceedings of the Annual Conference of JSAI, vol. JSAI2020, 34th Annual Conference, J2F6GS1302–2F6GS1302 (2020)

10. Ohnishi, T., et al.: Development of a remote consultation system using avatar technology. EEJ Trans. Electr. Inf. Syst. **129**(7), 1408–1415 (2009)
11. Shiraiwa, G., Sugaya. M.: Verification of learning motivation, improvement by voice of robot - quantitative evaluation by EEG analysis [Transrated from Japanese]. IEICE Technical Report, pp. 1–6. 119, 446, CNR2019–45 (2020)

E-learning Ergonomic Challenges During the Covid-19 Pandemic

Noemí Suárez Monzón[1(✉)], Janio Jadán-Guerrero[2], Reinaldo Requeiro Almeida[3], and Karina Elizabeth Delgado Valdivieso[4]

[1] Research Center for Human Sciences and Education, Human Sciences, Education and Social Development, Faculty. Universidad Tecnológica Indoamérica, Ambato, Ecuador
nsuarez@unibe.edu.ec

[2] Research Center in Mechatronics and Interactive Systems, Universidad Tecnológica Indoamérica, Quito, Ecuador
janiojadan@uti.edu.ec

[3] Applied Biological Sciences Department, Cienfuegos University, Cienfuegos, Cuba
ralmeida@ucf.edu.cu

[4] Human Sciences, Education and Social Development, Faculty. Universidad Tecnológica Indoamérica, Quito, Ecuador
karinadelgado@uti.edu.ec

Abstract. The studies continuity during COVID-19 pandemic has led to the widespread use of information technologies. They keep inadequate ergonomic positions, which lead to various health conditions due to the curricula huge demands. The current learning practices developed in conditions of social isolation and at home exacerbate ergonomic factors unfavorable to young students' health. The objective is to describe the problems associated with ergonomic factors affecting young Ecuadorian university students. The research is developed under a descriptive approach. It has been carried out based on the application of a survey to a non-probabilistic sample of 585 students who are being trained as teachers. The descriptive analysis of the data informs: inadequate conditions concerning to furniture, illumination, noise, temperature and chromatography. Most of the students showed difficulties in adapting to virtual learning. Finally, it is concluded that the ergonomic challenges described find a research in the immediate future between the disciplines of school health promotion pedagogy, technology and learning ergonomics.

Keywords: Ergonomics · E-learning · Wellbeing · Student · Covid-19

1 Introduction

The studies continuity within health conditions produced by COVID-19 pandemic has led to the widespread use of information and communication technologies. Numerous challenges for students and teachers' learning and health have appeared. In higher education, students regularly spend many hours sitting in front of the computer. They keep inadequate ergonomic positions, which lead to various health conditions due to the curricula huge demands.

S. Nazir et al. (Eds.): AHFE 2021, LNNS 269, pp. 324–330, 2021.
https://doi.org/10.1007/978-3-030-80000-0_38

Before Covid-19 appearance, the growing discussion regarding university development strategies had been resumed towards a research university model that articulates the learning processes linked to research processes that perfect them [1]. According to the same author, it is possible to find contradictions in the mission statements focused entirely on teaching, while others on research. This dilemma may have affected the speed with which universities have restructured their responses to be able to face the health crisis and investigate the best ways to face this problem.

The perceived learning experience in an online course with an inverted design in such circumstances has modified the beliefs and preconceptions that could stigmatize online learning in both undergraduate and graduate degrees [2]. The study contradicts the idea that master's level students have more difficulties to get adapted to the digital environment learning. The majority of students show a positive motivation to learn in a digital environment, regardless of the educational level that they are pursuing undergraduate or master's degrees, while their levels of productivity and participation are increasing [3].

The feasibility of online learning has also been validated [4]. The author verified the improvements in learning qualifications; however, they underline the convenience of promoting a systems mentality, in the process of competences conformation in information technologies aimed at improving work in virtual environments from an integrative and interdisciplinary approach, which favors the critical thinking development.

Such research results fail to perfectly assess the ergonomic challenges of virtual education and learning practices now exacerbated by conditions of social isolation and at home as the privileged site during the Covid-19 pandemic. The consequences of the intensive use of technology on ergonomic and mental health problems of those who participate in this type of professional training are not known, but especially for students who fall into the digital gap and do not have access to the study necessary facilities, to the students' networks on campus, a crucial aspect for their academic success [5].

Consequently, the current learning practices developed in conditions of social isolation and at home as the privileged place, exacerbate ergonomic factors unfavourable to young students' health. The objective of the study is to describe the problems associated with ergonomic factors that young Ecuadorian university students go through.

2 Theoretical Foundations

The long time spent studying through the computer, in view of the demands resulting from the different study programs, leads to inappropriate ergonomic positions that inevitably bring both physical and mental health conditions. Effects on the use of furniture, lighting, noise, temperature, chromatography and adaptation to the type of device used by the student during online learning, constitute elements to take into account, in the set of ergonomic challenges of virtual education during the Covid-19 pandemic. Therefore, they need to be prevented from an area of resilience, for which the conditions that lead to the awareness of an educational process carrying such quality must be taken into account.

In all homes, there are not the most appropriate environments to carry out an ergonomically satisfactory individual study, nor are there adequate conditions that enhance physical and mental health [6]. Drastic learning practices modifications, in conditions of social isolation at home, decisively influence learning. It should be remembered that, as homes became the place where the studies were to be carried out in an accelerated way, this new environment was not considered, nor was it amortized in advance. Family environments in this stage of Covid-19 have also been vulnerable due to economic and emotional concerns as a result of the pandemic prolongation.

A verifiable effect on home study environments is external noise levels, which continues to have an unfavorable impact on students' health. The perceptual learning occurs when there is asymmetric transfer between stimuli at different levels of external noise: learning with zero / low noise can be transferred significantly to the same stimulus with loud noise, but not vice versa [7].

The mechanisms underlying this asymmetric transfer have been investigated through psychophysical, neurophysiological, brain imaging, and computational modeling studies. It is to be assumed that, in the new home study environments, the unexpected, disorderly and unfamiliar noise levels for university students induce health problems. The urban spaces in which the dwellings are located may be less favored, in relation to the noise phenomenon, for this reason it is convenient in this case to study more specifically the permeability and social interaction existing in the neighborhood conglomerates as factors that are inevitably associated with the generation of noise. In addition, an element of ergonomic connotation incident in the learning quality in virtual environments [7].

The search for better spaces for virtual learning, such as silence, has led students to move to closed places with less lighting and ventilation in which they must make use of smaller devices such as the Cell Phone, Tablet and, consequently, be subject to eventual musculoskeletal damage. These devices are the most growing digital technology on the planet, but also one of the most damaging to the spine due to the bad postural habits it generates [8]. Meanwhile, the use of desktop and laptop computers by students experiences a downward trend, when they should be more widely accepted for their long-term well-being [9–11].

University students tend to adopt significantly higher degrees of cervical and upper thoracic flexion during cell phone use, thus, postural variables show significant correlations with neck and upper back pain. A significant part of university enrollment uses the cell phone to access virtual knowledge. The researchers have warned about the high level of persistent musculoskeletal pain and/or gastrointestinal symptoms, in university students, without an organic explanation. These elements could have non-ergonomic causes, but with psychological affectation [12].

Another of the most affected sensory analyzers is the vision, recent studies address the subject in a general way. Many people experience eye discomfort and vision problems when viewing digital screens for prolonged periods leading to computer vision syndrome, just because ergonomics is not taken into account of the computer use [13]. The authors emphasize that the most common cases of visual fatigue behavior due to prolonged use of the computer and inadequate machine-man distance. This is verifiable by monitoring the frequency of blinking, tearing, and headaches.

A cross-sectional study carried out in the United Kingdom contributes to reveal problems associated with other ergonomic factors that young university students present. This time associating sleep interruption and circadian rhythmicity, as a central characteristic of mood disorders and that could be associated with subjective well-being and as a consequence of cognitive function [14]. For the authors, poor sleep can exacerbate pre-existing diseases due, essentially, to the ineffective biological detoxification of the organ systems; this detoxification occurs, in an interrelated and continuous way, at different times of the night.

It is known that sleep has a restorative effect on the body and therefore in academic performance, its consequences are also noted in the quality of interpersonal relationships, in this way educational environments can become more tense, as a result of irritability that brings about the sleep deficit. The consequences of an inefficient sleep force teachers to make greater efforts to gain the attention of the affected students, without ultimately achieving a definitely satisfactory learning.

The factors associated with temperature, lighting, physical space and chromatography conditions are also closely related to learning. The investigations allude to the positive influence of spaces that express and communicate the student with the sociocultural environment to which he belongs. To achieve this, it is imperative to reinforce the aesthetic, such as the walls colors due to their influence on perceptions, memory, and effect on the expression of emotional states. For this reason, choosing the colors for the interior infrastructure should be oriented so that visibility is easy and natural, that the environment is comfortable, that provides a feeling of calm that facilitates concentration, stimulate performance and prevents negative emotional reactions [15]. All conditions now home transferred generate great challenges.

Despite the potential of e-learning to become an incentivizing process for resilience, however, however, the authors indicate as inescapable conditions to achieve this purpose, the university community support, leadership, adequate planning of teaching activities and promotion of healthy styles in the computers use [16]. Although universities in a greater or lesser extent have been agile to migrate to online education, one cannot speak of an action of their medical staff that takes into account the probable ergonomic challenges of virtual education during the Covid-pandemic and, consequently, appropriately issue adequate hygiene recommendations through the same online platforms.

3 Methodology

The research is developed under a descriptive approach. It has been carried out based on the application of a survey to a non-probabilistic sample of 585 students who are being trained as teachers. From these, 76.2% are female and prevail students between the ages of 25 and 40. The results are part of a wider study about the use of tics in learning and the students' well-being. From the web questionnaire only the values of the environmental ergonomic variables associated to the physical conditions that surround the student during the online learning were taken into account. In confined conditions they can influence in health problems and, by consequence in the academic performance. Specifically, the following variables were analyzed: noise, furniture, illumination, chromatography, temperature and the adaptation to the type of electronic device that the student uses during the online learning.

4 Analysis and Discussion of the Results

A reading of the data shows the current ergonomic challenges in the training of university students in conditions of social isolation, especially in Latin American countries such as Ecuador. In the sample under study, the data reports that 47.3% are exposed to external noises that interrupt learning. It is observable that 76.7% revealed their adaptation to electronic devices with greater emphasis on Phones and Tablets, a fact that may be associated with the possibilities that these free mobility devices offer to be used in different spaces, but not necessarily with the possibilities of concentrating on learning, or of adopting healthy postures.

Another data that reinforces the possibility of musculoskeletal disorders is that 70% have inadequate conditions in the furniture, 42.7% have little natural lighting, which can impair vision, 40.5% said that their study place is not painted with light colors, and this can generate lack of concentration, anxiety, animosity lack and negative predisposition to study.

Only 5.2% agreed that the temperatures in their learning environment at home are not very favorable (Fig. 1).

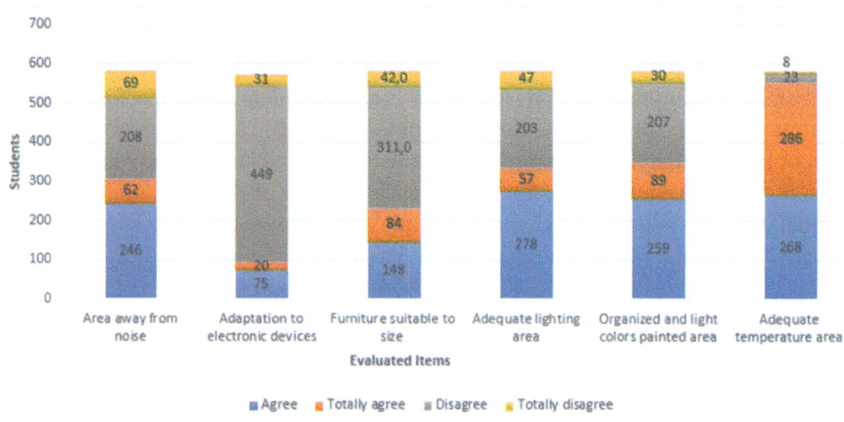

Fig. 1. Virtual learning conditions at home

The students' own appreciation of the health problems caused by the intensive use of the computer to carry out university studies in isolation conditions shows that, on a global level, various effects are seen. Figure 2 shows that approximately between 70% and 80% respond to agree and totally agree that the conditions surrounding their learning environment affect the musculoskeletal system, manifested in back discomfort, visual problems, auditory and anxiety levels that exceed 80%.

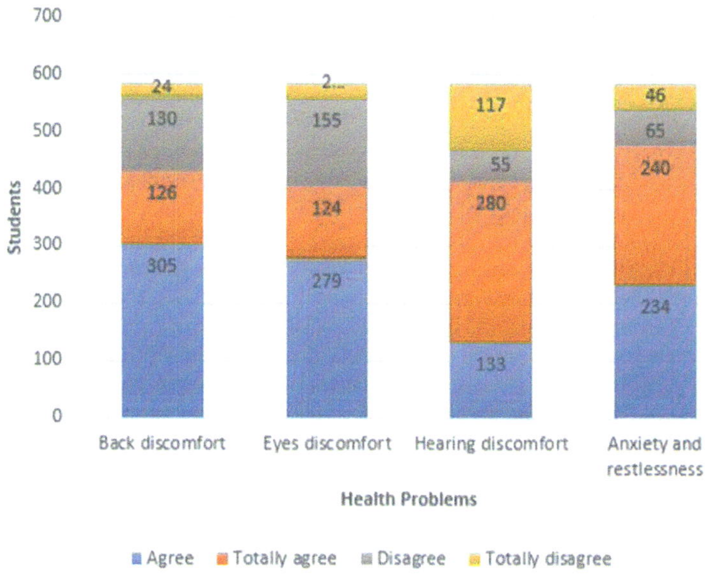

Fig. 2. Main health problems in university students

5 Conclusions

Universities have provided studies continuity, but have not promoted health to help students face the risks that this type of learning entails. Therefore, there are challenges of an ergonomic nature that affect health and it is necessary to develop health care programs for university students, in accordance with the demands that arise from the new ergonomic reality, imposed by the growing online learning and for which specific recommendations are not systematized to support students in taking care of their own health.

The outcomes found in the sample of young Ecuadorian university students who are trained as teachers indicate multiple problems related to the environment that surrounds the learning conditions in the new home context and that necessarily influence their health and learning conditions. Most related to musculoskeletal, vision and hearing problems.

The Covid-19 virus showed its potential to interrupt classroom teaching, which makes it necessary for each university to build personalized plans studying their vulnerabilities and areas for improvement, given the uncertainty that future crises may bring. In this way, the teaching-learning process will not only take place, but will take place in a resilient and therefore healthy way.

References

1. Anafinova, S.: The role of rankings in higher education policy: Coercive and normative isomorphism in Kazakhstani higher education. Int. J. Educ. Dev. **78**, 102246 (2020)

2. Sorokova, M.: Skepticism and learning difficulties in a digital environment at the Bachelor's and master's levels: are preconceptions valid? Heliyon **6**(11), e05335 (2020)
3. Xie, X., Zang, Z., Ponzoa, J.: The information impact of network media, the psychological reaction to the COVID-19 pandemic, and online knowledge acquisition: Evidence from Chinese college students. J. Innovat. Knowl. **5**, 297–305 (2020)
4. Neupokoeva, E., Chapaev, N., Suslova, I., Khokhlova, N., Sosnin, A.: The development of a systems mindset in the first-year students studying online at a vocational pedagogical university while mastering the basics of working with digital technologies. Think. Skills Creat. **39**, 100753 (2021)
5. AlAteeq, D., Aljhani, S., AlEesa, D.: Perceived stress among students in virtual classrooms during the COVID-19 outbreak in KSA. J. Taibah Univ. Med. Sci. **15**(5), 398–440 (2020)
6. Bennett, R., Uink, B., Cross, S.: Beyond the social: cumulative implications of COVID-19 for first nations university students in Australia. Soc. Sci. Hum. Open **2**(2), 100083 (2020)
7. Xie, X.-Y., Zhao, X.-N., Yu, C.: Perceptual learning of motion direction discrimination: location specificity and the uncertain roles of dorsal and ventral areas. Vision. Res. **175**, 51–57 (2020)
8. Zerouati, W., Bellal, T.: Evaluating the impact of mass housings' in-between spaces' spatial configuration on users' social interaction. Front. Architect. Res. **9**(1), 34–53 (2020)
9. Szeto, G., Tsang, S., Dai, J., Madeleine, P.: A field study on spinal postures and postural variations during smartphone use among university students. Appl. Ergon. **88**, 103183 (2020)
10. Khalid, M., Pedersen, M.: Digital exclusion in higher education contexts: a systematic literature review. Procedia. Soc. Behav. Sci. **228**, 614–621 (2016)
11. Taifa, I.W., Desai, D.A.: Quality function deployment integration with Kano model for ergonomic product improvement (classroom furniture)-a review. J. Multidiscipl. Eng. Sci. Tech. **2**(9), 2484–2491 (2015)
12. Andreasson, A., et al.: Properties of the sickness questionnaire in an Australian sample with chronic medically unexplained symptoms. Brain Behav. Immunity Health **3**, 100059 (2020)
13. Sharmila, T., Srinivasan, R., Nagarajan, K., Athithya, S.: Eye blink detection using back ground subtraction and gradient-based corner detection for preventing CVS. Procedia Comput. Sci. **165**, 781–789 (2019)
14. Lyall, L.M., et al.: Association of disrupted circadian rhythmicity with mood disorders, subjective wellbeing, and cognitive function: a crosssectional study of 91 105 participants from the UK Biobank. Lancet Psychiatr. **5**(6), 507–514 (2018)
15. Castro, M., Morales, M.E.: Los ambientes de aula que promueven el aprendizaje, desde la perspectiva de los niños y niñas escolares. Rev. Electr. Educ. **19**(3), 1–32 (2015)
16. Dohaney, J., Róiste, M., Salmon, R., Sutherland, K.: Benefits, barriers, and incentives for improved resilience to disruption in university teaching. Int. J. Dis. Risk Reduc. **50**, 101691 (2020)

E-learning and M-learning Benefits in the Learning Process

Omar Cóndor-Herrera[1], Mónica Bolaños-Pasquel[1], and Carlos Ramos-Galarza[1,2](✉)

[1] Centro de Investigación en Mecatrónica y Sistemas Interactivos, MIST y Carrera de Psicología, Universidad Tecnológica Indoamérica, Av. Machala y Sabanilla, Quito, Ecuador
{omarcondor,monicabolanos,carlosramos}@uti.edu.ec
[2] Facultad de Psicología, Pontificia Universidad Católica del Ecuador, Av. 12 de Octubre y Roca, Quito, Ecuador
caramos@puce.edu.ec

Abstract. This paper elaborates around the perception that students have regarding the learning process based on technological means: e-learning and m-learning combined. The experiment's sample consisted of 16 students between 9 and 12 years old (37.2% female and 62.5% male). The results showed that most students are in favor of using technology to improve their motivation to learn, which significantly improves their willingness to complete learning activities.

Keywords: E-learning · M-learning · VLE · ICT

1 Introduction

New technologies have generated a revolution, which affects not only the industry but also the educational field, resulting in a diversification of the sources of knowledge [1]. Thus, the development of technology is a central topic of interest in the educational context, as it has enabled a change in the traditional educational paradigm. Currently, most educational systems focus on the application of these technologies in the teaching-learning process, likewise, teachers have been forced to develop skills in the management of ICT applied to teach.

With the implementation of these resources and tools, different teaching methods have been developed, among which we will focus on e-learning and m-learning. These immerse the student in a virtual learning environment where the educational process takes place. The pillar of these methods is the use of technological resources to develop knowledge in students.

In the following sections, the characteristics of a virtual learning environment (VLE) and the benefits of using e-learning and m-learning in learning are presented. Afterward, the results of the perception of students regarding the learning process based on technological means are described.

S. Nazir et al. (Eds.): AHFE 2021, LNNS 269, pp. 331–336, 2021.
https://doi.org/10.1007/978-3-030-80000-0_39

2 Virtual Learning Environments

A Virtual Learning Environment (VLE) offers spaces to develop knowledge using digital technology. This development has tackled geographical, psychological, and pedagogical barriers, resulting in easier to connect between different continents and countries. Furthermore, it has made possible the appearance of learning management techniques known as Learning Management Systems (LMS) that complements the learning management tasks in the VLEs [2] such as administration, documentation, monitoring, reporting, automation, and delivery of educational courses, training programs and learning and development programs (Fig. 1).

Fig. 1. Learning management systems

Today, learning management systems in both desktop and mobile devices are widely adopted to help to learn and improve the development of competencies and skills in students. This is done in environments other than conventional classrooms, which implies the emergence of new teaching models, the terms designated for this type of education are known as e-learning and m-learning [3].

2.1 E-learning Benefits

E-learning can be defined as education that is totally or partially done through a web browser [4], students can take online courses and learning activities at their own pace. Another benefit is that teachers can manage learning content, activities, and assessments anytime and anywhere through cloud applications [5] or online education tools. Also, enabling gamification activities in a VLE (Fig. 2).

E-learning is an educational method that facilitates the application of educational programs to train students. This method provides the advantage of easy access regardless of time or place where students are using information and communication technology (ICT).

For this reason, different educational systems around the world have gradually adopted this type of teaching. However, the successful implementation of e-learning depends on the preparation of all educational actors. The implementation of this system

Fig. 2. Genially activity

without adequate preparation, it would probably fail [6]. Thus, it is needed that teachers develop skills in managing educational technologies, these will allow them to develop fruitful activities for their students and broaden their understanding of e-learning and the operation of a VLE. As a result, the learning experience in this modality can improve for students.

2.2 M-learning Benefits

Mobile learning is considered an evolution of e-learning. Although the latter allowed the development of virtual learning environments that could be accessed initially only by desktops, the appearance of smartphones enabled mobile access to VLEs giving rise to m-learning expanding access to this type of technology. As a consequence, contemporary learning processes are influenced by the intensive use of technologies [7]. Therefore, it is important to investigate what is the influence of smartphones in the dynamization of the pedagogical strategies of teaching and learning.

Smartphones use for pedagogical purposes is divided. Some teachers argue that the smartphone is only a distraction for the student. While others argue, that it can be a distractor only when the potential and tools are unknown. The latter suggests that it is essential to update teaching staff about digital pedagogies as recommended by UNESCO in previous years [8].

The smartphones undeniably have facilitated learning opportunities as it provides changes in the learning context, access to resources via web, and even in the same physical space, it is possible to move around various virtual spaces that can promote higher levels of learning. The massive use of smartphones and the great ease in which young people learn to interact with technology, help them develop the digital skills that adapt naturally to technology development [9, 10].

One of the m-learning advantages that we want to highlight is its adaptability to the individual learning needs. This helps to promote permanent learning in a personal and daily virtual environment, also, it allows the strengthening of digital literacy which represents an essential competence in the context of nowadays digital society [11].

2.3 Student's Motivation to Use Technology

According to UNESCO [8], students and teachers already use mobile technologies in various contexts for a wide range of teaching and learning purposes. In this sense, ICTs are factors of change in knowledge sorting, and when adopted by teachers it has allowed a teaching transformation through the construction of knowledge supported by technology, resulting in an attractive teaching process to students [12]. Nowadays, students have the advantage of been born in a digital age and enjoy learning with technology. Their technology skills make it natural to learn using technology and therefore increase their interest and motivation, which improves their predisposition to learn.

3 Methodology

A questionnaire was used to identify aspects such as the benefit, motivation, difficulties, and student favor in using technology for learning. The sample consisted of 16 students who consented their voluntary participation. All the students were in a range between 9 and 12 years old, 37.5% were female and 62.5% were male. All the participants belonged to the educational system of the city of Quito, Ecuador.

4 Results

Table 1 shows the participant's answer frequency.

Table 1. Questionnaire results.

Item	Answer frequency				
	Strongly agree	Somewhat agree	Neutral	Somewhat disagree	Strongly disagree
Q1. Do you think that the technology used encouraged your participation in the learning process?	12	4	0	0	0
Q2. Did you have difficulties managing the technologies applied in the activities?	2	7	0	1	6
Q3. Did you feel motivated by this learning methodology?	13	3	0	0	0
Q4. Did you like the game activities and quizzes that were used in this experience?	15	1	0	0	0
Q5. Would you like to receive other subjects with this methodology?	13	2	0	0	1

(continued)

Table 1. (*continued*)

Item	Answer frequency				
	Strongly agree	Somewhat agree	Neutral	Somewhat disagree	Strongly disagree
Q6. Do you think you could learn the topics studied in this way of working?	12	4	0	0	0
Q7. How would you rate this learning experience?	14	2	0	0	0

The reader is invited to review Fig. 3, which shows the percentage of the participants' responses.

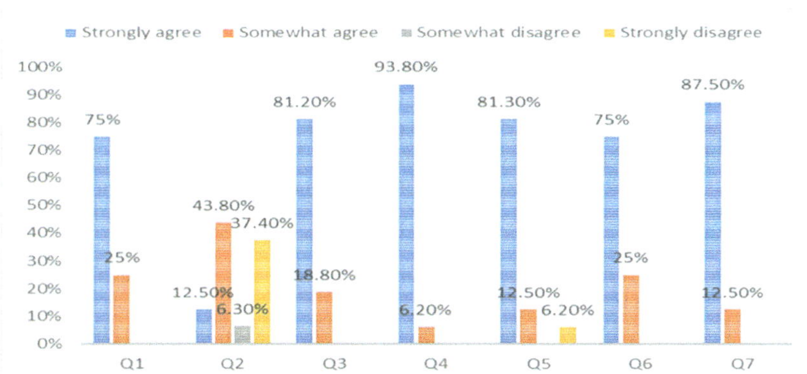

Fig. 3. Participants answers percentages.

5 Discussion

This article shows the results of analyzing the students' perceptions about the learning process based on technology.

The majority of students have been identified with favorable responses regarding the use of technology to improve their motivation to learn and increased willingness to perform learning activities.

This article results invite you to reflect on the learning process offered in settings such as Ecuador, as it has traditionally been characterized by a didactic method in which activities are done in the classroom without much use of technology. However, this study invites the use of other means in the teaching and learning process, such as the technological resources described.

Future research is the application of technological resources in experimental studies at the longitudinal level that will allow determining, through pre and post-test measurements, the effectiveness of this type of technology in the academic performance of a student.

References

1. Fernández, M.D., Álvarez, Q., Mariño, R.: E-learning: Otra manera de enseñar y aprender en una Universidad tradicionalmente presencial. Profesorado. Revista de Currículum y **17**(3), 273–291 (2013)
2. Guaña, E., Llumiquinga, S., Ortíz, K.: Caracterización de entornos virtuales de enseñanza aprendizaje (EVEA) en la. Ciencias Holguín **XXI**(4), 1–16 (2015)
3. Chen, H.-j.: Clarifying the impact of surprise in e-learning system design based on university students with multiple learning goals orientation. Educ. Inf. Technol. **25**, 1–20 (2020)
4. Riahi, G.: E-learning systems based on cloud computing: a review. Procedia Comput. Sci. **62**, 352–359 (2015)
5. Yee, L., Ling, S., Hoe, S., Chew, M.: Usability factors predicting continuance of intention to use cloud e-learning application. Heliyon **5**(6), 1–11 (2019)
6. Al-arabi, A., Mahrin, M., Yusoff, R.: Technological aspect factors of E-learning readiness in higher education institutions: Delphi technique. Educ. Inf. Technol. **24**, 567–590 (2019)
7. Rush, S.: Problematic use of smartphones in the workplace: an introductory study, Australia: tesis de grado, Central Queensland University (2011)
8. UNESCO: El futuro del aprendizaje móvil. Implicaciones para la planificación y la formulación de políticas. Organización de las Naciones Unidas para la Educación, la Ciencia y la Cultura (2013)
9. Silva, A., Martínez, D.: Influence of the Smartphone on learning and teaching processes. Suma de negocios **8**(17), 11–18 (2017)
10. Fajardo, I., Villalta, E., Salmerón, L.: ¿Son realmente tan buenos los nativos digitales? Relación entre las habilidades digitales y la lectura digital. Anales de Psicología **32**(1), 89–97 (2016)
11. Vilalonga, C., Lazo, M.: Modelo de integración educomunicativa de 'apps' móviles para la enseñanza y aprendizaje. Pixel-Bit. Revista de Medios y Educación (46), 137–153 (2015)
12. Paredes, J.: La motivación del uso de las TIC en la formación de profesorado en educación ambiental. Ciência Educação (Bauru) **18**(2), 353–368 (2012)

Evaluation of Achievement Effect on Visual Tools for Online Teaching-Learning Program in China During the Covid-19 Pandemic

Wenyu Wu(✉), Ziwei He(✉), Xinran Yan, Zijian Han, Yaning Li, and Chengqi Xue

Department of Industrial Design, School of Mechanical Engineering, Southeast University, Nanjing, China

Abstract. Online education has been promoted and applicated since the covid-19 pandemic spread world widely, meanwhile, big data and the vigorous development of online education brought the explosion of education data, unlike traditional offline education, online education has higher flexibility and universality, but at the same time, the way that teachers and learners how to deal with massive amounts of data of learning and visual interactive display needs to be analyzed and solved in the field of learning problems. Through the investigation and analysis of learners and teachers of the higher education platform, this paper finds out the pain points of the low interaction efficiency and incomplete information understanding of the existing online education APPs and puts forward some optimization strategies for the information visualization of online education APPs.

Keywords: Online education · Software tools · Interactive optimization · Information visualization · Teaching efficiency

1 Introduction

Since the outbreak of the epidemic at the beginning of the year 2020, colleges and universities have been actively carrying out online courses in various ways, making online education the largest in scale [1], with the largest number of online courses and the widest coverage and audience in history [2]. Moore summarized the interactive behaviors of online learning into three categories: Student-Content (SC) Relationship, which is the one-way interaction between learners and knowledge, courses and resource system, thus constructing individual knowledge network and acquiring new knowledge; Student-Student (SS) Relationship refers to the multi-point and mutually beneficial interaction mode between online learners, and Student-Teacher (ST) Relationship refers to the Teacher's guidance and evaluation of learners. Studies have found 70% of the teachers don't believe that the platform can help them solve some important problems of teaching [3], only 50% of the teachers can use the learning management system data to improve student performance, visible in the field of education, teachers don't know how to use the data to promote teaching [4]. Classroom data visualization has greatly improved the knowledge and understanding of classroom interaction and can set the transparency of classroom interaction independently to enhance the freedom of classroom interaction while fully respecting students [5].

S. Nazir et al. (Eds.): AHFE 2021, LNNS 269, pp. 337–342, 2021.
https://doi.org/10.1007/978-3-030-80000-0_40

2 Definition and Core Concept

Study analysis visualization refers to the use of information technology and other technical means, will get a lot of learning information before in the form of ICONS and graphics for learners and teachers as well as other stakeholders [6], which provide instant feedback for teachers and learners according to the different needs of users. Combined with online learning behavior, the common data visualization in online learning can be divided into three categories, namely, the visualization of knowledge learned, learning process, and learning effect [7].

3 Method

3.1 Questionnaire Design

The first thing of the questionnaire is to determine the identity information. The online education APP user's age, gender, education, the use of time, and purpose (learning/teaching) will divide the users into learners and teachers.

3.2 Social Network Analysis of Online Education APP

Social network analysis is used to analyze the communication mode of higher online education, and the communication process of online education is analyzed through data, so as to provide new ideas for the efficiency of an online education platform.

3.3 DEA

Data Envelopment Analysis (DEA) was first put forward in 1978 by A. Harners and W.W.Cooper. It is an effective system analysis method to solve the relative effectiveness evaluation between the departments with multiple inputs (inputs) and outputs (outputs).

1. Select Decision Units
   In the form of interviews, 10 students in Southeast University were selected, three of whom had some working experience.
2. Establish an Evaluation Index System
   In the research, the author sets up the input/output index system of e-learning (Table 1). During the interview, we quantified the corresponding indicators according to the students' answers. For indicators that cannot be directly quantified, we used Likert Scale.

4 Result

4.1 Questionnaire Result

The team designed a questionnaire at the end of January "online education technology APP information visualization design and efficiency of classroom research". Finally, a total of 110 valid questionnaires have been collected.

Table 1. Result of input and output

	1	2	3	4	5	6	7	8	9	10
Interest	6	2	6	2	2	8	5	5	6	8
duration	3	2.5	5	1	2	6	6	5	8	7
Test feedback	2	4	6	4	2	2	2	6	4	2
Interaction	2	2	4	2	2	4	2	4	6	8
Study plan	6	2	6	2	2	4	4	6	4	2
Academic performance	2	4	6	4	2	6	6	8	8	4
Ability to cooperate	6	4	6	6	8	6	4	6	4	4
Learning motivation	2	6	4	6	6	6	8	4	4	4

Among the respondents, 54 are male and 56 are female. The ages of them include 96 people from 20 to 25 and 6 people from 25 to 30. Most of the participants were undergraduate/junior college students and graduate students. The duration of users using the APP varies from three months to three years. Users generally use APP for the purpose of learning skills, learning in class, and daily teaching, among which, the purpose of learning skills is the most, accounting for 84.45%.

Due to the small number of respondents who participated in the survey as teachers, the survey did not have good reliability and validity. It can be seen from the survey results that most teachers can intuitively understand the learning evaluation situation, the change of ranking and the progress of most students in time. However, almost all teachers cannot fully understand each student's mastery of learning content, the opinions of students and parents on the course content and the problems of students.

According to the survey results, most of the users are satisfied with the help of software to improve learning efficiency and at any time it can indeed help understand the learning level. But the interaction density and efficiency are insufficient and various time nodes of the course are also a key point of later visual optimization design.

4.2 Social Network Analysis Result

In this example, a total of 21 users participated in the discussion of a certain topic, and 36 replies were received. The relationship matrix among the data was used to establish the relationship diagram of the user network structure (Fig. 1). It can be intuitively seen from the figure that there are "ring", "Y-shaped", "chain", "star" and other structures.

Fig. 1. Social network

According to Fig. 1, this user network has 21 nodes and 34 relationship lines in total. Through calculation, the network density of this network is 0.081. Therefore, the interaction between users of the higher online education platform is poor.

By calculation, the mean value of the distance between any two points in the user relationship structure is 1.897. It can be inferred that users of advanced online education mostly form a virtual team temporarily based on a common learning goal.

Closeness centrality is a measure of the dependence between individual users and other users in a relational network. It can be judged that the overall centrality of this case is low, that is, in the process of information transmission, nodes are not closely related to each other, and no "Circle" structure with the community has been formed.

4.3 Result of DEA

Frontier Analyst was used to input data and solve the C2R model (Fig. 2). It was concluded that learners 4, 5, 6, and 7 were relatively effective in learning ($\theta = 1$). The other 6 were not relatively effective ($\theta < 1$, and the learners with θ closer to 100% had higher learning efficiency compared with those with lower scores. In this evaluation, learner No. 3 has the lowest relative efficiency, which is 70.6%).

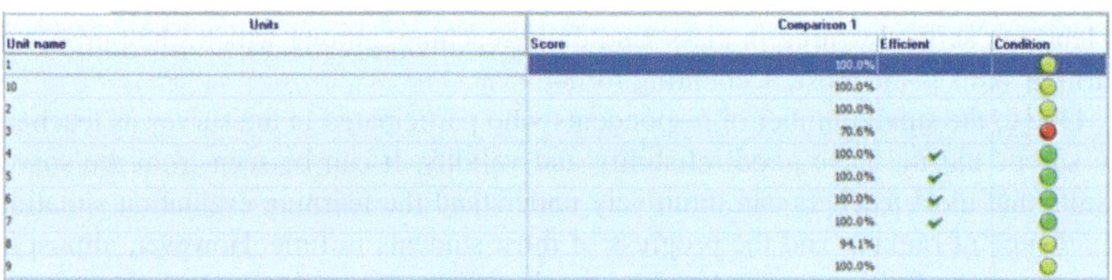

Units	Comparison 1		
Unit name	Score	Efficient	Condition
1	100.0%		
10	100.0%		
2	100.0%		
3	70.6%		
4	100.0%	✔	
5	100.0%	✔	
6	100.0%	✔	
7	100.0%	✔	
8	94.1%		
9	100.0%		

Fig. 2. C2R model

As can be seen from Fig. 3 distance autonomous learners have the least deficiency in "score" output. However, there is the largest deficiency in "motive" output, which is 34.18% to be improved. Secondly, "ability" is 22.61%. There is a large redundancy of "interest" in the input index, and the redundancy is 10.27%.

From the learning results of 10 learners, it can be seen that network-based learning is the most important thing. The second problem is cooperation and participation.

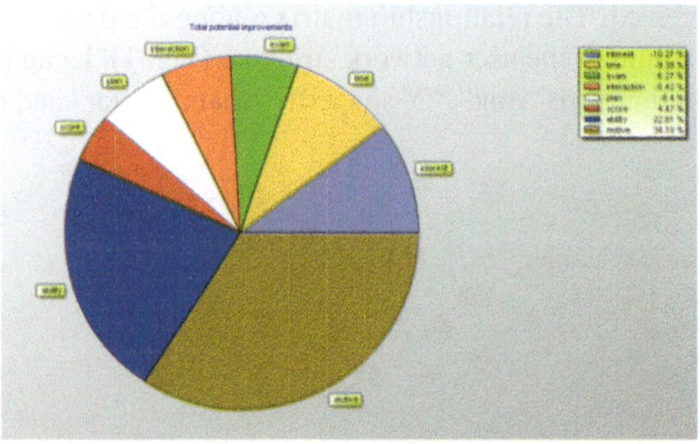

Fig. 3. Result of DEA

5 Discussion

Based on the current online education teaching situation investigation, it is concluded that learners pay more attention to their own degree of completion, mastery of knowledge points, homework submission, and other time nodes as well as their own behavior. On the basis of focusing on the performance of the whole class, teachers also attach great importance to individual knowledge mastery, students' learning progress, and the degree of interaction between the class. In view of the above situation, we put forward some visual design optimization strategies based on MOOC online education platform.

5.1 Learner-Oriented Learning Analytics Visualization Design

The personal report is mainly the comparison of students' personal learning activities with the average level of the class, including the comparison between the classroom performance and the feedback of homework results, as well as the comparison between the current progress that should be completed and the progress that has been completed in the course interface. As shown in Fig. 4, the green line for the students means the learning process has been completed. Yellow for the student's learning curve is the online performance as a function of the homework feedback. The blue dashed line represents the class average level. When the learning level below the class average, a red flag will remind it. When there is not much difference, the class average flag is blue. When higher than the average level, the symbol is green. In the column of course, there is a red dotted line to indicate the learning progress that should be completed. In front of the course name, there is a sign to indicate the degree that should be paid attention to. Red is very urgent.

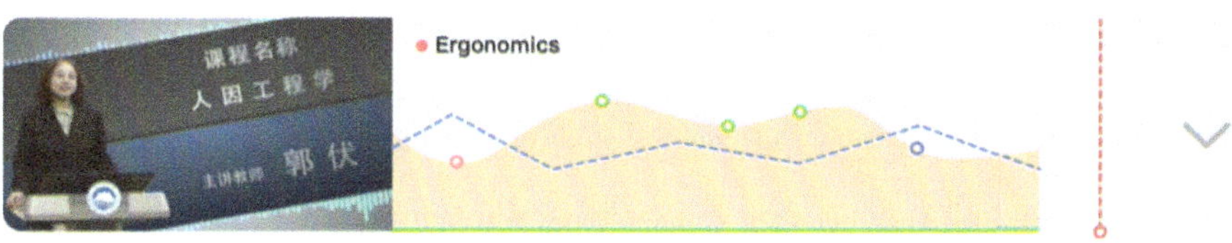

Fig. 4. Learner-oriented learning analytics visualization design

5.2 Visualized Design of Teacher-Oriented Learning Analytics

The teacher-oriented visualization of learning analysis mainly includes three aspects: first, the mastery of class progress, overall classroom effect, and classroom interaction. This function can discover the overall learning effect of the whole class, find out the difficulties under a certain topic, and provide targeted support and help for most students. Secondly, the mastery of learners' questioning situation can effectively solve the pain point that online learners cannot get timely feedback after raising questions. Thirdly, it is to master the learning progress and learning effect of individual students. Through visualization, teachers can intuitively know which students are lagging behind in what aspects and to what extent they lag behind, so as to intervene and help them. (Fig. 5).

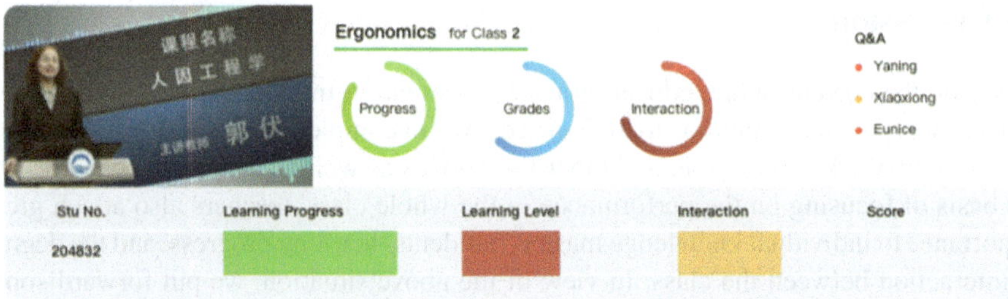

Fig. 5. Visualized design of teacher-oriented learning analytics

6 Conclusion

Online education APP is widely used today, online education has advantages in terms of convenience and data statistics, but the interaction efficiency has decreased a lot. Through the data statistics and visualization of teaching can strengthen the efficiency of information transfer between three kinds of communication mode, teachers oriented information visualization can also help teachers improve team cohesion of online education, the better study pays attention to the efficiency of data analysis and dynamic analysis of visualization, also will be more meet the demand of online education individualized learning.

References

1. Chen, T.G., Peng, L.J., Yin, X.H., Rong, J.T., Yang, J., Cong, G.D.: Analysis of user satisfaction with online education platforms in China during the COVID-19 pandemic. Healthcare **8**(3), 200 (2020)
2. Larionova, V., Brown, K., Bystrova, T., Sinitsyn, E.: Russian perspectives of online learning technologies in higher education: an empirical study of a MOOC. Res. Comp. Int. Educ. **13**(1), 70–91 (2018)
3. Brown, M.: Learning Analytics : The Coming Third Wave (2011)
4. Spykman, N.J.: The Social Theory of Georg Simmel. Taylor and Francis, New York (2017)
5. Siemens, G.: What are learning analytics. Nordic J. Digit. Liter. (2010)
6. Siemens, G.: 1st International Conference on Learning Analytics and Knowledge Consortium TNM. Learning Analytics. The Horizon Report 2011 Edition, pp. 28–30 (2011)
7. Siemens, G.: Learning analytics: the emergence of a discipline. Am. Behav. Sci. **57**(10), 1380–1400 (2013)

Competency and Skill Management in Education

Topic Mapping to Support Users in Literature Search

Ethan B. Trewhitt[1] (✉), Elizabeth T. Whitaker[1], Robert E. Wray[2], and Dawn Riddle[3]

[1] Georgia Tech Research Institute, 250 14th Street NW, Atlanta, GA, USA
{ethan.trewhitt,elizabeth.whitaker}@gtri.gatech.edu
[2] Soar Technology, 3600 Green Court Ste 600, Ann Arbor, MI, USA
wray@soartech.com
[3] Naval Air Warfare Center Training Systems Division, 12211 Science Dr., Orlando, FL, USA
dawn.riddle@navy.mil

Abstract. This paper describes the design and development of the Topic Map, a visualization and user interaction component of a cloud-based tool, Archimedes. Archimedes is designed to help individuals and teams examine and organize the results of a literature search and use them to understand the space that they are researching. The Topic Map is a document surrogate, designed to help the user visualize the topic space represented by the search corpus. It shows frequently occurring but generally uncommon topics in a user's workspace corpus. The Topic Mapper component generates the Topic Map automatically by extracting a list of topic phrases from the papers in the workspace, filtering and prioritizing what is displayed to the user based on a set of rules. It then visually distributes them in a two-dimensional space. This paper describes the motivation and design of the topic extraction implementation and its user interaction capabilities within Archimedes.

Keywords: Topic map · Visualization · t-SNE · TF-IDF · Human-machine interaction

1 Introduction

The development of topic map extraction, visualization and human-machine interaction to support research in support of a technological literature search capability to support project research is motivated by the users' need to understand the topics represented in a corpus of retrieved documents and the need to filter and prioritize the papers to be studied based on the concepts that they contain. This paper describes the design and development of the Topic Map, a visualization and user interaction component of a cloud-based tool, Archimedes, which is designed to help individuals and teams examine and organize the results of a literature search and use them to understand the space that they are researching. The Topic Map provides a surrogate view, i.e., a topic or concept summary of the document corpus, to help the user visualize the concept space represented

S. Nazir et al. (Eds.): AHFE 2021, LNNS 269, pp. 345–352, 2021.
https://doi.org/10.1007/978-3-030-80000-0_41

by the retrieved document corpus, showing important topics represented in the set of documents that the user has retrieved and stored in their workspace for literature search and understanding in Archimedes.

The Topic Map visualization shows the important topics in a user's workspace which is used to store the documents being used for the literature search for the research project in Archimedes. The Topic Map is particularly helpful when a domain ontology is not available. It is designed to allow the user to rapidly assess the concepts included in the document corpus. The human-machine interaction (HMI) approaches enable the user to select subsets of the document corpus and display and compare the topics contained in those documents by toggling between different selections. The user can use this surrogate view to understand the topics available in the overall corpus and in individual papers.

The Topic Map contains functionality to allow the user to remove or block concepts which are not relevant to the user's search (Block List) and to always display user chosen concepts if they appear in the corpus (Priority List).

The HMI allows the user to customize the display to include the hierarchical relationships between topics, i.e., assign a child to its parent topic and to define synonyms within the Topic Map display.

2 Topic Mapper Functionality and Implementation

The Topic Mapper component generates the Topic Map automatically by extracting a list of topic phrases from the papers in the workspace, filtering and prioritizing them based on a set of rules. It then selects the most salient topic phrases and visually distributes them in a two-dimensional space. This automated extraction and visualization process consists of two stages: Stage 1 is performed independently for each paper, while Stage 2 is performed across the results of Stage 1 for all papers in the workspace corpus (Fig. 1).

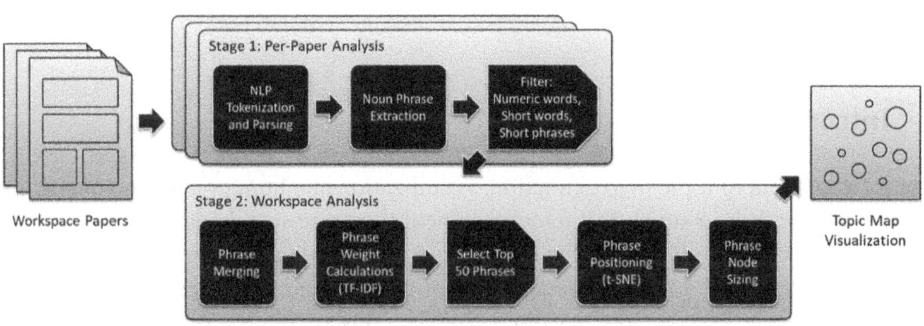

Fig. 1. Topic Mapper functionality in two stages

The objective of Stage 1 is to perform natural language processing (NLP) to extract topic phrases from each paper, independent of any other papers. This stage consists of the following steps, in which the Topic Mapper:

1. Performs standard NLP tokenization and parsing
2. Extracts all noun phrases
3. Filters (removes) stop words, short phrases, phrases containing too many/only short words, and phrases containing numbers or banned punctuation.

For each paper, the final output of Stage 1 is the filtered list of topic phrases that appear in that paper. This list is not de-duplicated, so phrases that appear more than once in a paper will appear multiple times in this list. Each phrase is represented as a pair: the original text representation of the phrase in the paper, plus a simplified version that has been lemmatized, converted to lowercase, and has standardized whitespace. These two representations enable additional reasoning in Stage 2.

Once all papers in a workspace corpus have been processed with Stage 1, Stage 2 takes place over the entire workspace. In Stage 2, the Topic Mapper:

1. Removes phrases that appear in the block list
2. Merges representations of phrases, regardless of capitalization, using phrase lemmas
3. Calculates phrase weights (TF-IDF)
4. Selects the best phrases to keep/show: include any externally specified priority phrases, then fill the remaining $n = 50$ slots using those with the highest TF-IDF weights.
5. Positions phrases in 2D space (t-SNE)
6. Determines phrase node sizes.

The phrase weight calculation uses the *term frequency, inverse document frequency* (TF-IDF) [1] scoring mechanism, which prioritizes terms that appear many times within a single document while simultaneously giving lower weight to terms that appear across many documents. In this application, each term is one of the simplified topic phrases. A final list of $n = 50$ (this is a customizable parameter that can be changed) topic phrases is produced by selecting any phrases in the user's *priority list*, plus the top-scoring automatically extracted phrases. The *t-distributed stochastic neighbor embedding* (t-SNE) [2] algorithm is used to calculate reasonable positions for the nodes in a two-dimensional space using paper co-appearance as the similarity metric. Thus, if any two topic phrases appear in the same document, t-SNE attempts to place those two phrases near each other. It also attempts to distribute the phrases evenly in space. Each topic's node size is determined by the number of documents in which the topic phrase appears at least once. Finally, the text/label of each topic phrase node is the most common original text representation of that topic phrase (capitalization, non-lemma version, etc.).

Because Stage 1 of the process is the most time-consuming part of Topic Map generation, the results of that stage are cached for each paper after they are computed. While Stage 1 may take several seconds per paper (longer papers need more processing time), this process only needs to be performed a single time after a paper is imported into Archimedes. All other user-accessible settings that affect Topic Map generation (block list, user-added list) involve changes that affect Stage 2 only.

The visual representation of the Topic Map is rendered in-browser using the popular D3.js visualization library [3], along with user interface widgets to select nodes and customize the visualization, as described in the following sections.

3 Customizing the Topics Displayed: Priority List and Block

Part of the customization enabled by the Topic Map is the ability for the user to specify a *priority list* of topics specified as labels in the user's work space. An important part of the Topic Map functionality is the human-machine interaction to provide the user the ability to customize the visualization. The Topic Map interface allows the user to block topics that have been automatically extracted (*block list*) but are not of interest in the user's research, and to include topics that are created as labels in the user workspace (*priority list*) whenever they occur in the selected papers. The user can specify these priority list and block labels or topics and can later undo or edit them (Fig. 2).

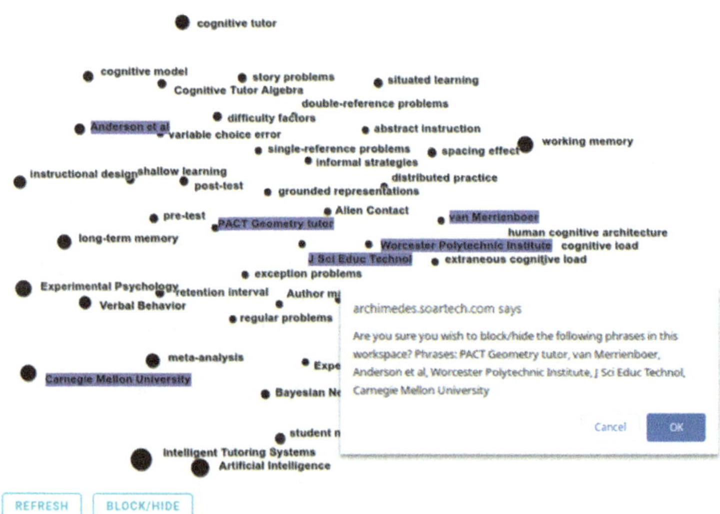

Fig. 2. Selecting terms to block/hide (add to the block list)

Additionally, it is possible to review and/or modify the phrases on the block list within the Workspace Settings page, as shown below (Fig. 3).

Fig. 3. Workspace Settings section for modifying the block list

Topic phrases are automatically added to the *Priority List* when they are created as labels by the user, as shown below. Machine-generated labels (those starting with "Source:" or "Type:") are not included (Fig. 4).

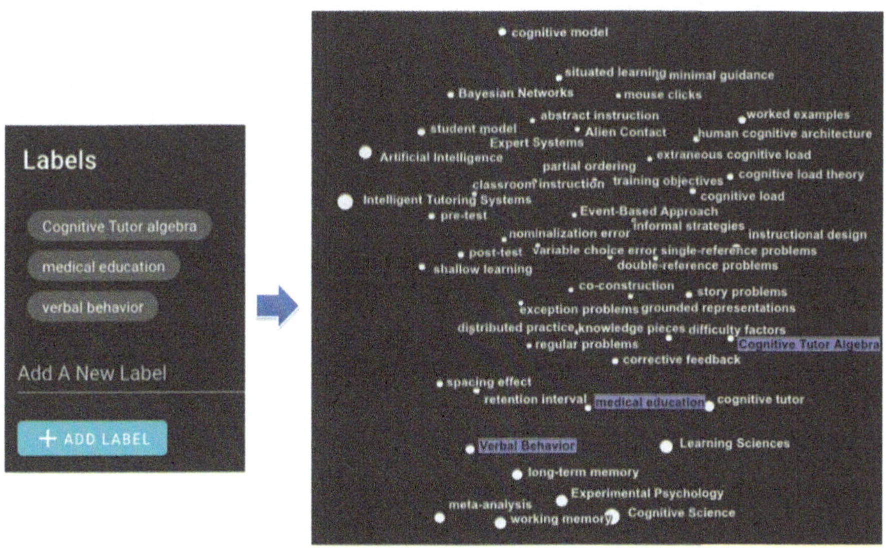

Fig. 4. Prioritized phrases appear, regardless of calculated weight, if they are created as labels by the user. Machine labels are not included in this process.

4 Displaying Topic Synonyms and Hierarchal Relationships

The user may also identify synonyms in the Topic Map and create hierarchical relationships among the topics to further customized the Topic Map's appearance. The user may identify synonyms in the Topic Map and create hierarchical relationships among the topics to further customize the Topic Map's representation (Figs. 5 and 6).

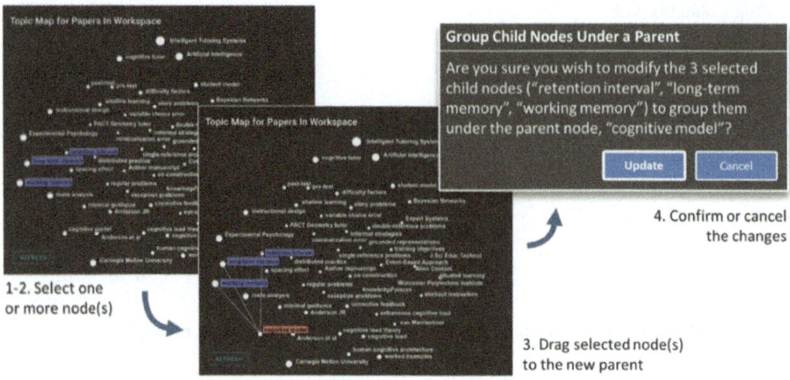

Fig. 5. Forming a hierarchical relationship by interacting with the Topic Map. This is a preliminary design for this feature that has not yet been fully implemented.

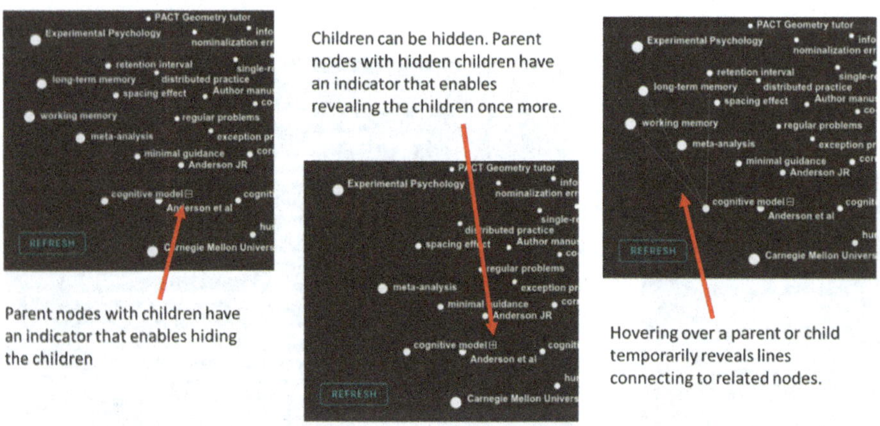

Fig. 6. Showing and hiding the children of a node after a hierarchical relationship has been created. This is a preliminary design for this feature that has not yet been fully implemented.

5 Topic Map to Explore the Corpus: Toggling Among Selected Documents

In order to provide tools to help the user select and prioritize the set of papers to read and analyze for a research project, the Topic Map provides the capability to explore the specific topics contained in any subset of papers in the corpus. The user can add papers to the selected set and see in the visualization which new topics were added (highlighted) and can delete papers from the selected set to see which topics are removed from the selected set if particular papers are de-selected. The goal is to enable the researcher to select for reading and studying a prioritized set of papers based on the concepts or topics that are needed for the research project.

When the user selects a subset of papers in the workspace, the Topic Map visualization highlights the topics that appear in those papers (Fig. 7).

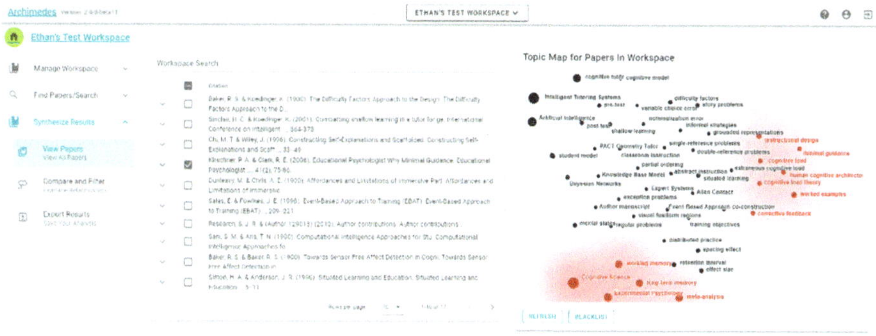

Fig. 7. The "View Papers" page of Archimedes, listing all pages in the workspace (center) and showing the Topic Map visualization (right)

The Topic Map is visible in two places within Archimedes: the *View Papers* page (where all papers within the workspace are visible) and the details page for an individual paper. When viewing the workspace paper list, the Topic Map shows all topics within the workspace. When one or more papers are selected, all topics that appear within the selected paper(s) are highlighted in the Topic Map. This serves to show, visually, the activated region of the workspace's conceptual topic space (Fig. 8).

Fig. 8. Topic Map for entire workspace (left), the same topic map with red indicating the topics present in the selected paper(s) (right). Node size proportional to topic popularity across the workspace.

6 Next Steps and Ongoing Improvements

Archimedes is under active development with new versions deployed regularly. The team is focused on improving the implementation of the Topic Map hierarchy and synonym capability, as well as improvements to the selection of high-quality topic phrases using natural language processing by rejecting additional types of spurious text. For example, the NLP named-entity recognizer can recognize, e.g., locations like "New York" that have appeared before and are likely undesirable as concept phrases. Topic Mapper will use this additional information to decide when to eliminate phrases, exposing these mechanisms to the users that the user can override the decisions if a workspace requires it. Continuing the previous example, if a workspace is being built to study "municipal infrastructure projects in Northeast US cities", then "New York" would likely be a topic that should not to be filtered.

7 Conclusions

The Topic Map provides a surrogate visual model of the workspace corpus contents, automatically summarizing the topics covered for easy understanding, with customizations available where the user prefers. It is being used by a small set of users participating in overall "beta tests" of the Archimedes tool. As we continue to receive comments and feedback from users, its function and user interface will be refined and improved.

Acknowledgments. The authors thank teammates and collaborators for contributions to concepts and implementations reported here including Tyree Cowell, Timothy Siggins, Dr. Jennifer Fowlkes, Kelly Neville, Caitlin Tenison, Austin Brehob, Dana Foley, and Ryan Yao. Opinions ex-pressed here are not necessarily those of the Department of Defense or the sponsor of this effort, the Naval Air Warfare Center Training Systems Division (NAWCTSD). This work funded under contract N68335-19-C-0004.

References

1. Jones, K.S.: A statistical interpretation of term specificity and its application in retrieval. J. Doc. **28**, 11–21 (1972)
2. Van Der Maaten, L., Hinton, G.: Visualizing data using t-SNE. J. Mach. Learn. Res. **9**, 2579–2605 (2008)
3. Myatt, G.J., Johnson, W.P.: Making Sense of Data III: A Practical Guide to Designing Interactive Data Visualizations, vol. 3. Wiley, Hoboken (2011)

An Analysis of the Relationship Between Goal Change and Skill Improvement in the Consumption of Education Services: A Case Study of the Management Game

Kazuhiro Fujimura(✉)

Faculty of Economics, Kagawa University, 2-1 Saiwai-cho, Takamatsu 760-8523, Kagawa, Japan
fujimura.kazuhiro@kagawa-u.ac.jp

Abstract. In the consumption of education service, the goals are subject to upward transformation after the accomplishment of the initial goal and/or various experiences in the course of learning. The goal upward transformations are also influenced by the learner's life goal—the matter of whom the learner intends to benefit by acquiring and using the targeted ability. We are advancing our study to expound and identify more factors that can affect the upward transformation of a goal as well as the effects of goal change on learning behavior. We began by carrying out explorative analysis using previous study results to identify these factors. An explorative analysis showed many interesting results. As one of the main results, we found a link between goal-raising and ability-improvement.

Keywords: Educational service · Goal change · Ability improvement · Benefit delay

1 Introduction

Even when a service is consumed at the same time it is delivered, the customer does not always receive the desired benefit from the service immediately. This is the case with educational and medical services; a time lag exists between the point when the service provider acts collaboratively with the customer to generate the desired benefit, and the point when the customer actually receives this benefit. In the case of education, this time lag arises because the desired benefit—an improvement in competency or ability—requires an accumulation of knowledge, skill, and experience. Defining this time gap as "benefit delay," we have been exploring, theoretically and empirically, the effects of benefit delay on customer satisfaction, participation, and goal change [1].

The present study focuses on educational services, a classic example of the kind of service in which benefit delay occurs. The purpose of the study is to identify the factors that influence goal change and to identify how goal change affects learning behavior and receiving benefit (ability-improvement). One reason for examining goal change is that users of educational services are likely to change their goal during the benefit delay, and such a goal change can affect the users' satisfaction and the way they engage in the

S. Nazir et al. (Eds.): AHFE 2021, LNNS 269, pp. 353–361, 2021.
https://doi.org/10.1007/978-3-030-80000-0_42

learning. Besides these short-term effects, goal change may affect, in the longer term, the profit and happiness of the learners themselves and of organizations or communities to which they belong. In this study, which forms the first step in the process of identifying the impact of goal change, we conducted an exploratory/inductive analysis on the results of a previously conducted questionnaire survey among players of the Management Game, a tool for training business managers.

2 Goal Change in the Learning Process and Its Effect on Ability-Improvement

2.1 Goal Change in the Learning Process

In the consumption of service which benefit delay occurs, the relatively lengthy service delivery process (compared to the cases where the user benefits immediately) makes the user more likely to change their initial goal. Insofar as the service requires the user to participate in its delivery, the longer the delivery takes, the more opportunities there will be for the user to encounter experiences or feedback that create positive or negative emotions. The user may downgrade the goal after experiencing negative emotions or difficulties in attaining the initial goal; alternatively, they may raise the bar higher after achieving the initial goal or after having encounters that positively affect their values or attitude. Furthermore, because the user receives no immediate or clear feedback regarding the outcomes of their participation, the user's motivation is likely to decrease, and this too could lead to the user downgrading the goal.

For learners and the organizations to which they belong to enjoy long-term happiness and profits, it would be better for the user to set the goal higher, not lower. If the user achieves a goal, sets an even higher goal, achieves this higher goal, and so forth, this positive spiral will draw out the user's potential and ultimately amplify the benefits for the user and the community. Accordingly, it is essential, both academically and socially, to identify the factors that encourage learners to upgrade their goals.

2.2 The Relationship Between Goal Change and Ability-Improvement

By setting a clear goal, learners can easily understand the gap between the goal and their present situation, making it easier to choose and execute the right steps for narrowing the gap. Clear goals also offer motivational benefits; the benefit delay can cause learners to lose motivation, but having a clear goal helps the learners stay motivated or even improves their motivation.

However, goal clarity is insufficient in itself to secure the long-term profit and happiness for learners and their communities; for this, the learners must engage in a continual cycle of achieving and then setting a higher goal. If the learner's goal in consuming educational services is to change their ability in a positive sense, then upgrading the goal would mean facilitating a positive change in level and direction. "Direction" refers to the type of goal, whereas "level" refers to the degree. Thus, learners could upgrade their goal in one of three ways: (1) They could keep the same direction as in the initial goal but aim for a higher level. (2) They may switch direction after considering the kinds of ability society requires and considering their own potential. (3) They may change both the level and direction.

The upgrading of goals is determined to a large extent by the learner's experiences and encounters in the service delivery process and by whether they achieved the initial goal. Achieving the initial goal—acquiring the desired ability—probably makes learners more likely to upgrade their goal in that it boosts their self-affirmation and self-efficacy. An upgraded goal would, in turn, make the user more likely to improve their ability. Thus, we have posited that there is a mutually reinforcing relationship between goal-raising and ability-improvement; the two create an upward, positive cycle.

2.3 The Importance of Altruism in the Relationship Between Goal Change and Ability-Improvement

If this positive cycle of goal-raising and ability-improvement does indeed occur, it would be shaped by the learner's life goal—the matter of whom the learner intends to benefit by acquiring and using the targeted ability. In other words, a positive spiral would be shaped by whether the learner's goal in life is to pursue their own profit and happiness or whether it is to contribute to the profit and happiness of their stakeholders or the wider community. When learners switch their life goal from pursuit of their own interests to the pursuit of others' (stakeholders or the wider community's) interests, the learning process starts to diversify in terms of learning perspective, learning content, and the kinds of experiences the learner encounters in the learning process. This diversification can encourage the learner to upgrade their goal and improve their ability.

3 Ability-Improvement and Goal Change in Management Game

3.1 Description of Management Game

Many business games have been developed on the assumption that such games can encourage active participation, and that learning could be easily established through that experience. One of them is Management Game (hereinafter "MG"), which was developed in 1976 by a subsidiary of Sony Corporation as an educational tool for studying management. MG is a face-to-face analogue simulation game in which gamers and instructors gather at a specific place, where a dedicated miniature game board is installed at the center of a table [2].

3.2 Modeling the Relationship Between Ability-Improvement and Goal Change in Management Game

As they would in any other learning process, MG participants will begin with an initial goal (to improve their management ability). Then, as they progress in the game, they will begin to notice improvements in their management ability and upgrade their goal. Accordingly, we plan to examine how an improvement in management ability leads to a change in goal direction and change in target level and a change in the intended beneficiary of the targeted ability, and how these changes, in turn, affect ability-improvement.

However, as this research remains at an early stage, the aim of the present study is limited to examining the efficacy of MG in training managers and the role of benefit delay in MG. To this end, the present study conducted an exploratory/inductive analysis upon the results of a previously conducted questionnaire survey.

Because the survey had just a few questions related to goal change, only three items were analyzed: clarification of self-directed growth goal (a goal to improve and benefit oneself, as opposed to benefiting the community), upgrading of self-directed growth goal, and altruization of goal (contribute to company). These items only cover a portion of the model for the relationships between ability-improvement and goal change. Therefore, we constructed the partial model shown in Fig. 1.

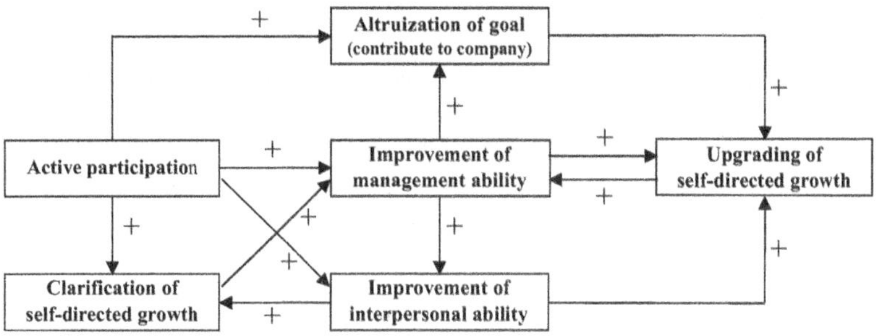

Fig. 1. Model of relationship between ability-improvement and goal change in MG

4 Methods

To test Fig. 1's model of the relationships between ability-improvement and goal change through MG participation, an analysis was conducted on the results of the questionnaire survey (effective respondents: 219), which itself was conducted among MG players between October 1 and November 24, 2018.

After the exploratory/inductive analysis, the model in Fig. 1 was tested using an analysis of covariance structures.

5 Results and Discussion

5.1 Specifying Question Items for Measuring Model's Construct Validity

A factor analysis for the extraction of MG-induced ability and customer participation was carried out using the maximum likelihood method/promax rotation on the question items to measure the model concepts shown in Fig. 1.

Table 1 shows the summarized results of extracted factors (concepts), the Cronbach's α calculated to examine internal consistency, and averages of question items (scores) showing the high factor loadings for each item. Two factors are extracted as MG-induced ability-improvement: improvement of management ability and improvement of interpersonal ability. The former can be interpreted as improvement of cognitive ability and the latter as enhancement of non-cognitive ability. Regarding customer participation, two forms were extracted: participation with a learning goal, which means active participation, and reluctant participation, suggesting passive participation. Because the present analysis is concerned with the effects of active participation in MG (as opposed to reluctant participation in MG), Table 2 only shows those question items that pertain to active participation. All question items consist of statements that respondents expressed agreement or disagreement with on a 5-point scale.

Table 1. Extracted elements (concepts) and descriptive statistics for question items.

	Concept	Question item	Each question item Average(SD)	Cronbach α	Score for each factor Average(SD)
Improvement of ability thorugh MG	Improvement of management ability	MG made me aware of the important role investment plays in business.	4.32 (0.90)	0.952	3.98 (0.74)
		Thanks to MG, I now understand how important future-oriented investment is.	4.45 (0.82)		
		MG gave me a fresh perspective on business.	4.20 (0.93)		
		MG has made it possible to plan from a long-term perspective.	3.88 (0.98)		
		MG helped me see business management in a broader context.	3.97 (1.01)		
		MG has helped me understand the right way to make profits.	3.77 (0.98)		
		Thanks to MG, I have acquired management skills.	3.79 (1.02)		
		MG prompted me to consider what must be done to ensure business continuity.	4.02 (0.97)		
		The accounts program helped me understand accounting procedures in business.	3.99 (0.97)		
		MG has made it possible to foresee the risks associated with management.	3.57 (0.97)		
		MG taught me how to communicate problems and the purpose of an action to colleagues and friends.	3.76 (0.97)		
		MG taught me how to price things.	3.89 (1.04)		
		MG helped me understand my organization's strengths and weaknesses.	3.85 (0.97)		
	Improvement of interpersonal ability	Thanks to MG, I can now face any situation with a smile.	3.46 (0.99)	0.885	3.52 (0.78)
		MG has enabled me to build a good relationship with those around us.	3.53 (0.94)		
		Thanks to MG, I can now strike up conversation with people I've never met before.	3.54 (1.07)		
		MG has taught me how to get my points across accurately.	3.45 (0.99)		
		MG has made me more empathetic to others.	3.67 (0.94)		
		MG has made me more aware of how managers and subordinates feel.	3.50 (0.97)		
Customer Participatio	Active participation (Participation with a learning goal)	Whenever I join MG, I set some goals regarding what I should learn.	3.93 (0.93)	0.811	4.20 (0.69)
		Whenever I participate in MG, I always try to learn something.	4.34 (0.82)		
		Whenever I participate in MG, I think about how I can use it for my company.	4.21 (0.86)		
		Through MG, I hope to master the strategic thinking skills of managers.	4.30 (0.86)		
Goal Change	Clarification of self-directed growth goal	MG has helped me clarify my personal goals.	3.71 (1.03)		
	Upgrading of self-directed growth goal	MG has encouraged me to set a higher personal goal.	3.79 (1.04)		
	Altruization of goal (contribute to company)	MG has encouraged me to prioritize the common good over personal gain.	4.02 (1.00)		

5.2 Relationships Between the Factors

Figure 2 shows the relationship between amount of MG experience (periods played) and perceived acquisition of cognitive and non-cognitive abilities. The results indicated that both cognitive and non-cognitive abilities only improved after an accumulation of MG experience, which is evidence of the presence of benefit delay. The results also indicated that the benefit delay is greater for non-cognitive ability, as represented by improvement of interpersonal ability.

Figure 3 shows how players' goals changed with an accumulation of MG experience (periods played). All three kinds of goal change became more likely to occur as

MG experience accumulated, but the increase was stepwise rather than straight and continuous. Specifically, goal change alternated between two phases: (i) a phase in which increased MG experience correlated with increased likelihood of goal change, and (ii) a phase in which increased MG experience did not affect the likelihood of goal change. A comparison of Figs. 2 and 3 reveals that this delay in goal change is greater than the delay in the improvement of management ability.

Of the three kinds of goal change, altruization of goal (contribute to company) was the most likely change; this finding might be attributable to the fact that many of the participants were playing Management Game as part of their job.

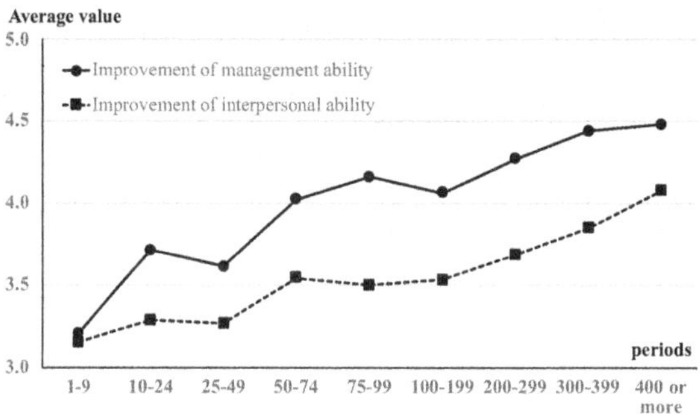

Fig. 2. Correlation between MG experience (periods played) and perceived ability-acquisition

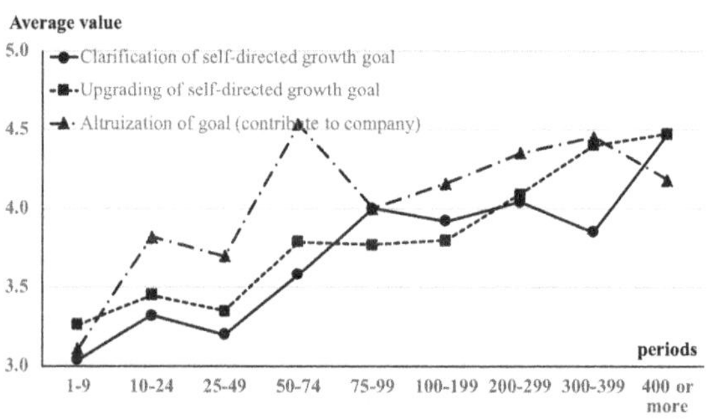

Fig. 3. Correlation between MG experience (periods played) and goal change

Given that an increase in MG experience is associated with an increase in perceived ability-improvement and a stronger tendency for of goal change, the analysis focused next upon the relationship between goal change and ability-improvement.

Table 2 shows how perceived ability-improvements (improvement of management ability and improvement of interpersonal ability) varied depending on which of the three goal-change items occurred. Eight different occurrence patterns were analyzed. For each of the three goal-change items, the goal change in question was deemed to have occurred if the score exceeded 3.0 and to not have occurred if the score was 3.0 or under. In the table, a check mark (✓) indicates that such a goal change occurred, while a cross (×) indicates that it did not occur.

Table 2. How perceived ability-improvement varies by goal-change pattern

Goal-change pattern	Goal-change			Improvement of management ability	Improvement of interpersonal ability	Frequency
	Clarification of persona-growth goal	Upgrading of persona-growth goal	Altruization of goal (contribute to company)	Average (SD)	Average (SD)	
1	✓	✓	✓	4.4 (0.45)	4.0 (0.63)	91
2	✓	×	✓	4.2 (0.43)	3.5 (0.46)	12
3	✓	✓	×	4.0 (0.34)	3.6 (0.53)	20
4	✓	×	×	3.6 —	3.2 —	1
5	×	✓	✓	4.0 (0.52)	3.5 (0.55)	18
6	×	×	✓	3.8 (0.44)	3.1 (0.45)	34
7	×	✓	×	3.7 (0.66)	3.4 (0.45)	11
8	×	×	×	2.9 (0.79)	2.6 (0.77)	32
Total				4.0 (0.74)	3.5 (0.78)	219
F-test amount				32.551***	22.952***	

1) A check mark (✓) indicates that the goal change in question occurred, while a cross (×) indicates that it did not occur.
2) *** $p < 0.001$

Of the eight patterns, Pattern 1, in which all three types of goal changes occurred, is associated with the highest perception of ability-improvement. The patterns associated with the next highest level of perception were Patterns 2, 3, and 6 in which two types of goal change occurred. The pattern associated with the lowest level of ability-improvement perception was Pattern 8, in which no goal change occurred. These results offer evidence that goal change impacts ability-improvement.

5.3 Testing the Model

To test the model of relationship between ability-improvement and goal change in MG, an analysis of covariance structures was performed. Figure 4 shows the results of testing.

Although the path from "active participation" to "improvement of interpersonal ability" in Fig. 1 was found to be statistically nonsignificant, all the other paths were found to be statistically significant. However, the analysis indicated that the path from "upgrading of self-directed growth goal" to "improvement of management ability" and the path from "altruization of goal" to "upgrading of self-directed growth goal" were significantly *negative*, rather than positive as the model posited.

The negative path from "upgrading of self-directed growth goal" to "improvement of management ability" is probably desirable. Inasmuch as learners use goals to measure learning outcomes, once they set a higher goal, they will at that point in time be aware of a gap between the goal and their current management ability. This gap awareness would then motivate the learner to improve their management ability and close the gap. Less desirable is the negative path from "altruization of goal (contribute to company)" to "upgrading of self-directed growth goal." This path implies that the more a learner prioritizes the interests of the whole community over their own, the less likely they are to upgrade their self-directed growth goal.

In Japan, there is an embedded idea of shūshin koyō ("lifetime employment")—the idea that graduates join a company and remain there for the rest of their working lives.

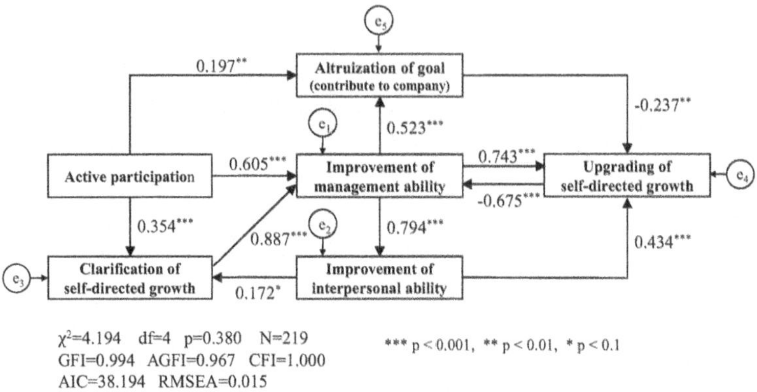

Fig. 4. Results of testing the model of relationship between ability-improvement and goal change in MG

Shūshin koyō does incentivize employees to acquire certain skills—the skills they need to play a role in the company and help it make a profit; by acquiring these skills, the employees can stay employed at the company and rise through the ranks. On the other hand, if it is true that employees can contribute to both their own interests and those of the wider community/society by fulfilling their inner potential, then it would be better for employees to try to raise their own market value and acquire more transferrable business skills that they can also use outside of organizations or communities they currently belong to. Thus, it is essential to find a way to change the path from "altruization of goal" to "upgrading of self-directed growth goal" into a positive path, without encouraging an exodus of talented employees who have improved their management ability.

6 Conclusion and Future Study

The results of this study revealed that active participation in MG leads to an improvement in the participants' management ability, and that an improvement in management ability facilitates improvement of interpersonal ability, upgrading of self-growth goal, and altruization of goal (contribute to company). The results also revealed that improvement of interpersonal ability facilitates clarification of self-directed growth goal, which in turn reinforces improvement of interpersonal ability through improvement of management ability. These findings imply that, in MG, ability-improvement and goal change have a mutually reinforcing relationship that leads to continuous improvement in ability. Thus, a talent management service like MG enables more effective and efficient talent management as it aids ability improvement and encourages employees to clarify their self-directed growth goals and set the bar higher.

However, a limitation of this study was that it only analyzed three items for goal change. It is necessary to explore goal change in a more multifaceted and detailed way in order to refine the model of the relationship between ability-improvement and goal change. It is also necessary to conduct follow-up research among MG participants to identify how the above relationship leads to higher ability at an individual level. Future research should address these issues.

References

1. Fujimura, K.: The negative effect that benefit delay in service consumption causes to a customer and the countermeasure against it. In: Freund, L.E., Cellary, W. (eds.) AHFE 2017. AISC, vol. 601, pp. 203–214. Springer, Cham (2018). https://doi.org/10.1007/978-3-319-60486-2_19
2. Fujimura, K., Ozaki, F.: Developing a model on the effects of management games for human resource development: from a benefit delay and acquisition perspective. In: Nazir, S., Ahram, T., Karwowski, W. (eds.) AHFE 2020. AISC, vol. 1211, pp. 204–211. Springer, Cham (2020). https://doi.org/10.1007/978-3-030-50896-8_31

Supporting Users in Review and Analysis of Academic Literature

Robert E. Wray[1](✉), Elizabeth T. Whitaker[2], Dawn Riddle[3], Tyree Cowell[1], Timothy Siggins[1], and Ethan B. Trewhitt[2]

[1] Soar Technology, 3600 Green Court Ste 600, Ann Arbor, MI, USA
{wray,tyree.cowell,timothy.siggins}@soartech.com
[2] Georgia Tech Research Institute, 400 10th St NW, Atlanta, GA, USA
{elizabeth.whitaker,ethan.trewhitt}@gtri.gatech.edu
[3] Naval Air Warfare Center Training Systems Division, 12211 Science Dr., Orlando, FL, USA
dawn.riddle@navy.mil

Abstract. Today, information search and retrieval tools are ubiquitous and powerful. However, there is comparably little investment in helping users with analysis of materials found via search. This paper introduces a software tool designed to help individuals and teams improve their ability to examine and to use the results of search. The existing tool focuses on literature review and analysis and emphasizes support for users who are building an initial mental model of a domain new to them. The paper describe rationales and goals for helping users perform literature searches, current implementation and features, preliminary user feedback, and future work.

Keywords: Information retrieval · Relevance ranking · Information extraction · Concept-based query and organization

1 Introduction

Today, it is faster and easier to find information relevant to a topic than at any time prior in human history. Search is ubiquitous and powerful. Many tools have been developed to support both general Internet search (e.g., Google, Bing) and targeted search (e.g., information retrieval systems). However, in comparison to search, there is comparably little investment in helping searchers analyze search results.

This paper describes on-going research targeted to address this limitation. We focus specifically on the design and development of technological support for *examination*, or evaluation of the information found via search [1]. The goal is to help users quickly prioritize what to closely read and to analyze from results of those searches, identify what information to extract, and help users apply and communicate their findings.

To support these goals, we describe an implemented, cloud-based software tool, **Archimedes**, designed to help individuals and teams improve their ability to examine and to use the results of search. The existing tool focuses on review and analysis of academic literature. It emphasizes support for users who are building an initial mental model of a domain new to them.

S. Nazir et al. (Eds.): AHFE 2021, LNNS 269, pp. 362–370, 2021.
https://doi.org/10.1007/978-3-030-80000-0_43

Archimedes is an implemented, functional software prototype. It is under current beta-testing and evaluation by an initial community of selected users. In what follows, we introduce the overall user-interaction and technical design rationales, summarize the current implementation and features (including contrast with other tools designed to facilitate examination), and outline some changes in approach that resulted from testing and user feedback.

2 Goals and Design

A key limitation of human search is poorly managed use of time examining search results [1]. Information seeking is a repetitive process of forming a search query, examining the results of a search, and reformulating the query. This process repeats until a satisfactory result set is found. Information sciences researchers have systematically evaluated the way people perform search and the tools that are designed to support that search [1–5]. This research also reveals that the way in which search results are displayed impacts the amount of time spent searching and shapes the direction of subsequent searches.

Archimedes draws design inspiration directly from the information-seeking framework described by Marchionini and White [1]. A high-level summary of the framework is illustrated in Fig. 1. The figure lists the stages of the framework and outlines the specific features within Archimedes that support each stage (further described in subsequent sections). Their analysis shows that the most time-intensive aspects of information seeking are examination and use, which are also comparatively less supported by today's search tools [1].

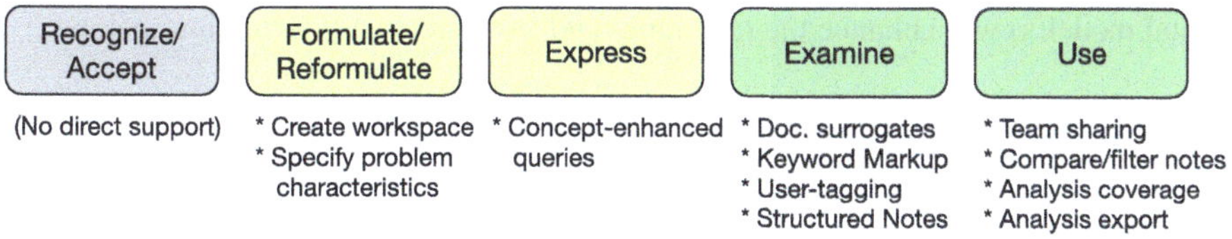

Fig. 1. Steps in information seeking (adapted from [1]) and how they are supported in Archimedes. Archimedes' primary focus is to support examination and use of search results (green) but also offers some support for problem formation and expressing search (yellow).

Even when supported by powerful Internet search tools such as Google and Bing, studies show that people generally are not effective searchers and they mismanage their time examining search results. For instance, one study found that the amount of time spent examining a series of search results reduced rapidly from the first to sixth responses; only 5% of the total search time was spent on the 7th thru 10th results [4]. If highly ranked matches are relevant, the searcher will find them. However, when relevant results are not near the top of the search results, they are easily overlooked because the examination strategy heavily invests in the first few items returned. Thus, the examination process is both time intensive and oftentimes ineffective.

These observations led us to focus Archimedes on direct support for efficient and effective examination and use of search results (the green boxes in Fig. 1). A searcher, especially one new or not well-versed in the domain of their search, may find many hits (1000s) to a general search query and may not get past the first page or two of results. By focusing on adaptable and targeted examination, we expect users will be able to find a greater number of relevant hits than by simply using general Internet search tools for formulation and expression.

Building tools to support the evaluation of results is complicated by the different goals (e.g., browsing, fact finding, information gathering) and search strategies (e.g., berry-picking, orienteering) that may be used throughout the search process and differ across users. A number of studies have found that individuals will often begin a search with a general query, then use the results of the initial search to break that query into specific sub-problems that can be answered through more directed queries and searches [2, 5]. If the initial search results presented are too specific to the query users may become fixated on their starting strategy and miss a potentially valuable set of resources [6]. In contrast, if the later search results are too general then the search process may take too long. As a consequence, Archimedes aims to present results for examination in a way that neither overly constrains nor overwhelms the user. It focuses especially on the use of *document surrogates* that can help convey substantive details of an individual source in readily interpretable ways.

Finally, because Archimedes is targeting analysts who may not have deep familiarity with existing literature, it has some support for helping those user(s) create a more nuanced mental model of the domain. Features that support this goal include the ability to define and refine labels, notes, and keywords associated with a problem and supplementing search with concepts relevant to the evolving mental models. These are the yellow boxes in the figure. While these features less directly support the goal of rapid but deep examination, we hypothesize that capturing and allowing expression of user mental model(s) will enhance the 'examine' and 'use' steps of information seeking.

3 Analysis Workflow and Archimedes Features

Archimedes was implemented on a cloud-based platform, using Vue.js for the web-based front-end, Java with a SQL database on the backend, and an integration with python that enables the use of several, existing machine learning and natural language processing modules. To provide structure for discussion of the features and capabilities of Archimedes, Fig. 2 shows the various activities that a user can do within the tool in terms of the information seeking framework introduced above. As suggested by the figure, Archimedes allows transitions between the various stages of information seeking. Different activities within those stages can also be combined to support various workflows.

Recognizing and accepting a problem is outside of the scope of the tool. We assume that users receive tasks (or, for students, assignments) that result in accepting the need to conduct a search. We now describe the other elements of the workflow in terms of the information seeking framework.

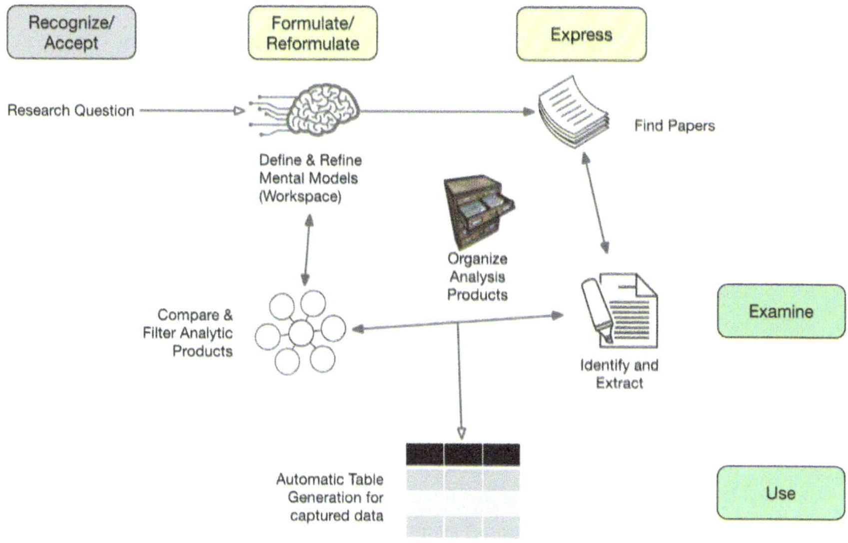

Fig. 2. Workflow(s) enabled by Archimedes.

3.1 Workspaces: Formulating the Problem Space

Archimedes offers support for formulation (further defining a search problem) and reformulation (iterative refinement) via its Workspace feature. Workspaces allow individual users and teams to identify specific questions or areas of interest. Within a workspace, a user (or a team of users working on the same analysis problem) define both characteristics of the problem (important terms, concepts, etc.) and various steps anticipated in the analysis process (e.g., categorizing the relative importance or contribution of a particular source document).

Users can define "tags" or labels, note templates, and keywords to highlight within papers. These items signify topics and process within the workspace. Labels identify relevant topics ("augmented reality"), specific steps in the analysis process ("paper to read") and outcomes ("foundational paper"). Archimedes allows users to create their own labels. As we discuss further below, once labels are defined, they can be "tagged" to individual papers and used to find and filter papers with various properties.

Within a workspace, users can also define templates for the type of notes they wish to capture. These templates include user-defined fields within the template. Figure 3 illustrates an example. In this case, a user has defined a note that allows for information capture about the components within an intelligent tutoring system. The user can specify the kinds of fields (a defined list of options, numeric data, short phrases, or text boxes) in the note and provides a description of what should be entered. The note templates provide structure for analysis as individual papers are read and notes are captured, specifying what information should be captured for each kind of note. We hypothesize that note templates will be especially beneficial for teams of analysts working together; note templates codify the team's shared understanding of what information needs to be captured during the shared analysis.

Fig. 3. A screenshot showing the interface for defining a note template in a workspace.

A final feature within the workspace is the ability to specify categories of terms/keywords that the user to have highlighted in a paper. All the identified terms are automatically highlighted in the user's chosen color for each category and a simple syntax for regular expressions allows the user to specify similar words together (e.g., "tutor", "tutors", "tutoring"). These "markups" are defined for the workspace (but can be customized and extended for individual papers). The consistent markup across the workspace enables users to share their understanding of key terms that appear in papers.

3.2 Document Surrogates: Speeding Initial Review and Examination

Because there are so many powerful search tools available already, Archimedes offers only limited support for expressing a search problem. One way it does support enhanced expression is that it can automatically customize a series of search steps (a "search plan" [7]) using the problem characteristics identified in the formulation process. Archimedes interfaces to a local repository of papers as well as several internet repositories that offer public APIs: Microsoft Academic, CORE, and arXive.

Once search results are returned Archimedes employs a number of abstract representations of the content of a paper, or *document surrogates*, to support rapid evaluation. Archimedes uses various text analysis, extraction algorithms, and a machine learning library to enable users to quickly assess an individual paper. Authors, keywords, publication, abstract, and a summary graph of citations are extracted to support fast review and assessment. Figure 4 presents examples of one of these document surrogates, the automatic extraction of citation and abstract data.

These document surrogates use backend processing of the paper to extract information from the papers to provide more information to the user than would be available from a typical internet search. Publication meta-data is extracted from an open-source machine learning library (Grobid). We have tuned Grobid's performance to improve extraction from conference papers and explored tuning it for use in patent and technical report documents. A topic map presents a summary of topics in a paper in the context of a collection of topics in the workspace, using TF-IDF and t-distributed stochastic neighbor embedding (t-SNE) algorithms. Topics that occur closer to one another appear more often together in papers in the workspace and the size of the circle represents the overall frequency of terms in the workspace. (Another paper in this conference presents a more detailed discussion of the topic map feature.)

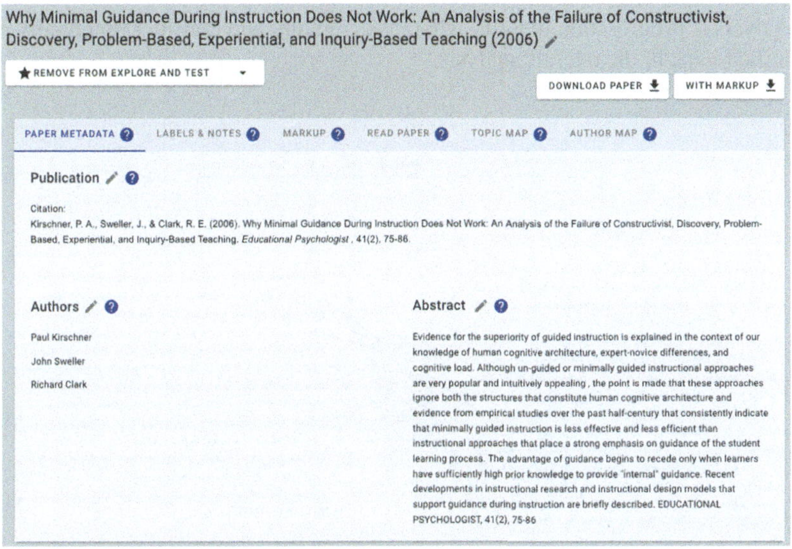

Fig. 4. Archimedes document surrogates. Automatic text extraction provides a summary of authors, publication, and abstract. Topic and author-citation maps are also generated for papers.

3.3 Capturing and Comparing Notes: Supporting Aggregate Analysis

As suggested above, users can add tags to an individual paper and capture notes within the paper. Archimedes allows users to automatically associate an excerpt from the paper with a note, providing additional context for captured information. All notes can be exported for use in reports and briefs. The note and labeling feature allows users to capture analysis resulting from individual papers; however, Archimedes also includes aggregate-level analysis tools for all the papers in a workspace. The topic map discussed above is one example, enabling a user to see and to evaluate topics covered in the workspace.

Archimedes also introduces a novel (to our knowledge) capability that allows a user to view the progress of analysis by comparing and filtering papers in the workspace based on user-captured labels and notes. The user interface for this feature is shown in Fig. 5. The user chooses labels and notes, and even fields within notes, to see connections, overlaps, and potentially missing items within the papers in the workspace. In this specific example, the interface shows links between all papers that have a label, indicating that they remain to be fully read and processed ("paper to read"), all papers with some empirical results, and all papers that outline a theory. The light blue lines show the connections for the selected node (empirical results). Any comparison nodes dragged to the "Filters" box will result in only papers that match the comparison node. For example, if the "Empirical Results" node was made a filter, only the three papers with empirical results (light blue lines) would be displayed. This comparison and filter interface is meant to support rapid assessment of the state of an analysis. By mixing workflow labels ("paper to read"), topic labels ("ITS"), and outcome labels ("Foundational"), users can readily determine how analysis is progressing, identify what tasks remain, and consider potentially as yet unidentified gaps in the overall analysis.

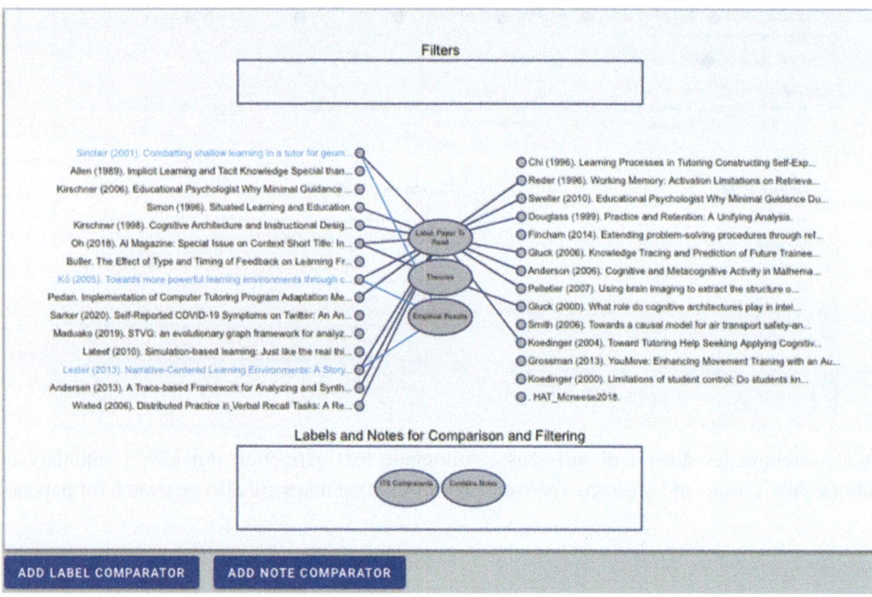

Fig. 5. The Archimedes comparison and filter interface.

3.4 Exporting Analyses: Enabling Use and Sharing of Analyses

A final feature of Archimedes is the ability to export the results of a user's analysis. All structured notes and labels can be exported for use in reports and briefs, including the paper excerpts associated with those notes. In the current implementation, Archimedes allows a user to explore all the notes of a particular type (e.g., all notes capturing "empirical results"), as well as all notes associated with the workspace. The data is exported in a Microsoft Excel-compatible format, allowing straightforward incorporation of the output tables into common office productivity tools.

We have also explored more customized analysis formats, such as the ability to create a matrix of various dependent and independent variables used in experiments, in which cells identify whether there is evidence of a positive or negative relationship between the variables. Many such specialized interfaces to support analysis could be supported in the future. However, in the current tool, we focus on text export which is relatively straightforward to implement and ensures that users can use and further transform their analyses after capturing it in Archimedes.

4 Testing, Iterative Refinement, and Status

Archimedes was designed and initially implemented as a "minimum viable product" (MVP) to provide useful and compelling capability for users, while also providing opportunity for initial users to provide feedback on the implementation and identify/suggest new features. Early feedback from testers resulted in two significant changes. First, rather than a desktop application, users requested a cloud-based application. This request was often motivated by technical constraints (e.g., some users lacked permissions to install or update software on organization-controlled computers). Shifting to a cloud application, however, also facilitated achieving a second recurring request, to support multiple, shared workspaces. Second, in the MVP release, there was no explicit workspace concept. Now users can create multiple workspaces and workspaces can be shared. We did envision these new features in the long-term plan. However, user feedback motivated introducing these features much earlier and shifting to the cloud.

While these changes have led to additional requirements in backend data management and a more complex software environment, initial feedback from on-going testing with these new features indicates that users find the tool much more valuable in this new configuration.

Acknowledgments. The authors thank teammates and collaborators for contributions to concepts and implementations reported here including Dr. Jennifer Fowlkes, J.T. Folsom-Kovarik, Kelly Neville, Caitlin Tenison, Austin Brehob, Dana Foley, and Ryan Yao. Opinions expressed here are not necessarily those of the Department of Defense or the sponsor of this effort, the Naval Air Warfare Center Training Systems Division (NAWCTSD). This work funded under contract N68335-19-C-0004.

References

1. Marchionini, G., White, R.: Find what you need, understand what you find. Int. J. Hum. Comput. Interact. **23**, 205–237 (2007)
2. Anick, P.G.: Adapting a full-text information retrieval system to the computer troubleshooting domain. In: Croft, B.W., van Rijsbergen, C.J. (eds.) Proceedings of the 17th Annual International ACM SIGIR Conference on Research and Development in Information Retrieval, pp. 349–358. Springer, New York (1994). Doi: https://doi.org/10.1007/978-1-4471-2099-5_36
3. Hearst, M.: Search User Interfaces. Cambridge University Press, Cambridge (2009)
4. Joachims, T., Granka, L., Pan, B., Hembrooke, H., Gay, G.: Accurately interpreting clickthrough data as implicit feedback. In: Proceedings of the 28th Annual International ACM SIGIR Conference on Research and Development in Information Retrieval. ACM (2005)

5. Teevan, J., Alvarado, C., Ackerman, M.S., Karger, D.R.: The perfect search engine is not enough: a study of orienteering behavior in directed search. In: Proceedings of the SIGCHI Conference on Human Factors in Computing Systems, pp. 415–422. ACM (2004)
6. Hertzum, M., Frøkjær, E.: Browsing and querying in online documentation: a study of user interfaces and the interaction process. ACM Trans. Comput. Hum. Interact. (TOCHI) **3**, 136–161 (1996)
7. Whitaker, E., Simpson, R., Burkhart, L., MacTavish, R., Lobb, C.: Cognitive factors in homeland defense: reusing intelligence analysts' search plans. In: Proceedings of Human Factors and Ergonomics 2004, New Orleans (2004)

Application of Virtual Simulation Experimental System of Curtain Wall Based on Educational Transformation and Upgrading

Chenguang Ma(✉) and Manhai Li

Chongqing University of Posts and Telecommunications, Chongqing 400065, China

Abstract. The development of experimental course on curtain wall design in universities faced with such difficulties as high construction cost of teaching laboratory, lack of teaching security and poor experiment efficiency. In order to improve the teaching reform of experimental course on curtain wall design and strengthen the cultivation of students' specialty and Information Literacy, the team applied virtual simulation technology to curtain wall experiment teaching. Through the "1+3+1" step-by-step innovative teaching model, a progressive experimental system integrating learning, practice and examination is formed. The results show that the virtual simulation experiment system is superior to the traditional curtain wall design laboratory in terms of experiment efficiency, field scope, experiment completion rate, experiment security and cost control. The combination of Virtual Reality Technology and curtain wall experiment course can improve the teaching effect significantly.

Keywords: Curtain wall design · Virtual simulation experiment · Module teaching · Effect analysis

1 Introduction

In the field of architecture, curtain walls have gradually become the preferred decoration scheme for building façades because of their beauty, energy saving and easy maintenance. Under the situation of increasing market demand, the output of curtain wall design talents has not kept up in time, this is due in large part to the limitation of curtain wall teaching practice. In the curtain wall practice teaching in universities, many factors restrict the curtain wall laboratory construction, which makes the curtain wall experiment course in colleges and universities face great difficulties. Such as low safety factor and high test threshold, high cost and low efficiency. Therefore, the research team designed and made virtual simulation experimental system for curtain wall, made an attempt to improve the teaching quality and efficiency.

2 Functional Framework of Virtual Simulation Experimental System for Curtain Wall

With the continuous progress of modern information technology, experimental teaching is no longer limited to the laboratory, more students can carry out experimental learning

S. Nazir et al. (Eds.): AHFE 2021, LNNS 269, pp. 371–378, 2021.
https://doi.org/10.1007/978-3-030-80000-0_44

through the way of network platform [1]. In order to promote the teaching reform of curtain wall design course and improve students' professional and information literacy, the educational circles have to consider upgrading the existing curtain wall laboratory and upgrading the previous experimental methods to a more intelligent and digital experimental system. So, in order to solve many problems in curtain wall laboratory teaching, the Chongqing University of Posts and Telecommunication's virtual simulation team is aiming at the much needed informationization of experimental teaching, transforming complex engineering problems and teaching and research results into a virtual simulation experimental system with multiple contents and complete types [2].

The system is based on the Internet, including login section, project introduction section, course learning section, creative section, personal space section and examination section.

3 The "1+3+1" Experimental Model for Virtual Simulation Teaching System

In order to expand the scope of application of the teaching system and better serve the users in the design level of various curtain walls, we designed the curtain wall simulation experiment module as "1+3+1", The first one represents the curtain wall cognitive experiment, number three represents the unit component experiment, the curtain wall installation experiment, the template design experiment, and the second one represents the free creation experiment, detailed screen is shown in Fig. 1. The design results follow the step by step learning order from easy to difficult, after the initial -advanced-higher order design practice, the experimental form is gradually extended from the basic operation in the system to the support of external resources. It accords with the basic law of knowledge acquisition.

Fig. 1. Virtual simulation experimental system of curtain wall

3.1 Curtain Wall Cognitive Experiment Module

Curtain wall cognitive module is the basic cognitive module of curtain wall design experiment. The theoretical part explains the basic knowledge of curtain wall (including glass curtain wall, stone curtain wall, metal curtain wall) from the definition, classification, material, development history and application environment of three kinds of common curtain walls. In the practice part, five kinds of explosion pictures are preset to study the detailed structure of curtain wall, including bright frame glass curtain wall, point-supported glass curtain wall, unit glass curtain wall, stone curtain wall and metal curtain wall, detailed screen is shown in Fig. 2. When students approach the building with a cursor, they are guided to the learning pages of various curtain walls and choose their own learning order. Students can learn relevant knowledge in detail through curtain wall appearance, details, explosion drawings, and auxiliary graphics and so on.

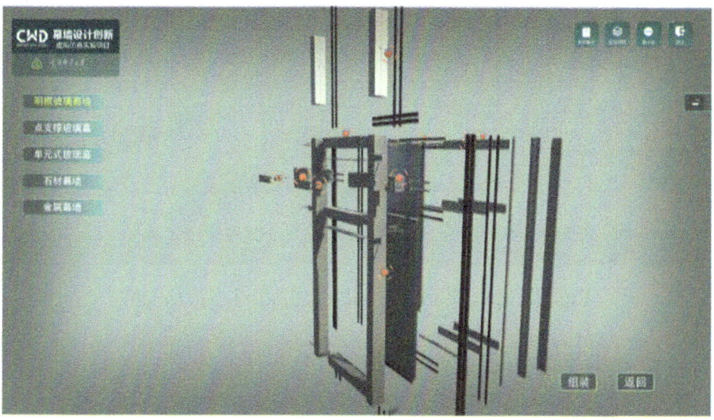

Fig. 2. Bright frame glass curtain wall cognition experiment

3.2 Experimental Module of Unit Component, Experimental Module of Curtain Wall Installation, Experimental Module of Template Design

This part is the advanced module of curtain wall design experiment and the key point of curtain wall design. They include the above three kinds of curtain wall design methods, material characteristics, the installation sequence of each component, the relationship with the wall and other experimental contents. At the same time, this part is injected into the curtain wall material and color adjustment engine, so that students can complete the curtain wall appearance design on the basis of completing the curtain wall unit assembly experiment and the installation step experiment. Finally, students can generate personalized curtain wall works based on existing architectural models in the system. Through this part of learning, students can experience the design and installation process of common curtain walls in the market in a variety of virtual building models, and make a good foundation for free creation experiments.

During the experiment, the active pop-up instructions can help students learn how to use the system quickly. The skeleton materials, plates, sealing materials and structural adhesive materials used in common curtain wall engineering are integrated in the alternative column on the left side of the operating interface. Students can complete the curtain wall unit assembly experiment by clicking on the curtain wall assembly sequence and its corresponding items. In learning mode, users can also click on the "novice guide" to get the tips of the experimental operation interface and get the correct step guide, which is much faster than waiting for the teacher's answer, detailed screen is shown in Fig. 3.

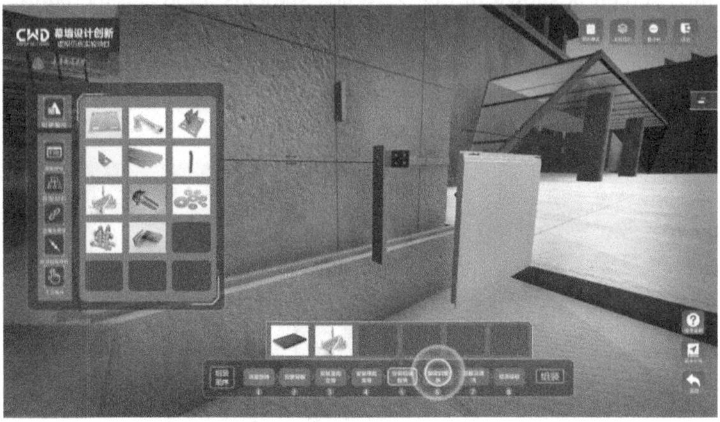

Fig. 3. Metal curtain wall installation experiment

3.3 Free creation Experiment Module

Free creation experiment is the advanced stage of curtain wall design experiment. This part of the experiment is separated from the system built-in data, all modeling, attached material, mapping, rendering and other processes are designed by the students themselves. Then the students upload the design scheme to the system in the form of pictures, documents, animation, models and so on, as the basis of communication learning and performance evaluation.

In addition, this module experiment also based on the construction safety code of building curtain wall construction set the knowledge of construction safety code, curtain wall wind-resistant animation and other experiments. Learning construction safety is a compulsory part of curtain wall experiment. Students should learn safety standards from the aspects of personnel wearing, equipment inspection and carrying tools. The experimental process of curtain wall wind-resistant animation integrates the real-time detection function of curtain wall structure and the real-time monitoring function of each force point, which can help students to detect the safety of curtain wall structure design and installation to a certain extent. Students only need to convert the independently designed model to the prescribed format of the system and upload it according to the specified path, and the calibration data of curtain wall strength can be obtained according to the built-in algorithm of the system.

4 Application Effect Analysis of Virtual Simulation System for Curtain Wall Design

The data inside the virtual simulation system of curtain wall design can clearly record the process of teaching experiment, so as to facilitate the analysis of classroom experiment. By reading the stored data, we can see that compared with the traditional laboratory, the system has great advantages in experimental efficiency, experimental site, experimental completion rate, safety, cost control and so on.

4.1 Experimental Efficiency Analysis

The efficiency data as the Table 1 show can be obtained by extracting the data records in the system and comparing the actual operation time of the students with the experimental class hours specified in the syllabus.

Table 1. Statistical table of experimental efficiency.

Experiment name	Academic hour plan	Academic hour used	Efficiency
Module of unit component			
Curtain wall cognitive	2	1	200%
Unit combination	4	3	133.3%
Curtain wall installation			
Assembly sequence	4	2	200%
Skeleton and panel	4	3	133.3%
Connection, sealing and process	4	3	133.3%
Template design	4	3	133.3%
Free design	4	2	200%
Develop learning	2	1	200%

It can be seen from the data in the table that after the students study in the virtual simulation teaching system of curtain wall, the actual time of completing the experiment will be greatly shortened, and the efficiency of the experiment will be improved obviously. The main reasons are:

The virtual experiment environment is convenient and time-consuming. Teachers can convert laboratory preparation time into tutoring or answering questions and other related teaching activities.

The time for students to enter the experimental environment is not limited by the working hours of the laboratory. They can use their spare time to prepare for class and reduce the probability of misoperation.

The virtual simulation experiment teaching system of curtain wall has the advantage of multi-party cooperation. It can realize remote experiment, cooperative development, cooperative research and so on through virtual curtain wall environment technology. In the virtual laboratory, we can integrate the media, skills, teaching, industry, materials and other related chains to become a multifunctional curtain wall design laboratory. It can enable learners to acquire more curtain wall knowledge and master more curtain wall design skills in a short time.

The powerful data processing function and interactive function of the system help to improve the teaching progress and improve the teaching quality. Through virtual simulation technology, the key construction nodes in the teaching process are "enhanced and reproduced", which can make students overcome the cognitive obstacles in learning and help to improve the learning effect [3]. Relying on the system data, the teacher can synchronously view the experimental process of each student, and can also understand the mastery of each student through the statistical query function of the experimental results. From the point of view of data value, the data recorded by the system can also help teachers to accurately find out the students' mastery and weakness of curtain wall experiments, accurately judge the heavy and difficult points of the course, and optimize and update the teaching contents in time.

4.2 Experimental Site and Completion Rate Analysis

In order to accurately analyze the influence of virtual simulation experiment system on students' selection of experimental site and the completion rate of experiment, we extract the activity data recorded in the system and count the data of students' experimental site, as the Table 2 shows.

Table 2. Site analysis and completion rate analysis table.

Teaching week	The number of people in the lab	Number of people in virtual simulation system	Completion rate on the teach day	Completion rate for the rest of the time
1	16	75	93.41	6.59
2	10	81	92.31	7.69
3	12	79	93.13	6.87
4	28	63	86.82	13.18
5	20	71	87.92	12.08
6	18	73	94.51	5.49
7	11	80	96.71	3.29
8	9	82	97.81	2.19

Statistics show that the laboratory is not the only test site for students. On the premise of free choice of the site, most students show their preference for new technology and choose to do the experiment in the virtual simulation system. The practice of decentralized development not only reduces the space demand and hardware pressure of the laboratory, but also brings students the opportunity to advance synchronously and solve problems together in the process of experiment. In addition, virtual reality environment has great potential to improve students' participation, interaction and enthusiasm for learning. Virtual environment can help students to identify relevant information effectively in detail, improve students' spatial understanding ability and operation ability [4], highlight the key points of experiments, and reduce the difficulties of experiments. As a result, most complete the experiment on the day of class, and a few students are free to arrange time to continue their study according to their own situation.

4.3 Experimental Safety Analysis

The construction of virtual simulation system can solve the security problems caused by the experiment site. In the virtual building scene, students can not be restricted by the experimental site, the curtain wall material knowledge, organizational components, assembly technology and other experimental links are safe and reliable. Enterprises do not have to stop cooperating with schools because of safety risks, and schools may expand their professional orientation and increase the number of professionals by dispelling safety concerns. Besides, students' mistakes in the course of the experiment will not lead to personal injury and environmental harm.

4.4 Experimental Cost Control Analysis

On the one hand, the virtual simulation system of curtain wall reduces the cost of laboratory construction. The curtain wall simulation system can virtual the real experimental environment, which has the characteristics of short development period, low construction cost, fast upgrade speed and convenient maintenance in the later stage. It not only builds the process of green environmental protection, but also can with the upgrading of curtain wall industry technology rapid response iteration, to ensure the integration of industry and education.

On the other hand, the use of curtain wall virtual simulation system reduces the cost of use and controls carbon emissions. In the virtual simulation system, the material loss and the output of construction waste are almost zero, which saves the material purchase cost, the equipment wear depreciation fee and the garbage disposal fee, and reduces the experimental cost while avoiding the environmental pollution.

5 Conclusion

The feedback of virtual simulation curtain wall design experiment in practical teaching is as follows: strong interest in learning, strong sense of reality of scene substitution, and impressive operation of experiment. The virtual simulation system of this kind of simulation laboratory reconstructs the physical space of learning, improves the perception effect of teachers and students, provides multi-modal training situation for learners, and promotes the deep interaction between teachers and students and the learning environment and learning content [5]. In the process of application, the deficiency of curtain wall simulation experimental system also exists, such as insufficient innovation, low compatibility between model and system from outside the system, insufficient training of field operation ability and so on. Therefore, in the future, we need to strengthen the integration with the traditional laboratory "virtual" and "real", develop the virtual reality online learning resource system, and promote the wide application of virtual reality online learning resources. At the same time, it is necessary to integrate information resources and integrate new information technology to perfect the system itself.

References

1. Chunhua, S., Yuxu, D.: Construction of experimental online teaching simulation platform based on virtual technology. Mod. Electron. Tech. 156–159+164 (2020)
2. Haibo, L., et al.: Construction of virtual simulation experimental teaching resource platform based on engineering education. Exp. Technol. Manage. 19–22+35 (2019)
3. Xuehong, G., Wei, W., Chenhui, J.: Research on the application of AR technology in the construction teaching of assembly integral structure. Zhejiang Constr. 59–62 (2018)
4. Wang, P., et al.: A critical review of the use of virtual reality in construction engineering education and training. Int. J. Env. Res. Public Health 1204 (2018)
5. Ying, L.: An analysis of the modal reconstruction in virtualized foreign language teaching and learning. Technol. Enhanced Foreign Lang. Educ. 24–30+4 (2020)

Innovative Forms of Social Non-formal Youth Education

Nataliia Levchenko(✉), Yuliya Klymenko, Olena Baldyniuk, Svitlana Roienko, and Nataliia Koliada

Pavlo Tychyna Uman State Pedagogical University, Uman, Ukraine
koliada_n@ukr.net

Abstract. The article reveals the features of non-formal education, its connection with formal education; various definitions of the concept of "non-formal education" are presented; This article analyzes scientific sources of researchers who have dealt with the issue of non-formal education, the main forms of non-formal education, which are known in Ukraine. The purpose of the article is to carry out a theoretical analysis of the scientific problem and highlight the features of the introduction of non-formal education into the process of preparing social workers to work with young people. In this research used methods of analysis, comparison and generalization of sources from the researched problem.

Keywords: Training of youth workers · Work with youth · Youth · Non-formal education · Forms of work with youth

1 Introduction

Youth is an active subject of social reproduction, the main innovative potential of society and a significant guarantor of its development. It is the generation of young people that is an important condition for further socio-economic change. One of the effective means of successfully solving problems with young people is social work with this category of population. This type of social work affects the nature of life of young people and as a result determines its qualitative characteristics. Therefore, social guarantees and support for young people in Ukraine should become one of the priorities of the state, as well as provision of high-quality education - the main goal of institutions and organizations that provide it.

In our opinion, this can be done at the international, state, regional and local levels. The main tools for achieving this goal are formal, non-formal and informal education.

In the process of organizing the professional training course of a social worker to work with young people, it is necessary to focus on the formation of a high level of professional culture, development of a need for continuous professional self-improvement, which guarantees success and efficiency in the modern educational environment.

In Ukraine, the main aspects of youth policy and professional youth work training are regulated by: the draft Law of Ukraine "On Youth" (2015), the Law of Ukraine "On Promoting Social Standing and Development of Youth in Ukraine" (1993, with

S. Nazir et al. (Eds.): AHFE 2021, LNNS 269, pp. 379–386, 2021.
https://doi.org/10.1007/978-3-030-80000-0_45

changes), the Resolution of the Cabinet of Ministers of Ukraine "On approval of the State targeted social program "Youth of Ukraine" for 2016–2020" (2016), the Law of Ukraine "On Higher Education" (2014), the Program "Youth Worker" (2014), the Personal and professional development of youth "The Statesman" (2018), Resolution of the Cabinet of Ministers of Ukraine "On approval of standard regulations on the youth center and the expert council at the youth center" (2017), draft regulation "On training of youth workers" (2017) and others.

The current state of socio-pedagogical work in Ukraine is characterized by a variety of professional roles of social workers in various fields of their activities. The Ukrainian education system considers formal, non-formal, informal education, each of which plays an important role in preparing future social workers to work with young people.

Article 8 of the Law of Ukraine "On Education" (2017) stipulates that "formal education is education that is obtained through educational programs in accordance with the levels of education, fields of knowledge, specialties (professions) defined by law and provides for the achievement of level of education defined by educational standards of learning outcomes of the appropriate level of education and acquisition of qualifications recognized by the state. Non-formal education is education that is usually obtained through educational programs and does not involve the award of state-recognized educational qualifications by level of education, but may result in the award of professional and/or partial educational qualifications. Informal education (self-education) is education that involves a person's self-organized acquisition of certain competencies, in particular during daily activities related to professional, social or other activities, family or leisure" [3].

The normative document states that learning outcomes obtained through non-formal or informal education are defined in the formal education system in the manner prescribed by law, so all three types of education are interrelated.

At the state level, the most common forms and types of non-formal education, namely in the training of specialists to work with young people, are conferences, programs, courses, internships, etc.

2 Materials and Methods

In the process of determining the role of a youth worker in the field of innovative development of social work and substantiation of theoretical and methodological principles of preparation of future social workers for youth work, a series of training courses were conducted to provide theoretical knowledge of practical skills in the implementation of youth policy.

We covered 24 participants of the program of personal and professional development of youth "The Statesman" in Uman.

The study used a set of interrelated theoretical and empirical methods: analysis, comparison and generalization of scientific sources on the researched problem, study of the theory and practice of non-formal education and the effectiveness of the program of personal and professional development "The Statesman"; observations, online interviews and surveys were used to diagnose the level of effectiveness of the system of personal and professional development of young people in Ukraine aged 18 to 25, the development of responsible leadership and increasing the level of social activity of young people; formative experiment - a series of lectures, seminars, trainings aimed at providing theoretical knowledge of the development of practical skills in the field of public, private and public sectors of the state formation.

In 2014, in order to disseminate and mainstream youth work in Ukraine, the "Youth Worker" program was launched with the support of the "United Nations Development Program of Ukraine" and in cooperation with the Ministry of Youth and Sports of Ukraine, the State Institute for Family and Youth Policy and the UN Children's Fund (UNICEF), which is based on the principles of non-formal education. The "Youth Worker" program is a part of the "Resuscitation package of youth policy reforms in Ukraine" and one of the priorities of the "State target program" Youth of Ukraine 2016–2020", for the implementation of which funds were provided in state and local budgets [6].

The implementation of the "Youth Worker" program in Ukraine is a unique opportunity to raise the professional level of representatives of civil services and public associations working with youth, which will increase the number of young people involved in the youth policy-making and strengthen national cooperation between the state and youth public institutions in the field of social formation of youth [6].

According to the program, for the first time in Ukraine, a professional group of specialists working with youth has been singled out, and youth work has been recognized at the state level as a separate area of professional training of such specialists.

To participate in the program, you need to have experience and desire to work with young people [6].

The program is aimed at leaders and activists of public and charity organizations working with youth; civil servants responsible for the implementation of the youth policy at the local and regional levels; employees of other organizations involved in working with young people, regardless of the type of ownership. The purpose of the program is to improve skills of youth leaders and specialists in the youth field, in the context of decentralization, in order to increase the efficiency of development and implementation of substantiated youth policy at the local and regional levels.

The main objectives of the program "Youth Worker": improving theoretical knowledge and practical skills of youth leaders and professionals in the youth field; formation of a database of youth specialists and best practices in the implementation of youth policy at the local and regional levels; providing consultations on the formation of local youth policy, etc. [6].

The program provides 3 levels of training: basic (first level), specialized (second level), training for trainers (third level).

Each type of training is conducted in accordance with the curriculum. Each training course has its own purpose, tasks, curriculum, training modules.

The basic training lasts for three days and includes three training modules (24 teaching hours), in particular - the first module: "Youth policy and youth work. Key topics: Who are the youth? What is youth policy? Roadmap for reforming youth policy in Ukraine. Youth participation. Cross-sectoral approach. Youth work and youth policy. European approach to youth policy; second: Community and youth work in the community. Key topics: Community. Activation of the community. Youth work in the community. Cross-sectoral approach. Youth participation; third: Competences of the youth worker. Key topics: Youth worker. Competence approach. Competences of a youth worker. Portfolio of youth work of the Council of Europe.

The purpose of the basic training is to form the participants' basic theoretical knowledge and practical skills for the implementation of regional youth policy. The tasks of the training include: providing participants with the necessary basic level for the implementation of a successful youth policy; establishing a dialogue between different participants in youth policy; giving participants the opportunity to create their own identity as a youth worker.

Areas of activity of the youth worker: implementation of the state youth policy at the regional and local levels, provision and development of interrelation of the state bodies of the regional and local levels during the youth policy of the state, assistance in coordination of work of youth associations, development and realization of socially significant youth projects; counseling programs for adolescents, youth, young families (legal, social, psychological, career guidance, health, etc.), collection and analysis of information about the situation of youth in the region.

Upon completion of the training, participants receive a certificate of advanced training [6].

Graduates of the basic training can participate in a specialized training course. Such training consists of three modules and home assignments and are designed for three days. Each training has their own theme, for example: "Civic education for youth workers", "Engagement and coordination of volunteers", "Inclusiveness in youth centers" and others.

The course "Civic Education for Youth Workers" aims to help young people to become and to be active citizens, understand human rights and their own identity, be interested in social processes, critically evaluate them, as well as be the engine of these processes. The course "Engagement and coordination of volunteers" is designed for young people who have experience of volunteer work, are participants in volunteer projects and seek to further train volunteers in their organizations.

The course "Inclusion in youth centers" provides practical competencies for the implementation of inclusive approaches in youth centers, teaches how to take into account interests and needs of all groups of young people who may be potential visitors of the center [6].

Studying under the program "Youth Worker" for youth workers who have experience in training activities, youth work and are able to conduct training, is a training course for trainers.

The program also provides an international level, namely: the best graduates of the program "Youth Worker" from different regions of Ukraine have the opportunity to participate in an exchange program with an educational visit to the European Youth Center in Weimar (Germany). The purpose of this experience is to implement in Ukraine a system of training for civil servants and leaders of public institutions, whose activities are aimed at working with young people. The curriculum includes thematic blocks: active participation of young people at the local level; finding resources and successfully convincing others that your organization deserves support; definition of the concept of "youth worker" as a profession [6].

A good example of the development of non-formal education at the state level at the initiative of the Ministry of Youth and Sports of Ukraine and the Ukrainian Academy of Leadership is also the program of personal and professional development of youth "The Statesman". This program has been operating since 2018, its goal is to create in Ukraine a system of personal and professional development of young people aged 18 to 25, development of responsible leadership and increase the level of civic activity of young people. In 2020, 8 regions of Ukraine (Chernivtsi, Ternopil, Sumy, Kharkiv, Lviv, Cherkasy (Uman), Rivne, Kirovohrad) are involved in this program, it is aimed at students who are motivated; have desire to implement projects for their community, participate in training programs and internships, community activities, social and volunteer projects, etc.

The implementation of the "The Statesman" program in 2020 under the conditions of the COVID-19 epidemic takes place with the provision of security for the people of Ukraine by an online 10-day course.

Tasks of the program "The Statesman": to promote personal, professional and social development of youth; to raise the level of awareness of young people about the activities of the public, private and public sectors; to acquaint young people with the basics and importance of the state formation; to promote the development of leadership skills of program participants; involvement of young people in internships and entering the civil service, service in local governments; provide Ukrainian youth with information on anti-corruption mechanisms and familiarize them with anti-corruption tools in Ukraine; to visit anti-corruption institutions of Ukraine, in particular NAPC, NABU and others; to promote values of integrity and transparency among the youth of Ukrainian. The program consists of three components: educational (lectures, training courses, seminars); practical (educational visits to local governments, media, private companies and public associations, creation of project teams, writing and pitching projects before the jury); formational (reflection, team building, mentoring).

One of the important highlights of this year's program is the involvement in mentoring of graduates of the "The Statesman" of previous years, who have chosen a career in public service, international institutions, mass media, and public organizations. The curriculum of the program in 2020 consists of three modules "Creating ourselves and forming a team", "Exploring the community", "Building the state" with a total of 60 h. work with youth; emotional intelligence; active listening; goal setting; the art of decision making; personal reflection; building a career in the civil service; team building and roles, team dynamics, team motivation; the art of the feedback; reflection in the team; games as team exercises and tasks; main functions and tasks of local self-government; project management; effective attraction of financial resources for projects in the community: cooperation with business and local governments; creation and management of local communities; combating corruption and educating the values of integrity and transparency in Ukrainian youth; writing draft applications for social action projects; pitching of social action projects on key topics (education and upbringing; national culture and identity; healthy lifestyle; state formation).

According to the results of the program, the best graduates are offered internships and internships in public service during 2020–2021.

During the implementation of the program, a study was conducted, which consisted of two stages and included a survey: at the first stage, a survey was proposed for self-assessment of participants of their knowledge at the beginning of the program; at the second stage - surveys to assess their knowledge at the end of each of the modules; and the third stage - summarizing the results of the effectiveness of the program.

At the first stage, during the program "The Statesman", participants were offered a survey, which included questions: 1. Do you have a clear idea of state formation; 2. Are you familiar with writing social projects; 3. Your feedback.

At the second stage of the research (at the end of each of the modules) the participants were offered a survey that helped to assess the knowledge of the participants of the module.

In the third stage of the study, they found out, according to young people, whether this program is effective.

Thus, the state-level programs "Youth Worker", "The Statesman" are striking examples of innovative forms of non-formal education for young people. Formal and non-formal education play an important role in youth development. In our opinion, these two types of education should function in interplay, complementing each other, helping to form the modern young generation, which can influence the success of the whole country with their personal and professional achievements.

3 Results

At the first stage of the study, it was found that participants at the beginning of the program have a 50% clear idea of state formation; 50% are familiar with the process of writing social projects, but 25% are not aware of this topic at all.

In the second stage of the survey, it was found that after three modules, the results improved, namely: 89% of respondents are well aware of state formation, and 11% have a clear idea of public policy. To the question "Do you know the tools of social projects" 60% of respondents answered to know perfectly well, and 40% have a clear idea about it.

A comparison of the results of the survey of the first and second stages of the study led to the conclusion that the theoretical and practical material that was laid down in the program "The Statesman" showed sufficient results and made a positive difference.

The third stage of our study showed that the result was that the program is an effective form of non-formal education, which is implemented at the regional level.

4 Discussion

Our study confirmed that some of the effective innovative forms of non-formal education are the state-level programs "Youth Worker" and "The Statesman", which are designed for youth development, to train professionals who work with young people.

Screening of the scientific literature has shown that non-formal education is important for the professional development of young people.

Having analyzed the scientific literature, we single out the definition of non-formal education.

V. Onushkin, E. Ogarev believe that non-formal education is "... education, upon completion of which it is not expected to receive diplomas...", but E. Ogarev complements the definition by the fact that "... is a part of the educational component; programs, courses, after which there are no legal consequences...; education, which is focused on the replenishment of knowledge and skills in the field of amateur knowledge, entertainment in order to expand the cultural horizons and acquire knowledge and skills needed in everyday life, in the field of interpersonal communication" [8].

A. Honcharuk believes that non-formal education is an organized, structured and purposeful educational activity outside formal education institutions, aimed at meeting a variety of educational needs of different, including age (from early childhood to old age), groups of the population, however, does not provide a legalized diploma" [2].

A similar opinion is expressed by O. Lazarenko: "The educational process, which mainly takes place outside the educational institution, in the workplace, in the premises of providers of such education, which does not lead to the assignment of certain qualifications, but has certain time limits and is structured to use resources..." [4].

N. Pavlyk formed her own definition of non-formal education as a process of "additional purposeful dialogical training, education and development of youth, organized outside the content, forms and methods of educational institutions and state institutions" [9].

In our opinion, non-formal education is education based on the principles of voluntariness, activity, participation, self-organization and which can be obtained at the state, regional and local levels through various forms of training (programs, training courses, public education, online education, internship, etc.) without the award of educational qualifications, but with a document confirming its acquisition.

Non-formal education as a purposeful process of education, upbringing and development of young people is organized in a higher education institution or outside of it. For its effectiveness the following forms are used: programs, training courses, seminars, online education (MEP), professional internship, etc.

Thus, many researches have revealed important aspects of professional training of social workers in general and social workers in particular, but research on the theoretical foundations of training social workers to work with young people in terms of non-formal education has not been conducted.

Problems of development of various types of non-formal education in Ukraine and abroad have become the object of study of Ukrainian scientists, namely: the basics of formal vocational education - N. Vasylenko [1], N. Nychkalo [7]; non-formal adult education as a problem of theory and practice of educational activities, adult education and its informal forms abroad, the development of non-formal education as a component of lifelong learning - A. Honcharuk [2], L. Lukyanova [5]; theoretical and practical aspects of the organization of non-formal education in Ukraine - N. Pavlyk [9], N. Terekhina [10], T. Tkach [11].

5 Conclusions

The results of our study has shown that the participants of the program of personal and professional development of youth "The Statesman" - increased their awareness of the activities of the public, private and public sectors; developed leadership skills; received tools for managing social action projects for the community.

In Ukraine, youth work can be carried out both in higher education institutions and abroad, using formal and non-formal education. At the local level, higher education institutions provide formal education in accordance with higher education standards, educational programs, curriculum and other educational and methodological support, used various forms of non-formal education - training courses, seminars, camps, sections, conferences, readings, etc. At the national level, effective innovative forms of non-formal education include programs such as "Youth Worker", "The Statesman" and others. In our opinion, each education type has a common goal - to promote personal, professional development of young people by acquiring appropriate competence; to raise their level of awareness and assist their upbringing.

References

1. Vasylenko, N.G.: Status and problems of introduction of the system of recognition of results of informal professional training in Ukraine. Actual problems of professional orientation and professional training of the population in the conditions of social and economic instability. K.: IPK DSZU, S. 130–138 (2014) (in Ukrainian)
2. Honcharuk, A.: Non-formal adult education in EU countries. Pedagogical Sci. **54**, 31–36 (2012). (in Ukrainian)
3. Law of Ukraine "On Education" (2017) (in Ukrainian). https://zakon.rada.gov.ua/laws/show/2145-19#Text
4. Lazarenko, O.: Philosophy of adult education in the context of the practical paradigm of lifelong learning: a European example for Ukraine. Philos. Educ. **10**(1-2), 255–265 (2011) (in Ukrainian)
5. Lukyanova, L.B.: Conceptual bases of formation of the legislation in the field of adult education. Current problems of vocational guidance and vocational training of the population in conditions of socio-economic instability. K.: IPK DSZU, pp. 52–59 (2014) (in Ukrainian)
6. Training program "Youth Worker" (in Ukrainian). https://youth-worker.org.ua/about/. Accessed 20 Oct 2020
7. Nychkalo, N.G.: Continuing professional education as a world trend. Vocational Education in Foreign Countries: A Comparative Analysis. CHOICE, Cherkasy, p. 322 (2000) (in Ukrainian)
8. Onushkin, V.G., Ogarev, E.I.: Adult education: an interdisciplinary dictionary of terminology. SPb., RAO IOV, Voronezh, p. 232 (1995) (in Ukrainian)
9. Pavlyk, N.P.: Theory and practice of organization of non-formal education of future social teachers: monograph. ZhSU Publishing House. I. Franko, Zhytomyr, p. 350 (2018) (in Ukrainian)
10. Terekhina, N.O.: Features of development of non-formal education in Ukraine. Pedagogical sciences: theory, history, innovative technologies. Sumy State Pedagogical University named after AS Makarenko, Sumy, pp. 79–87 (2013) (in Ukrainian)
11. Tkach, T.V.: Development of personality by means of non-formal education (in Ukrainian). http://ecopsy.com.ua/data/zbirki/2008_15/sb15_72.pdf. Accessed 20 Oct 2020

An Examination of the Effect of Common Language on the Creation of a Sense of Unity Between Organization Members in a Management Game

Fumiyo Ozaki[1] and Kazuhiro Fujimura[2(✉)]

[1] CARTEQUE.LTD, 629-16 Yamakita-cho, Marugame 763-0072, Kagawa, Japan
ozaki@carteque.jp
[2] Faculty of Economics, Kagawa University,
2-1 Saiwai-cho, Takamatsu 760-8523, Kagawa, Japan
fujimura.kazuhiro@kagawa-u.ac.jp

Abstract. We studied the effects of the Management Game on individual members and organizations, which is an educational tool in training management human resources, from the viewpoint of "benefit delay". In this study, we found that personal growth leads to the growth of the organization. MG not only contributes to the improvement of individual participants' abilities, but also has an effect on their autonomous actions and forging a sense of unity between the members of an organization. The members' improvement of management ability plays a key role in their autonomous actions and creation of a common language, and they contribute to fostering a sense of unity in the organization.

Keywords: Common language · Common purpose · Sense of unity · Autonomous action · Improvement of management ability

1 Introduction

Training management personnel entails not only the acquisition of skills required to perform core work tasks, but also requires accumulating a myriad of social experiences from interactions with a variety of people; this can take time. We consider this time to be a "benefit delay" in the development of human resources, and our research focuses on reducing the length of this delay while also improving how we train personnel to be more effective and efficient.

We investigated teaching effectiveness and the factors that affect it for individuals when using the Management Game (MG) developed as an educational tool in training management human resources [1]. Furthermore, results of an interview revealed that, in companies where MG was adopted and management engaged in the game with employees, individuals familiar with MG interacted with each other. This led to a sharing of perspectives that caused individuals to act both autonomously and in cohesion with the entire company. We discovered cases of management innovation resulting from this.

S. Nazir et al. (Eds.): AHFE 2021, LNNS 269, pp. 387–395, 2021.
https://doi.org/10.1007/978-3-030-80000-0_46

Foremost in these cases, members of the organization whose management ability was improved through MG engaged in proactive communication with other members who had started the same process concerning the challenges they experienced and so on. As a result, shared empathy within the group deepened. We inferred that this kind of group phenomenon was the result of the "common language" cited by many respondents to the survey on the effects of MG. While considering what this common language meant, we confirmed how it came out of MG, and constructed an effect model related to the improved sense of unity in the organization.

2 Effects of MG on Individual Participants and Their Organizations

2.1 Characteristics of MG

A number of business games have been developed using the results of simulations. Developed in 1976 as an educational tool for learning management skills at a subsidiary of Sony Corporation, MG is a representative example. Presently, this MG is implemented by several research institutes and we conducted a survey of participants at Nishi Research Institute Co., Ltd., headed by Junichiro Nishi, the developer of MG.

For MG, instructors and players gather at a set place, where a dedicated miniature game board is installed in the center of a table. There are five to six people per table, and two or more tables are used. By competing over capital, players learn about management procedures and systems. In the game's simulation, each player adopts the role of a company president in the manufacturing industry, carrying out actions, such as financing, capital investment, recruitment, procurement, training, research and development, advertising, bidding, sales, and settlement of accounts. In MG, the number of games played is expressed as "number of periods," with the accounting period from the beginning to the end of the period considered as one unit. All bookkeeping and settlement calculations are performed manually using a pencil and a special sheet, and a table showing the settlement results, as well as a graph showing the transition in the results as periods proceed, is created. MG is characterized by the challenging burden of carrying out the management process as a single person, while dealing with the rules of severe bidding and time limits, along with keeping an accurate inventory and carrying out an original settlement program that adheres to business accounting standards.

In this way, MG encourages participants to be proactive and positive to allow for a more realistic simulated experience, which creates a challenge for players. However, the game's characteristics also allow players to communicate frequently with each other and instructors in the trial-and-error process, so that they can enjoy the learning experience. MG players are evaluated based on the amount of capital gained over the course of five periods (rounds) in the game. Players are ranked in descending order with awards given to those who are the most successful. Additionally, players receive awards for participating in 100, 200, and 300 game periods in total. This system implicitly assumes that there is a delay in effects of MG's human resource development strategy, and it is presumed that this system is intended to encourage an ongoing process of learning.

2.2 The Common Language of MG

MG was developed as a practical training tool and has been devised in such a way that the settlement of accounts can be completed quickly without knowledge of accounting. Its developer achieved this by breaking down the accounting system and simplifying it. However, this settlement of accounts differs greatly from the calculation methods used in everyday practice and is not automatically calculated using a computer. The calculations have to be done by hand; however, the purpose of this is to make participants understand the meaning behind the numbers being calculated. Each factor and formula of this original settlement program and accounting figures used in MG is referred to as "MQ accounting." MQ accounting terms coined to refer to its elements include P (price), V (variable cost), M (margin), Q (quantity), F (fixed cost), G (gain), PQ (P × Q), VQ (V × Q), and MQ (M × Q).

Based on the survey of MG participants, it was surmised that understanding of these MQ accounting terms is one aspect of the common language shared by those who participate in MG. Through successive experiences of the MG periods, participants come to understand the meaning of the MQ accounting terms and interact by using them to communicate with other participants in the activity. Additionally, those who participate as a part of company training come to use the MQ accounting terms to communicate at their company. Language standardization is an important contributing factor in promoting intercultural and multilingual trade and exchange. Being able to communicate directly in the absence of an interpreter or translator is not only a matter of convenience and efficiency, but it reinforces the value of the training, forming a positive attitude toward participation and boosting individual growth.

In addition, by fostering communication and shared understanding, participants are more likely to feel positive. Positive emotions can increase their level of learning enjoyment and propel individual growth. This development as individuals allows human resources to become active autonomously under the umbrella of a shared understanding, which is thought to also help their organization to further grow.

Through the common language that comes from MG, participants not only learn MQ accounting terms and their meanings but there is also a sharing of the joys and struggles involved in the course of learning and application that allows them to imagine the interests and benefits of others, along with becoming better able to empathize.

2.3 Sense of Unity and Autonomous Action Among Members in Organizations

Improvement of interpersonal ability produces enjoyment of learning with others. Fujimura [1] makes it clear that emotional benefit and perspective benefit create motivation at work and contribute to its improvement [2]. This not only occurs in the interactions between service providers and their users in the service delivery process but can also occur within an organization.

When each member comprehends the purpose and goals of the organization as a whole and understands the importance of their role in achieving them, it becomes easier to elicit a positive outlook. Whereas the members previously did not see beyond themselves and overlooked the bigger picture, this new attitude allows them to grow into a long-term, systematic perspective, which promotes self-motivated, voluntary action. When each member of the organization takes on a management perspective, they will become able to engage in autonomous action. Workers grow from having a passive attitude where they merely carry out the orders of upper management into maintaining a balance on their own, where they become more active in their positions and strive to improve the profits of the company as a whole, take other departments into consideration, and have pride in their work. Behavior that takes the interests of others into consideration also builds mutual trust. While each member works independently, it is thought that having a common language makes it possible to create a sense of unity and shared purpose.

In this way, the sense of unity among the members of the organization created by MG means that, while as individuals they become more autonomous and less controlled, they share a purpose as a part of the organization as a whole.

2.4 Modeling the Relationship of the Common Language Created by MG and the Organization's Sense of Unity

Based on the results of interviews with MG participants and results of behavior observation, Fig. 1 illustrates the proposed model of how a sense of unity among organization members arises from a common language as discussed in Sects. 2.3 and2.4.

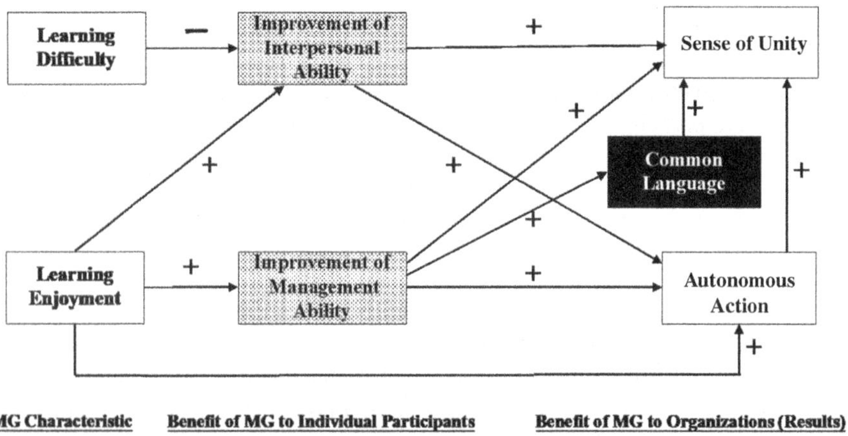

Fig. 1. Model of the common language created from MG and sense of unity of an organization

Two major characteristics of MG that were postulated was that it would be fun to learn with other participants and difficult to learn an original accounting system, and so on. Another premise was that, through repeated simulated experiences of MG, participants improve their management ability through what they discover and notice, and interpersonal ability also improves through communication opportunities, such as the participants teaching each other. These improvements in management and interpersonal abilities were also expected to have effects within the participants' companies, encouraging autonamous action and fostering a sense of unity. Furthermore, individual improvement of management ability was thought to contribute to a stronger sense of unity by encouraging the use of a shared language. In addition, the shared enjoyment of learning with others was thought to foster mutual recognition, empathy, trust, and respect, which would encourage autonomous action. This autonomy was thought to also contribute to enhancing the sense of unity within an organization.

3 Methods

To assess the effects on individual members and organizations as a whole when they adopt MG, we interviewed participants (managers and employees), and conducted an observational study on the building of relationships among members. In addition, we formed a hypothesis about the effects of MG on organizations and the factors that influence them based on listening to an online discussion about MG between MG developers and participants. Using the results of the questionnaire (sample #219) conducted from September 29 to November 29, 2018, we reanalyzed the effects of MG on organizations. Structural equation modeling was used to analyze the survey data.

4 Results and Discussion

4.1 Specifying Question Items for Measuring Concepts in the Model

Factor analysis to extract factors from MG brought out the benefits for participants (improvement of ability) and benefits for organizations (results), and MG characteristics were carried out using the maximum likelihood method/promax rotation on the question items for measuring the model concepts shown in Fig. 1.

Table 1 presents the extracted factors (concepts), the calculations for Cronbach's α to determine internal consistency, and the high load amount for each factor that is the compiling of the average values (score). As benefits of MG for individuals, there was improvement of management and interpersonal abilities, autonomous action and sense of unity were specified and as benefits of MG for organizations. With regard to common language, at the time the survey was conducted, the main purpose was to clarify the educational effects of MG on individual participants, so only one item was set.

Table 2 displays the relationship coefficients calculated using the score of each factor (concept) extracted in Table 1. Since the autonomous action of organizations' members, resulting from MG, highly correlates with the improvement of management and interpersonal abilities, it was speculated that, through the benefits that MG brings to individual participants (improvement of ability), there was also an effect on autonomous actions within organizations. Nevertheless, it was also speculated that autonomous action would lead to the improvement of management ability as a result of accumulated experience. In addition, while improvement of management ability, improvement of interpersonal ability, and autonomous action influenced each other, it was inferred they played an important role in shared organizational goals, or the sense of unity. Further, it seems that improvement of management ability contributes more to reducing the learning difficulty than improvement of interpersonal ability. Due to the few questions and ambiguous definition regarding common language, few factors showed a high correlation, but a relatively high correlation was found in connection with sense of unity.

Table 1. Extracted elements (concepts) and descriptive statistics for question.

	Concept	Question item	Each question item Average(SD)	Cronbach α	Score for each factor Average(SD)
Benefits of MG to Individual Participants (Improvement of Ability)	Improvement of Management Ability	MG made me aware of the important role investment plays in business.	4.32 (0.90)	0.947	4.00 (0.73)
		There are many discoveries and learnings from participating in MG.	4.45 (0.82)		
		MG gave me a fresh perspective on business.	4.20 (0.93)		
		MG has allowed me to think about the future more than the present.	4.25 (0.90)		
		MG has made it possible to plan from a long-term perspective.	3.88 (0.98)		
		The accounts program helped me understand accounting procedures in business.	3.99 (0.96)		
		MG helped me see business management in a broader context.	3.97 (1.01)		
		MG prompted me to consider what must be done to ensure business continuity.	4.02 (0.97)		
		Thanks to MG, I have acquired management skills.	3.79 (1.02)		
		MG has helped me understand the right way to make profits.	4.03 (0.97)		
		MG prompted me to proactively try various approaches to improve business operations.	3.88 (0.92)		
		MG has made it possible to foresee the risks associated with management.	3.57 (0.97)		
		MG taught me how to price things.	3.89 (1.04)		
		MG helped me understand my organization's strengths and weaknesses.	3.85 (0.97)		
	Improvement of Interpersonal Ability	MG has enabled me to build a good relationship with those around us.	3.53 (0.94)	0.922	3.50 (0.77)
		Thanks to MG, I can now face any situation with a smile.	3.46 (0.99)		
		MG has changed the way I interact with people around us.	3.68 (0.96)		
		MG eliminates anxiety about interpersonal relationships.	3.16 (0.97)		
		Thanks to MG, I can now strike up conversation with people I've never met before.	3.54 (1.07)		
		MG has taught me how to get my points across accurately.	3.45 (0.99)		
		MG has made me more empathetic to others.	3.67 (0.94)		
		MG changed my mind about the importance of my work.	3.56 (1.07)		
		MG has made me more aware of how managers and subordinates feel.	3.50 (0.97)		
Benefits of MG to Organizations (Results)	Autonomous Action	MG made me think for myself and start my work.	4.00 (1.02)	0.935	3.90 (0.78)
		MG has prompted me to proactively try new things.	4.00 (0.96)		
		MG has made me feel more positively about my work.	4.02 (0.98)		
		MG has taught me to admit my mistakes good-naturedly and improve work operations efficiently.	3.87 (1.00)		
		MG has prompted me to proactively try various things to improve my activities at work.	3.88 (0.92)		
		MG has allowed me to accept failure in a positive light and use it as an opportunity to move forward.	3.98 (0.97)		
		MG has given me the outlook to think for myself when facing problems on site.	3.91 (0.98)		
		I put into practice the experiences and realizations I had through MG.	4.06 (0.94)		
		MG helped me understand the role of myself and other departments in the company.	3.58 (0.99)		
		MG helped me better understand my company's vision and goals.	3.69 (1.05)		
	Sense of Unity	MG allowed me to share my visions and thoughts with top management and coworkers.	3.63 (1.02)	0.861	3.58 (0.88)
		MG has made it easier to communicate about accounting and goals in the company.	3.61 (1.08)		
		MG strengthened the unity between positions and teams in the company.	3.31 (1.11)		
		MG allowed me to discuss problems and actions within the company and with colleagues.	3.76 (0.97)		
	Common Language	MG terminology has become a common language in the company.	3.71 (1.32)		
MG Characteristics	Learning Difficulty	It is difficult to settle accounts manually.	2.44 (1.39)	0.847	2.38 (1.00)
		I experience a great deal of stress while performing the accounting program manually.	1.97 (1.20)		
		MG's financial statement preparation program is esoteric.	2.31 (1.13)		
		MG is difficult because I must think about many things that are not a part of my daily work.	2.82 (1.35)		
		MG is difficult because I must think about things with no relation to my position in the company.	2.37 (1.28)		
	Learning Enjoyment	In MG, I am happy I can experience things that I do not in my daily work.	4.20 (0.86)	0.770	4.30 (0.64)
		In MG, good relationships are formed where participants teach other about rules and settlement programs.	4.39 (0.78)		
		In MG, you can experiment with your own strategies.	4.28 (0.88)		
		In MG, you can try different strategies while observing the strategies of others.	4.32 (0.81)		

Table 2. Relationship between elements (concepts) in the model.

	Benefit of MG to Individual Participants (Improvement of Ability)		Benefit of MG to Organizations (Results)			MG Characteristics	
	Improvement of Management Ability	Improvement of Interpersonal Ability	Autonomous Action	Sense of Unity	Common Language	Learning Difficulty	Learning Enjoyment
Improvement of Management Ability	-	0.75 ***	0.90 ***	0.78 ***	0.36 ***	-0.45 ***	0.74 ***
Improvement of Interpersonal Ability		-	0.85 ***	0.75 ***	0.27 ***	-0.28 ***	0.60 ***
Autonomous Action			-	0.77 ***	0.32 ***	-0.39 ***	0.71 ***
Sense of Unity				-	0.60 ***	-0.29 ***	0.60 ***
Common Language					-	-0.16 *	0.25 ***
Learning Difficulty						-	-0.37 ***
Learning Enjoyment							-

***, **,* = P < 0.001, P < 0.01, P < 0.1, respectively.

4.2 Verification of the Model Relating the Common Language Created from MG and Sense of Unity of an Organization

Based on the extracted factors and the correlations between them, the model describing the relationship of common language from MG and an organization's sense of unity changed from Fig. 1 to Fig. 2, and verification was performed using covariance structure analysis.

There was a direct positive impact on sense of unity from improvement of management ability and improvement of interpersonal ability. Additionally, improvement of management ability had an indirect positive effect on unity through autonomous action and common language. In addition, due to the effect of improvement of management ability on improvement of interpersonal ability through autonomous action, it is thought that the improvement of management ability for each member of an organization plays a crucial role in creating the sense of unity within an organization. Learning enjoyment plays an important role in the improvement of this factor, and since this also has a positive effect on autonomous action, it is inferred that MG characteristically is effectively equipped with factors that benefit individuals and organizations.

Furthermore, since learning difficulty is reduced by improvement of management ability, it will presumably be reduced via the process of improving management ability by increasing the number of MG periods experienced.

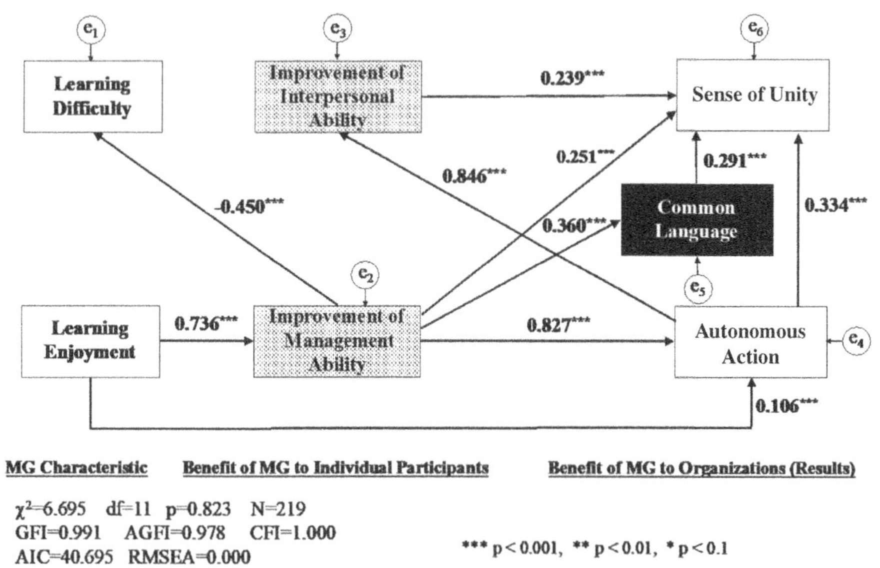

Fig. 2. Verification results of model of the common language created from MG and sense of unity of an organization

5 Conclusion and Future Study

5.1 Specifying Question Items for Measuring Concepts in the Model

From the verification results and interview results pertaining to the model for the relationship between the common language of MG and an organization's sense of unity, many things were made clear, but the following four items are especially important.

(1) MG as an educational tool for management human resources not only contributes to the improvement of individual participants' abilities, but also has an effect on their autonomous actions and forging a sense of unity between the members of an organization.
(2) The members' improvement of management ability plays a key role in their autonomous actions and creation of a common language, and they contribute to fostering a sense of unity in the organization. Therefore, in the development of an organization, it is indispensable to first improve the abilities of its members.
(3) Not only does making MQ accounting terms into a common language lead to an accounting-centric shared understanding that benefits business, but there is also a sharing of goals through active communication and interaction among members who know it.
(4) The autonomous actions of individual members promote the improvement of interpersonal ability, and this further contributes to a sense of unity in the organization.

Autonomous action has a greater direct, positive impact on sense of unity than on improvement of management ability or improvement of interpersonal ability. However, since the key factor that encourages autonomous action is improvement of management ability, it is clear that personal growth leads to the growth of the organization. Since there is a benefit delay in the improvement of individual abilities, this means that there is greater benefit delay in developing a sense of unity in an organization; this indicates there is a need to consider the growth of members and organizations from a long-term perspective. At present, the COVID-19 crisis limits in-person communication at organizations, but if autonomous action and sense of unity are fostered through the growth of each member, it is thought that organizational problems will be minimized.

However, since we used the results of a questionnaire conducted in the past to verify our model, there were restrictions on the measurement of the concepts of autonomous action, common language, and sense of unity. Hence, a more multidimensional and rigorous measurement of these concepts will need to be conducted in the future to better refine the model. Furthermore, there is also a necessity to consider the means for reducing the benefit delay in the development of the sense of unity and autonomous action in the organization. Presently, we are conducting an ongoing survey of MG participants, and we want to resolve these issues by using this research study.

Acknowledgments. This work was supported by JSPS KAKENHI Grant Number 16H03672.

References

1. Fujimura, K., Ozaki, F.: Developing a model on the effects of management games for human resource development: from a benefit delay and acquisition Perspective. In: Nazir, S., Ahram, T., Karwowski, W. (eds.) AHFE 2020. AISC, vol. 1211, pp. 204–211. Springer, Cham (2020). https://doi.org/10.1007/978-3-030-50896-8_31
2. Fujimura, K.: Constructing a work motivation model based on "benefit delay" and "benefit enjoyment." In: Kantola, J.I., Nazir, S. (eds.) AHFE 2019. AISC, vol. 961, pp. 251–262. Springer, Cham (2020). https://doi.org/10.1007/978-3-030-20154-8_24

Author Index

S. Nazir et al. (Eds.): AHFE 2021, LNNS 269, pp. 397–399, 2021.
https://doi.org/10.1007/978-3-030-80000-0

The manufacturer's authorised representative in the EU is Springer Nature Customer Service Centre GmbH, Europaplatz 3, 69115 Heidelberg, Germany. If you have any concerns regarding our products, please contact ProductSafety@springernature.com

Printed and bound by CPI Group (UK) Ltd, Croydon, CR0 4YY
30/09/2026
02225292-0008